Enzyme Engineering

Future Directions

Th… …ore

Enzyme Engineering

Future Directions

Edited by

Lemuel B. Wingard, Jr.
Department of Pharmacology
School of Medicine
University of Pittsburgh
Pittsburgh, Pennsylvania, U.S.A.

Ilia V. Berezin
and
Anatole A. Klyosov
Department of Chemistry
Moscow State University
Moscow, U.S.S.R.

PLENUM PRESS · NEW YORK AND LONDON

Library of Congress Cataloging in Publication Data

Conference on the Future of Enzyme Engineering Development, Tbilisi, 1978.
Enzyme engineering.

"Based on the proceedings of the Conference on the Future of Enzyme Engineering Development, held in Tbilisi, Georgian S.S.R., June 20–24, 1978."
"Conference cosponsors, State Committee for Science and Technology of the Council of Ministers of the USSR . . . [et al.]."
Includes index.
1. Enzymes–Industrial applications–Congresses. I. Wingard, Jr. Lemuel B. II. Berezin, Il'ia Vasil'evich. III. Klyosov Anatole Alekseevich. IV. Russia (1923- U.S.S.R.). Gosudarstvennyi komitet po nauke i tekhnike. V. Title. [DNLM: 1. Chemical engineering–Congresses. 2. Enzymes–Congresses. QU135 C749e 1978]
QP248.E5C68 1978 660'.63 80-12061
ISBN 0-306-40442-7

ACKNOWLEDGMENTS

The editors wish to thank the following for permission to reproduce a number of illustrations in this volume:

Academic Press, New York
DECHEMA (Deutsche Gesellschaft für chemisches Apparatewesen), Frankfurt/Main, FRG
Elsevier/North-Holland Biomedical Press, Amsterdam
IFIAS (The International Federation of Institutes for Advanced Study), Solna, Sweden
IPC Science and Technology Press, Guildford, Surrey, England
Pergamon Press, Elmsford, N.Y.
VAAP (The Copyright Agency of the USSR), Moscow
John Wiley & Sons, New York

Based on the Proceedings of the Conference on the Future of Enzyme Engineering Development, held in Tbilisi, Georgian S.S.R., June 20–24, 1978.

A Division of Plenum Publishing Corporation
227 West 17th Street, New York, N.Y. 10011

Printed in the United States of America

CONFERENCE ORGANIZING COMMITTEE

I. V. Berezin . . . . Chairman

S. V. Durmishidze . . . . Vice Chairman

E. I. Chazov . . . . Vice Chairman

V. F. Zurikov . . . . Vice Chairman

G. I. Kvesitadze. . . . . Science Secretary

A. M. Egorov. . . . . Science Secretary

N. S. Egorov

N. V. Kakhniashvili

M. G. Kohin

S. Nilsson

N. I. Serdujk

G. K. Skryabin

P. M. Skirda

S. D. Varfolomeev

CONFERENCE COSPONSORS

State Committee for Science and Technology of the Council of Ministers of the USSR, Moscow, USSR

International Federation of Institutes for Advanced Study, Solna, Sweden

Ministry of Higher Education of the USSR, Moscow, USSR

Academy of Sciences of the USSR, Moscow, USSR

Academy of Sciences of the Georgian SSR, Tbilisi, USSR

Institute for Plant Biochemistry of the Georgian SSR, Tbilisi, USSR

Moscow State University, Moscow, USSR

PREFACE

The Soviet Union has had an active research and development program in the study and application of soluble and immobilized enzymes since about 1970. Therefore, it was a natural consequence that an international conference should be held in the Soviet Union to focus on some of the developments that may lead to new and exciting practical applications of enzymatic catalysts in a variety of areas, especially carbohydrate conversions, medicine, energy transduction, and photochemistry.

The International Federation of Institutes for Advanced Study, with its goal of focusing international scientific and technological expertise on world problems, also has been very active in the area of enzyme engineering, with a major effort during 1972-1977 in several of the subareas covered in this volume.

The conference was held June 20-24, 1978 in the city of Tbilisi in the Georgian SSR of the USSR, under the title "The Future of Enzyme Engineering Development." The participants expressed a high sense of gratitude to the State Committee for Science and Technology and especially to the Georgian Academy of Sciences for their hospitality. The services of the Soviet linguists in providing simultaneous translation between Russian and English, and the brass band that played for the departing participants at the Tbilisi airport helped to make both the scientific and cultural parts of the conference memorable.

This volume contains most of the papers from the conference. The editors appreciate the efforts of the authors in submitting all of the manuscripts for this volume in English and in updating their papers to compensate for the later (1980) publication date. The initial editorial review of the papers was done at Moscow, with the final editing, mainly for clarity and conciseness, carried out at Pittsburgh. The efforts of Ms. Hall of the Department of Pharmacology in typing all of the manuscripts in final form ready for photoprinting is greatly appreciated.

The Editors

November 1979

CONTENTS

I. GENERAL PROBLEMS OF ENZYME ENGINEERING

Stabilization and Reactivation of Enzymes 3
K. Martinek, V. V. Mozhaev and I. V. Berezin

Theory of Reliability and Kinetics of Inactivation of Biocatalytical Systems 55
S. D. Varfolomeev

II. CARBOHYDRATE CONVERSIONS WITH ENZYMES

Enzymatic Conversion of Cellulose to Glucose: Present State of the Art and Potential 83
A. A. Klyosov and M. L. Rabinowitch

Immobilized Amyloglucosidase: Preparation, Properties, and Application for Starch Hydrolysis 167
L. A. Nakhapetyan and I. I. Menyailova

Substrate Stabilization of Soluble and Immobilized Glucoamylase Against Heating 197
A. A. Klyosov, V. B. Gerasimas and A. P. Sinitsyn

III. BIOMEDICAL POSSIBILITIES OF ENZYME ENGINEERING

Chemical Aspects of Enzyme Stabilization and Modification for Use in Therapy 219
V. P. Torchilin, A V. Mazaev, E. V. Il'ina, V. S. Goldmacher, V. N. Smirnov and E. I. Chazov

Modification of Trypsin Pancreatic Inhibitor by Polysaccharides for Prolongation of Therapeutic Effect 241
N. I. Larionova, I. Y. Sakharov, N. F. Kazanskaya, A. G. Zhuravlyov, V. G. Vladimirov and P. I. Tolstich

Enzymatic Modification of β-Lactam Antibiotics: Problems and Perspectives 257
V. K. Svedas, A. L. Margolin and I. V. Berezin

Modification of Enzymes with Water Soluble Polymers 295
I. M. Tereshin and B. V. Moskvichev

Immobilized Enzymes and Other Materials for the Study of Mammalian Cell Surfaces 313
L. B. Wingard, Jr.

IV. ENZYME ENGINEERING IN ENERGY TRANSFER, PHOTOGRAPHY, AND FINE CHEMICAL PROCESSING

Microorganisms as Hydrogen and Hydrogenase Producers 321
I. N. Gogotov

Spatially Structured Enzyme Support Arrangements in Electrochemical Systems 339
L. B. Wingard, Jr.

Application of Immobilized Enzyme Systems in Nonsilver Photography 357
I. V. Berezin, N. F. Kazanskaya and K. Martinek

Immobilized Enzymes: A Breakthrough in Fine Chemicals Processing . . . 405
E. Cernia

Problems of Efficiency and Optimization in Enzyme Engineering 415
A. Köstner and E. Siimer

V. ENZYMES IN FOOD AND NUTRITION

Novel Enzymatic Production of L-Malic Acid as an Alternative Acidulant to Citric Acid 439
F. Giacobbe, A. Iasonna, W. Marconi, F. Morisi and G. Prosperi

Application of Plant Phenol Oxidases in Biotechnological Processes . . 453
G. N. Pruidze

Immobilized Enzymes in Nutritional Applications 465
W. Marconi

VI. FUTURE PROSPECTS

Impact of Enzyme Engineering on Science Policy 487
B. Adams, C. G. Hedén and S. Nilsson

Some Thoughts on the Future 499
L. B. Wingard, Jr. and
A. A. Klyosov

Adresses of Authors 507

Subject Index 511

SECTION I
GENERAL PROBLEMS OF ENZYME ENGINEERING

STABILIZATION AND REACTIVATION OF ENZYMES

K. Martinek, V. V. Mozhaev and
I. V. Berezin

Moscow State University
Moscow, USSR

Prevention of the denaturation of enzymes is frequently indispensable to their technological application. This is exemplified by the following five operational factors of importance in the practical use of enzymes:

1. Enzymes isolated from their *in vivo* environment usually become labile, and their lifetime sometimes does not exceed minutes (1-4). For example, the stability of immobilized aspartase proved insufficient for technological use of this enzyme (5).

2. For many processes, elevated temperature is desirable. The rates of chemical reactions, including enzymatic, increase with temperature; so that at higher temperatures, the required conversion of the substrate will be achieved in a shorter time or with a smaller amount of immobilized enzyme. This is very important for technological processes where the cost of the enzyme is a limiting factor, e.g., glucoamylase (6). Also, higher temperatures allow germ-free conditions to be maintained (7), which are indispensable for, say, the food industry (8,9). On the other hand, denaturation of biocatalysts intensifies (1-3) at elevated temperatures.

3. Reaction equilibrium may be such that the required products are obtained only if the reaction is carried out in an aqueous-organic mixture with a high proportion of the organic component.

Under such conditions enzymes as a rule lose their catalytic activity or specificity (2,10; see also references in 11).

4. Sometimes the pH optimum of an enzymatic reaction and the pH range within which the enzyme is stable do not coincide. A good example is synthesis of penicillin antibiotics under the action of penicillin amidase; equilibrium in the synthesis is shifted toward the product in an acidic medium, where the enzyme is rather unstable (12).

5. It may not be convenient to use an enzyme immediately upon harvesting or isolation. Hence there is a need for stabilization against inactivation during prolonged storage (13).

To cope with the problem of inactivation of enzymes, several approaches have been suggested. First, intact (or partially degraded) cells can be used (14-19), and the enzyme remains in its natural environment. For instance, because of the low stability of free aspartase, continuous production of L-aspartic acid can be done using immobilized cells of *Escherichia coli* (5). Partially degraded cells may retain catalytic properties even in organic solvents (20). Second, an enzyme can be isolated from highly stable (e.g., thermophilic) strains (3,21) of microorganisms. However, each of these approaches has a drawback. The use of intact cells is hampered by diffusion of substrates and reaction products through cell walls; furthermore, metabolic pathways other than those necessary for the required catalyzed reaction can lead to the formation of undesirable side products. This is a problem especially in the production of drugs. Another drawback is that the choice of immobilization method is limited. With enzymes from thermophilic organisms it is always troublesome (and sometimes difficult) to screen specific strains for catalysis of a specific reaction.

Therefore, in this review we confine ourselves to the case where the enzyme molecule (globule) is stabilized against denaturing actions in an artificial way. This stabilization could not be solved merely by immobilization, because the stability of enzymes against inactivating actions sometimes increases, sometimes decreases, and sometimes remains unaltered upon immobilization. (16). Therefore, highly stable enzyme preparations often are the exception rather than the rule (17,22); and this problem of enzyme stabilization is barely touched on in a major volume on enzyme immobilization (19).

This chapter begins with a discussion of how thermal inactivation, due to protein unfolding, can be minimized. This is followed by a review of stabilization against extremes of pH. A brief discussion of enzymatic reactions in organic-aqueous media and the possibility for reactivation of denatured enzymes leads to concluding remarks on general principles of enzyme stabilization.

I. STABILIZATION OF ENZYMES AGAINST THERMO-INACTIVATION DUE TO UNFOLDING OF PROTEIN GLOBULE

Modification of enzymes by chemical reagents or their attachment to a water-insoluble support often entails alteration (increase or decrease) in the thermostability. This change in stability often is accounted for by the following phenomena (15-17,22,23):

1. Change in the conformation of an enzyme molecule compared to its native structure, on the assumption that the thermostability of a protein globule depends on its conformational state.
2. Change in the microenvironment of an enzyme molecule, on the assumption that the microenvironment of a protein globule affects the intramolecular linkages that maintain the enzyme in its native shape.

These views, even if entirely true, cannot be the sole foundation for elaborating general methods of stabilization of enzymes. It is not known how the stability of an enzyme is dependent upon its conformation or microenvironment; moreover, if this dependency exists at all, it should, in all probability, be unique for each enzyme. In addition it was demonstrated recently by direct physical methods that on binding between proteins and supports the protein macromolecules undergo hardly any conformational change (24-29).

Another reason why it is difficult to define the general ways of increasing enzyme stability is that the mechanisms of enzyme inactivation are still obscure (1,4); although intensive studies in this area have been carried out for years. It is likely that no general mechanism of denaturation of proteins exists, except that inactivation of enzymes under the action of heating or denaturing agents entails significant conformational changes, i.e. unfolding (1-4,30), in protein molecules (Fig. 1).

The thermostability of enzymes can be increased by rigidifying (fixing) the native conformation of the protein globule. In fact, if unfolding is recognized as being an indispensable step of enzyme inactivation, then it is clear that the more firmly the protein globule of the enzyme is fixed, the more difficult it is to unfold it and consequently to inactivate its catalytic center. An enzyme globule can be fixed to prevent its unfolding by several methods: by applying

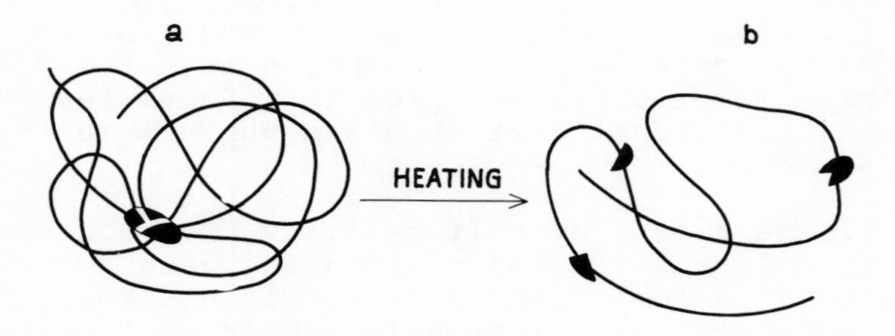

Fig. 1. Schematic description of unfolding of proteins.

intramolecular crosslinkages (Fig. 2a), by covalent attachment to support (Fig. 2b,c), and by mechanical entrapment into "tight" pores of an inert support (Fig. 2d). Each of these approaches is discussed separately.

A. Intramolecular Crosslinkages

The literature contains abundant data on the methods of chemical modification of enzymes and on the properties of the resulting preparations. But in no case can one predict how the thermostability of the enzyme will change as a result of its modification. This is associated with the extreme complexity of the molecular structure of biocatalysts. A good example of a rather complex dependence of the thermostability of the enzyme on the degree of modification of its surface layer is shown by the behavior of chymotrypsin with alkylated NH_2 groups (31). The enzyme was modified by treatment with acrolein followed by reduction of Shiff's bases by sodium borohydride; the procedures were described by Feeney et al. (32). As a result, the enzyme retained almost all of its activity. The degree of stabilization can be defined as the

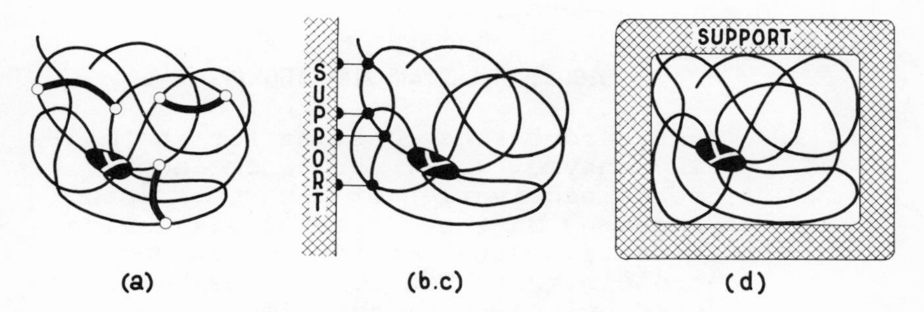

Fig. 2. Fixing of protein native conformation by intramolecular crosslinking (a), attachment to a support (b,c), or entrapment (d).

ratio of the first order rate constants of inactivation for the native enzyme as compared to the modified enzyme. With stabilization the denominator should decrease and the ratio increase. This is shown in Fig. 3, where alkylation of the NH_2 groups that were the first to undergo modification did not affect the thermostability of the enzyme. However, alkylation of the less reactive or less accessible NH_2 groups caused a drastic increase in

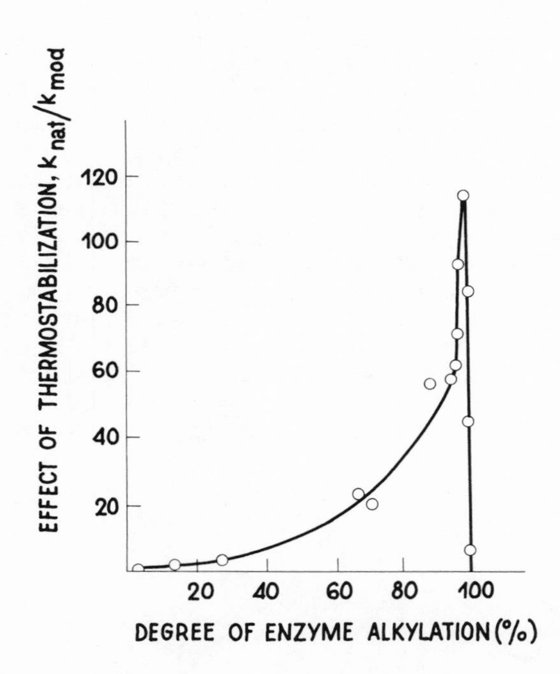

Fig. 3. First-order rate constant for the modified enzyme, k_{mod}, for monomolecular thermoinactivation of α-chymotrypsin depending on the degree of alkylation (titration data) of the NH_2 groups by acrolein, with Schiff's bases being reduced by sodium borohydride. The value of k_{nat} obtained for the native (unmodified) enzyme is assumed to be unity. Conditions: 50°C, pH 8, 5×10^{-3}M tris-HCl, 0.1 M KCl (31, 150).

the stability. Then, with complete modification of all NH_2 groups titratable with picryl sulfonic acid, the thermostability of the enzyme preparation decreased sharply. The peak point of thermostabilization of the enzyme came when only a few of the 15 titratable NH_2 groups had not been modified.

The situation is different if an enzyme is modified by bifunctional reagents which crosslink the protein globule in a covalent fashion. Here the conformational stability will continue to increase (33). Artificial crosslinking in protein molecules can be achieved by treating them with bifunctional reagents (34,35). The reagents commonly used are dialdehydes, diimidoesters, diisocyanates, bisdiazonium salts, etc. Analysis of the literature (33,36-45) shows that in most cases the effects of stabilization are not high. Sometimes they are due to simple (single-point) chemical modification of the enzyme (36) rather than to crosslinking; this was the case with glutardialdehyde (43). Sometimes the effects of stabilization proved to be more pronounced (33,42); such that the increase in stability correlated with greater rigidity of the protein molecule (33).

It seems that when an attempt is made to enhance the stability of an enzyme by treating it with bifunctional reagents, success or failure may depend on whether the length of the bifunctional molecule or "bracket" corresponds to the distance between the possible centers of attachment on the protein globule. Apparently, for any protein, there is an optimal size of the intramolecular crosslinking agent. Therefore, studies on determining the optimal size of bifunctional agents are of great interest.

1. *Variation in the Length of the Intramolecular Crosslinkage.* Studies on intramolecular crosslinkage length are rather rare, the reason being that many of the bifunctional agents that have been used as "brackets" are not commercially available and are rather difficult to synthesize.

For this reason it is very tempting to use bifunctional agents that normally are not suitable for intramolecular crosslinking of enzymes but become suitable after the enzyme is modified in a certain way. An example is some commercially available and relatively cheap aliphatic diamines (46) of 0 to 12 methylene groups in length. Diamines can be used for crosslinking the carboxyl groups of the enzymes after preliminary activation by carbodiimide. As a result, between the amino groups of the modifier (diamine) and carboxyl group of protein a strong amide bond is formed (47) (Fig. 4).

In conditions where intermolecular crosslinking, as shown in Fig. 4b, is absent, the thermostability of an enzyme crosslinked in an intramolecular fashion increases considerably. This was the case with α-chymotrypsin, where the thermoinactivation rate constant decreased more than fortyfold compared to that of a native preparation. This stabilization effect could not be accounted for by a single-point modification of the enzyme (Fig. 4c), since modification of chymotrypsin by 1-aminopropanol-3 decreased the thermostability. The effect of stabilization depended on the number

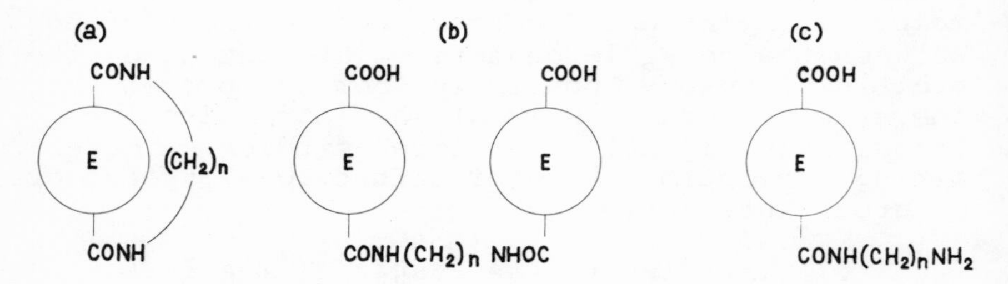

Fig. 4. Aliphatic diamine coupling to enzyme carboxyl groups. Intramolecular crosslinking (a), intermolecular crosslinking (b), and single point attachment (c) (46).

of methylene groups (n) in the molecule of the modifier, $NH_2(CH_2)_nNH_2$. Figure 5A shows that the dependence of the first order rate constant for thermoinactivation of chymotrypsin crosslinked by various diamines had a sharp minimum at n = 4, although at n = 2,5, and 6 the enzyme was also stabilized. The fact that treatment by tetramethylene diamine induced the greatest stabilization of the enzyme seems to indicate that the length of this agent best fit the distance between the carboxyl groups of the protein molecule. This in turn, could have given rise to a greater number of intramolecular linkages than with other diamines.

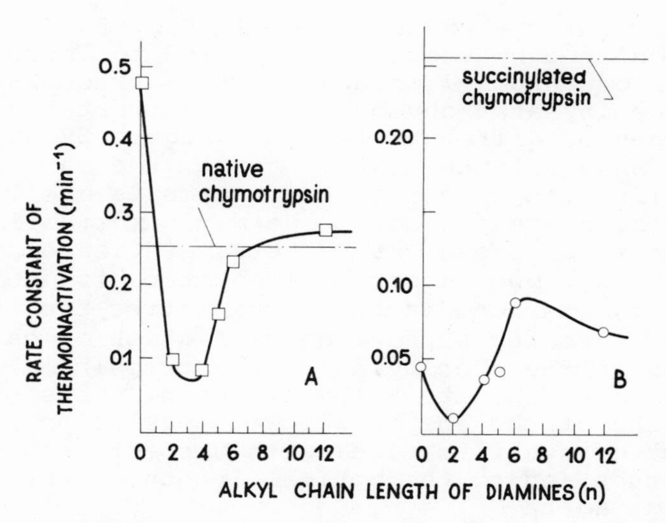

Fig. 5. First-order rate constant for monomolecular thermoinactivation of α-chymotrypsin (A) and succinylated α-chymotrypsin (B) intramolecularly crosslinked with diamines, $NH_2(CH_2)_nNH_2$. The broken line shows the level of thermostability of the native and succinylated enzymes. Conditions: 50°C, pH 7, 0.02 M phosphate (46).

2. *Ability of Enzyme to Interact with Bifunctional (Polyfunctional) Reagents.* It should be expected that the more "brackets" that are imposed on a protein molecule, the more stable the molecule will be against unfolding and, as a consequence, against inactivation. The number of "brackets" is, in turn, determined by the quantity and relative position of the functional groups on the surface of the protein globule that interact with the crosslinking agent. It is clear that the number of the functional groups can be increased by premodification of the protein. For this purpose, Reiner et al. (48) suggested that SH groups should be introduced with the help of N-acetyl-homocysteine thiolactone for subsequent use with bis-iodacetamides of different aliphatic diamines.

In the system described in Fig. 5, the enzyme was modified by succinic anhydride to increase the number of carboxyl groups in its surface layer (98). The succinylated chymotrypsin was treated with diamines of different length. Figure 5B shows the dependence obtained (46): first, the maximal stabilization effect of the enzyme is greater than for Fig. 5A; and second, the maximum stabilizing effect is produced not by tetramethylenediamine, as in Fig. 5A, but rather by a shorter bifunctional reagent, ethylenediamine. The latter fact suggests a greater surface concentration of carboxyl groups in the succinylated chymotrypsin molecule as compared to native chymotrypsin. Thus premodification of the enzyme allowed regulation of the effect of stabilization, with respect both to the magnitude and to the optimal length of the cross-linkage agent.

B. Multipoint Covalent Attachment of the Enzyme to the Complementary Surface of the Support

The methods of covalent attachment of enzymes to various supports are described in detail in a number of reviews (15-19,50). Such methods of immobilization (Fig. 2b) can, in principle, increase the conformational stability of the protein molecule and hence prevent it from unfolding and undergoing inactivation. This was evidenced by the experimental study of protein-support binding, and showed that the molecular structure of the immobilized proteins became more rigid or frozen (24,25,27). The same conclusion followed from abundant indirect data obtained by both physical methods and enzymatic activity measurements (25, 27,51-53). Also, Gabel in his pioneering work (54) showed that the greater the number of linkages by which trypsin was attached to Sephadex, the more stable was the protein against inactivation, i.e., unfolding (55), by urea.

So, the above considerations give grounds for believing (56) that a general principle of stabilization of enzymes is that of providing multipoint binding of the catalyst molecule to the support. This binding serves to make the conformation of the molecule more rigid but without necessarily altering the conformation, and hence more stable against unfolding and, ultimately, inactivation. This principle is difficult to realize methodologically for steric reasons as both the surface of the support and that of the protein have reliefs of their own that are, generally speaking, not congruent. In addition, it is obvious that even in the case of multipoint interaction, when an enzyme is attached to the support over only a small portion of the surface, one should hardly expect that all of the enzyme molecule will be affected the same. This seems to have been overlooked by some authors who attempted unsuccessfully to stabilize enzymes (24,57,58). The problem is actually one of providing a support with a surface complementary to that of the enzyme molecule; only then can the multipoint protein-support interaction favorable to stabilization of the enzyme be realized.

1. Preparation of a Polymeric Support Complementary to the Molecule of Immobilized Enzyme. Our approach (56,59,60) to the preparation of a complementary polymeric support is based on modifying the enzyme by an analogue of the monomer and then copolymerizing the resulting preparation with the monomer. This gives an enzyme chemically incorporated into the three-dimensional lattice of the polymer gel, the points of the enzyme-support binding being the centers of premodification of the enzyme molecule (Fig. 6). It is obvious that, because of the principle on which the method is based, the microsurface of the gel around the entrapped protein molecule should be complementary to the surface of the latter.

The method for preparation of copolymerization-immobilized enzymes was elaborated by Jaworek et al. (61). Their goal was to prevent immobilized enzymes from being washed out from gel supports. We have independently suggested this method (56,59,60) for stabilization of enzymes. As analogues of the monomer, acylating and alkylating agents with a double bond capable of undergoing copolymerizing may be used. Two examples include acryloylchloride, which acylates NH_2 and OH groups of proteins, and acrolein (56), which interacts with NH_2 groups with subsequent reduction of Schiff's bases by sodium borohydride to give alkylated proteins. There also is a method of vinylation of the carboxyl groups of proteins (62). Acrylamide, sodium methacrylate, and 2-hydroxyethylmethacrylate can be used as comonomers

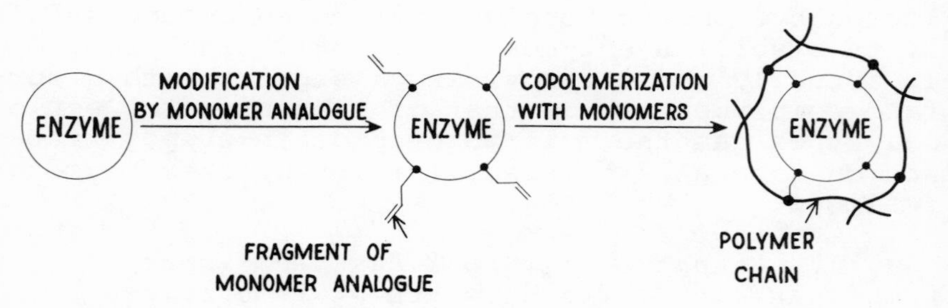

Fig. 6. Approach for preparation of complementary enzyme-polymer gels (56).

and N,N,-methylenebisacrylamide as crosslinking agents.

2. *The Scale of the Stabilization Effect.* A good example of the scale of the stabilization effect is shown in Fig. 7. The rate of inactivation of α-chymotrypsin entrapped in polymethacrylate and polyacrylamide gel is presented in Arrhenius coordinates (56). The thermostability of the immobilized enzyme preparations in Fig. 7 was so much higher than that of the free native enzyme that experimental comparisons could not be carried out at the same temperature. Instead comparisons were made of the values of thermoinactivation rate constants determined by extrapolation, e.g., at 60°C (the vertical broken line in Fig. 7). At 60°C acryloylated chymotrypsin

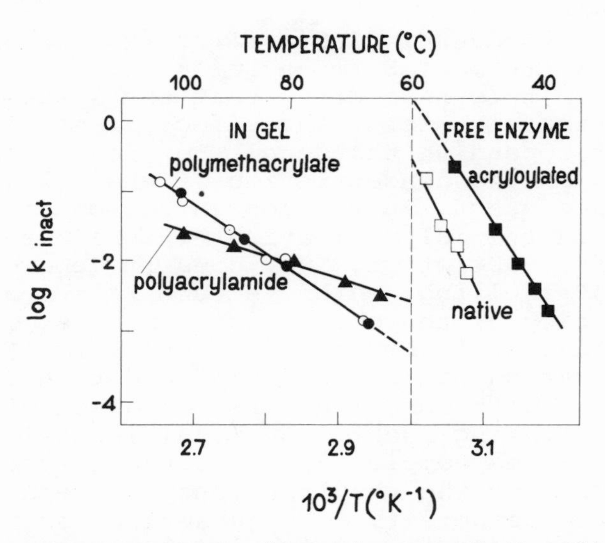

Fig. 7. Temperature dependence of the first-order rate constant (min^{-1}) for monomolecular thermoinactivation of free, native and acrylcylated, and immobilized (chemically entrapped in polymeric gel) α-chymotrypsin. Conditions: pH 8, 0.005 M tris-HCl, 3 M KCl. Water insoluble ●, water soluble O (56).

entrapped in methacrylate and polyacrylamide gels is 1000 and 200 times as stable as the free native enzyme, respectively. The difference is still more striking if comparison is made with free acryloylated chymotrypsin.

Fig. 7 shows that the effective values of activation energies of thermoinactivation for acryloylated chymotrypsin entrapped in both polymethacrylate and polyacrylamide gels (35 and 15 kcal/mole, respectively) were much lower than that of the native enzyme (110 kcal/mole). This means that, as the temperature went higher, the effect of stabilization on the immobilized enzyme increased as compared to the native enzyme. For example, extrapolation of Fig. 7 shows that at 102°C the gel-entrapped enzyme would be 10^8 times as stable as the native enzyme.

3. *Dependence of Stabilization Effect on the Number of Linkages Between Immobilized Enzyme and Support.* It is noteworthy that for systems such as described in Fig. 7 the effect of stabilization does not depend on the density of the gel support, that is, on the concentration of the polymer or, to be more exact, on the concentration of the monomer in the polymerization mixture over the range of 10-50% wt/wt. Nor does it depend on the degree of swelling, which in turn is a function of the quantity of the crosslinking agents, i.e., N,N′-methylenebisacrylamide. Moreover, if copolymerization of an acryloylated derivative of the enzyme with comonomers is performed without the crosslinking agent, the resulting gellike polymer can be completely dissolved in water; and most important the resulting enzyme preparation has a thermostability as high as that of the enzyme immobilized in a water-insoluble manner (Fig. 7) (56). This seems to be due to the polymeric structure of the gel in the vicinity of the enzyme molecule being congruent to the enzyme surface and thus primarily determined by the number of crosslinks with the protein.

The thermostability of the copolymerized enzyme is, in turn, affected by the degree of its covalent binding with the support; or, the greater the number of linkages between the immobilized enzyme and the support, the greater the thermostability of catalytic activity. For example, Fig. 8 shows that the effective thermoinactivation rate constant of α-chymotrypsin at 60°C decreases more than a thousand-fold if the degree of attachment of the enzyme to polymethacrylate gel is increased by about 75% (56).

4. *Mechanism of Stabilization.* The effect of stabilization cannot be ascribed to alteration of the thermostability of the enzyme resulting from its having been chemically modified. For example, free acryloylated chymotrypsin is even less thermostable than the free native enzyme (see Fig. 7). Neither can the stabilization be attributed to the effect of the gel microenvironment on the enzyme,

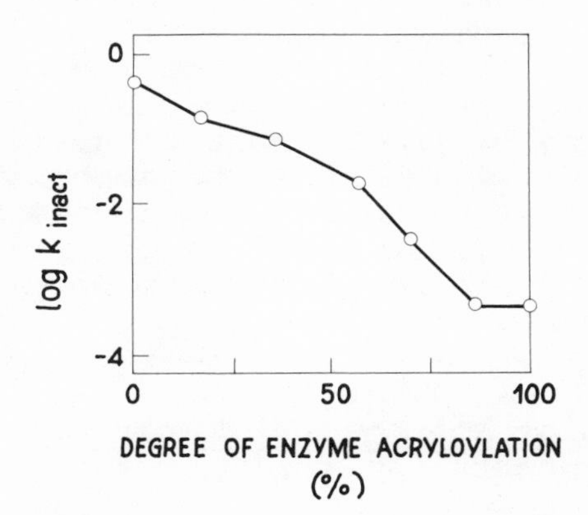

Fig. 8. Dependence of the first-order rate constant (min^{-1}) characterizing the monomolecular thermoinactivation of α-chymotrypsin entrapped in polymethacrylate gel on the degree of the enzyme-support covalent binding. Conditions: pH8, 0.005 M tris-HCl, 3 M KCl, 60°C (56).

as the thermostability of chymotrypsin physically entrapped in polymethacrylate gel is practically the same as that of the same concentration of free enzyme (63).

The nature of the gel is actually not decisive in enzyme stabilization. Increase in the thermostability was observed if the enzyme immobilized by copolymerization was chemically entrapped in polyelectrolyte polymethacrylate (56), electroneutral polyacrylamide (56), or poly (2-hydroxyethyl) methacrylate (64) gels (e.g., see Fig. 7).

The effect of stabilization occurs with various kinds of premodification, i.e., alkylation of NH_2 groups by acrolein or acylation of protein with acryloyl chloride (56) or vinylation of the carboxyl groups (62). Finally, stabilization was achieved with various enzymes, i.e., α-chymotrypsin and trypsin (56), glucose oxidase (64), and penicillin amidase (65).

All this means that the key role most likely belongs to the entrapment of the enzyme molecule into the three-dimensional lattice of the gel rather than to the premodification method or to the nature of the gel. When the protein is fixed, the unfolding of the globule on heating becomes more difficult. It should be emphasized that the native, catalytically active, conformation of the active center probably can be maintained only if the enzyme is attached to the support in a multipoint fashion (Fig. 8).

C. Multipoint Noncovalent Interaction of Enzyme with Support

Multipoint interaction of an enzyme with a support can in principle ensure the stabilization of a biocatalyst even if the enzyme has been immobilized by physical methods, (15,19,50), such as mechanical entrapment in gels (66), microcapsulation (67), and adsorption on supports (68,69). It is only recently, however, that the thermodynamic causes of enzyme thermostabilization have been discussed, and recommendations have been given for preparation of stabilized enzymes (63).

By way of example the thermostability of α-chymotrypsin mechanically entrapped in polymethacrylate gel is discussed in greater detail. With this polyelectrolyte as the support, the enzyme can form a great number of weak electrostatic or hydrogen linkages. Fig. 9 shows the dependence of the logarithm of the effective rate constant of thermoinactivation of chymotrypsin on the concentration of polymethacrylate gel, up to the limit of solubility of the monomer in water. One can see that if the concentration of the gel in water

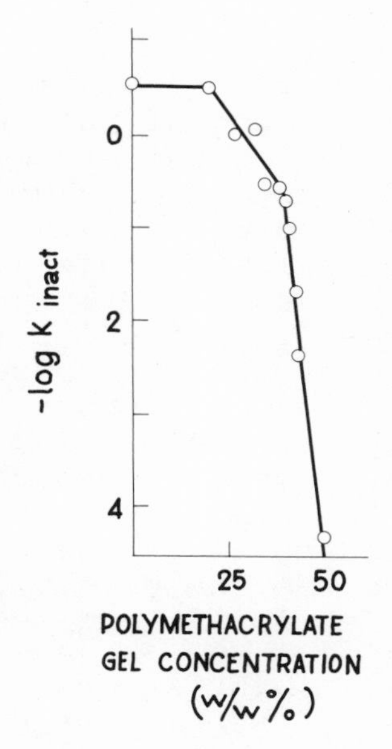

Fig. 9. Dependence on gel concentration, for the first-order rate constant (min^{-1}) characterizing the monomolecular thermoinactivation of α-chymotrypsin mechanically entrapped in polymethacrylate gel. Conditions: pH 8, 60°C (63).

changes from 0 to 30% wt/wt, the rate of inactivation of the enzyme at 60°C is almost constant but sharply drops if the concentration of the gel is further increased. With the gel concentration of 50% wt/wt, the scope of the stabilization effect amounts to 10^5 times. Moreover, extrapolation of the straight line portion of the Arrhenium dependence (log k_{inact} vs. 1/T) gave an apparent lifetime for the enzyme immobilized in 45% wt/wt polymethacrylate gel of hundreds of millions of years (63).

The mechanism of stabilization is conditioned by the interaction of the enzyme molecules with the polymeric support. In principle, the enzyme molecules can form electrostatic and hydrogen bonds with the carboxyl groups of polymethacrylate. However, these bonds are relatively weak and can hardly be realized in a solution of the polymer or in diluted gels; the formation of such an enzyme-support complex should produce a large loss in entropy due to quenching of the tranlaitional and rotational movements of the enzyme molecules. The situation is different in concentrated gel, since the tranlaitional and possibly rotational movement of the enzyme molecules is to a great extent already reduced because of steric hindrances caused by the dense three-dimensional lattice of the polymer. This means that in concentrated gels, formation of the protein-support complex should be thermodynamically more favorable, as there are fewer free energy losses due to changes in transitional or rotational entropy when the enzyme globule becomes adsorbed on the polymeric support. This does not exclude the possibility that in concentrated gels the mobility of the polymer chains becomes more limited. Hence, compared to diluted gels or polymer in solution, the entropy expenditure of free energy necessary for the support to be "frozen" as an enzyme-support complex is much lower. Finally, with concentrated gels it should be expected that enzyme-support interactions will be multipoint, at least for purely steric reasons, as the polymer chains adhere to the enzyme globule from all sides (see Fig. 10). And, as has been

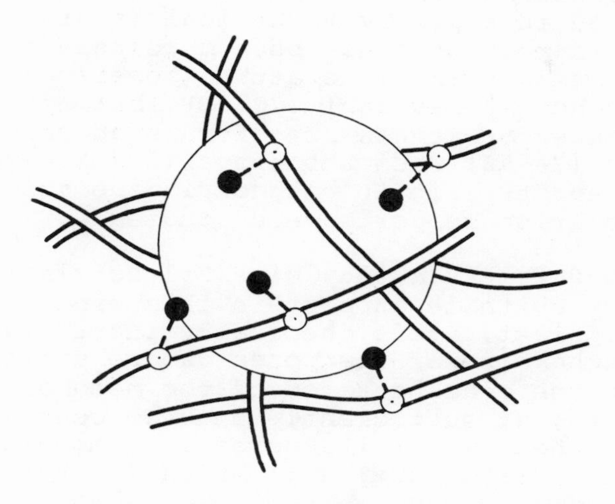

Fig. 10. Schematic of multipoint formation of enzyme-complex binding.

exemplified above with covalent enzyme-support binding, such multipoint interaction may lead to a sharp increase in the thermostability of an immobilized enzyme.

All of these conclusions were drawn from investigation of the thermostability and the tranlational and rotational mobility of enzyme molecules as a function of the concentration of gel (63).

D. Mechanical Entrapment of Enzymes in "Tight" Pores of Support

In principle protein globules can be prevented from unfolding if placed in a certain cell that does not interact with the molecule either chemically or sorptionally but that is sufficiently "tight" to prevent formation of an unfolded conformation for steric reasons. In this case, the native conformation of a protein globule could be maintained in a purely mechanical fashion (Fig. 2d). This mechanism has been more than once hypothesized in the literature (sometimes in a very incoherent fashion) (70-73), but experimental evidence for it has been furnished only recently (74-76) when the properties of α-chymotrypsin and trypsin in polyacrylamide gel or in/on polysaccharide supports were studied.

1. *Polyacrylamide Gel.* Polyacrylamide gel is a very suitable support in this kind of study because, first, it is chemically inert (77), second, the size of its pores can be varied by changing the concentration of the monomer (78), and, third, at sufficiently high concentrations of the gel, the size of its pores is commensurate with the sizes of enzyme globules (79,80).

Two methods were used to achieve mechanical entrapment of the enzyme into the gel. In one method polymerization of acrylamide of a certain concentration was carried out directly in a solution of the enzyme, as described by Martinek et al. (56,63). In the second method gel particles were taken without the enzyme and thoroughly washed from low molecular weight admixtures; the washed gels then were impregnated with the solution of the enzyme and finally lyophilized until the required concentration of polyacrylamide was achieved (75,76). In both cases the enzyme displayed the same thermostability (Fig. 11). At a moderately high concentration of the gel, but less than 45% wt/wt, there was no appreciable stabilization of the enzymes against irreversible monomolecular thermoinactivation; whereas in a more dense support, more than 50% wt/wt of polyacrylamide, the thermostability of the enzyme increased sharply.

2. *Scale of the Stabilization Effect.* The thermostability of the enzyme mechanically incorporated into highly concentrated polyacrylamide gel (lyophilized powder) is so much higher than

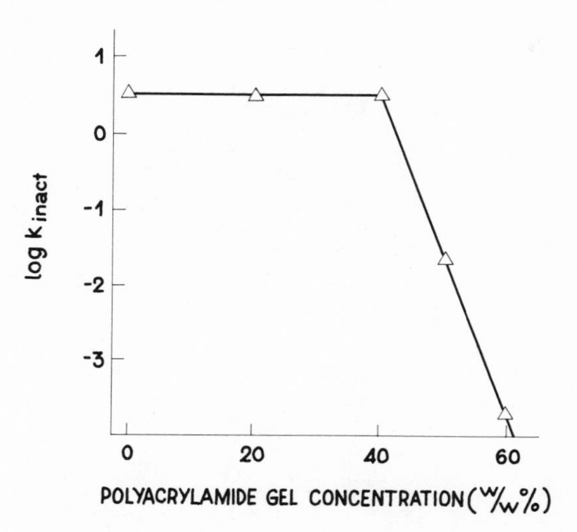

Fig. 11. Dependence on gel concentration for the first-order rate constant (min^{-1}) characterizing the monomolecular thermoinactivation of α-chymotrypsin mechanically entrapped in polyacrylamide gel. Conditions: 60°C, pH 8, 0.02 M tris-HCl, 0.1 M KCl (134, 76).

that of the enzyme in solution that these values cannot be determined experimentally at the same temperature. To find this out, we studied the temperature dependence of the rate of monomolecular thermoinactivation of chymotrypsin. Linear extrapolation of the data in Fig. 12 showed that at the midpoint of the temperature range, i.e., 120°C (vertical broken line) the effect of stabilization in fully dried gel was 10^{13} times as high as for the native enzyme in solution. For gels of lower concentration, the effect of stabilization was, of course, lower.

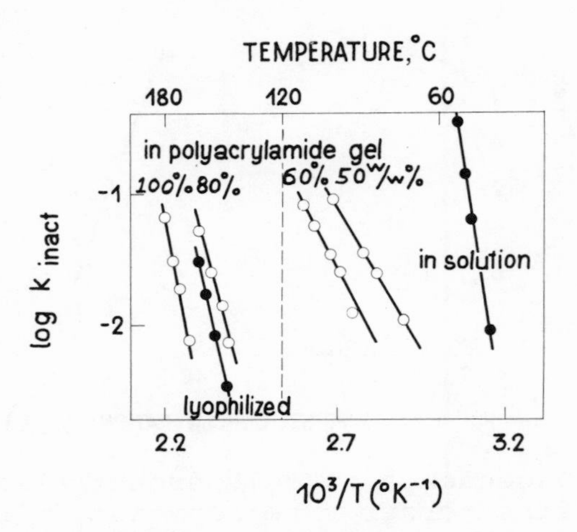

Fig. 12. Temperature dependence of the first-order rate constant (min^{-1}) characterizing the monomolecular thermoinactivation of α-chymotrypsin mechanically entrapped in 50%, 60%, 80%, and 100% wt/wt polyacrylamide gel. The concentrations of the polymer were determined after washed 60-μm gel granules had been lyophilized. For comparison the thermostabilities of the enzyme in an aqueous solution (ph 5-8, $5x10^{-3}$M tris) and in a lyophilized state are given (134, 76).

3. *Mechanism of Stabilization.* The mechanism of stabilization is not related to the non-polymerized acrylamide (81), as this material was washed completely from the powdered gel (75,76). Neither can the stabilization be explained by chemical attachment of the enzyme to the gel, as both chymotrypsin and trypsin could be washed completely from the swollen gel (75,76). The enzymes hardly ever interact with the support in a sorptional way up to a gel concentration of 50% wt/wt, at which concentration pronounced thermostabilization is observed. This indicates that at low temperature, 5-25°C, the protein globule retains a high rotational mobility (63,75,76).

One has then grounds for believing that as the concentration of the gel increases, the size of the pores decreases, and the enzyme globule becomes fixed so firmly that it cannot unfold (denature) because of the "cage effect".

One could assume a different mechanism of stabilization, i.e., dehydration of protein, believing that on dehydration the mobility of the polypeptide chains becomes lower and the unfolding of the globule is less pronounced. One cannot choose between these two possibilities by studying the thermostability of a lyophilized preparation of a free enzyme. Therefore, the reason for the extremely high thermostability of dried (lyophilized) enzyme preparations (70,72,73) e.g., chymotrypsin; (see Fig. 12) can be due both to "dehydration of polypeptide chains" and to the "cage effect". This problem was resolved when polysaccharide supports were used.

4. *Polysaccharide Supports.* Polysaccharide supports are inert and do not interact with proteins. This is evidenced by the fact that although the tranlational diffusion of protein molecules (lactoglobulin, ribonuclease, etc.) is almost arrested in a 40% solution of dextran, the rotational diffusion remains almost the same as in an aqueous solution containing no polysaccharide (82).

To resolve this question, the thermostability of chymotrypsin covalently bound to two polysaccharides, microcrystalline cellulose and Sephadex, was studied (76). These two supports are not greatly different in chemical composition; but structurally, they are very dissimilar; Sephadex, unlike cellulose, is porous. This means that if the enzyme is attached to the surface for cellulose, then the protein molecule will be entrapped in the pores for Sephadex. That is why on lyophilization in Sephadex, and not in cellulose, the "cage effect" can be realized. Table 1 shows that covalent immobilization only slightly affected the thermostability of chymotrypsin on both polysaccharide supports, as compared to the free enzyme. However, another result seems to be more important, i.e., additional lyophilization of the enzyme attached to the surface of microcrystalline cellulose did not greatly affect the thermostability of the enzyme preparation (see Table 1).

The situation is entirely different in the case of the enzyme attached to Sephadex G-150; this lyophilized enzyme preparation was not thermoinactivated at all under comparable conditions.

Since the only significant difference between these polysaccharide supports was that in Sephadex the enzyme after lyophilization was firmly entrapped in a pore and in microcrystalline cellulose it was attached to the surface of the particle, one can conclude that such an enormous increase in the thermostability of the enzyme in concentrated porus supports (Sephadex, see Table 1, and polyacrylamide, see Fig. 12) should be ascribed to the "cage effect" rather than to "dehydration of protein".

5. *Perspectives*. Enzyme preparations with such a high thermostability can find many applications if swelling of the support in water is eliminated. This can be achieved by using highly crosslinked polymeric supports or by providing organic solvents immiscible with water for enzymatic reactions (see below). Immobilization of

TABLE 1

FIRST-ORDER RATE CONSTANTS FOR IRREVERSIBLE MONO-MOLECULAR THERMOINACTIVATION OF NATIVE AND IMMOBILIZED α-CHYMOTRYPSIN [(a)]

Conditions	Rate Constant for Inactivation (min^{-1})		
	Free enzyme	Enzyme Attached to the Surface of Microcrystalline Cellulose	Enzyme Chemically Entrapped in Pores of Sephadex
Solution or suspension at pH 8 (2×10^{-2}M tris + 0.1 M KCl)	3	0.1	0.08
Lyophilized	0 [(b)]	0.05	0 [(b)]

(a) Thermoinactivation at 60°C.; references (74, 76).

(b) No thermoinactivation was detected for 30 min even at a higher temperature, 100°C.

protein globules in tight pores of an inert support opens up new perspectives for stabilization (owing to the cage effect) of even very labile enzymes, which usually, on being immobilized, become inactivated by interacting with the support.

E. Approaches Based on the pH Dependence of the Rate of Thermoinactivation

The rate of thermoinactivation sometimes greatly depends on pH (e.g., see Refs. 16,49). In this case the denaturation of the enzyme can be inhibited using the procedure based on the physicochemical principles described below.

II. STABILIZATION OF ENZYMES AGAINST INACTIVATION UNDER THE ACTION OF EXTREME pH VALUES

Many enzymes rapidly and irreversibly lose their catalytic activity under the action of extreme pH values (i.e., at pH values of the medium that are far from the pH range of the enzymatic action) (1). A good and well-studied case in point is alkaline inactivation of porcine pepsin, which starts even at pH 5, or acidic inactivation of chicken lactate dehydrogenase, which is effective even at pH 4.

In principle, on being immobilized, the enzyme can change its pH stability, as well as its thermostability. This has been observed experimentally with many enzymes (15,16,22,23). This is usually explained (e.g., see Ref. 16) by alteration of the microenvironment of the enzyme due to immobilization. Another possible explanation is alteration of the conformation of the enzyme on immobilization induced, for example, by its chemical modification or interaction with the support.

However, similarly to stabilization of the enzymes against heating, we must state that, although these considerations may be absolutely correct, they cannot be used as a foundation for elaborating the general principles of enhancement of the pH stability of the enzymes, because it is not known how the latter is associated with the microenvironment of the enzymes and their conformation. On the other hand, we can now single out several general physicochemical approaches with the help of which the pH stability of the enzyme will invariably be increased as a result of immobilization.

A. Rigidification (Fixation) of the Native Conformation of Enzyme Globule

As in thermoinactivation, pH inactivation of enzymes includes unfolding of the protein globule (2) caused by the alteration of the balance of electrostatic and hydrogen bonds. The latter changes result from pH-induced alteration of the ionization state of ionogenic groups on the protein. If this is the case, all the recommendations which were given when discussing stabilization of enzymes against thermoinactivation should be valid.

By way of example, let us consider the data on the pH dependence of the rate of thermoinactivation of glucose oxidase (84). Figure 13 shows that the rate of inactivation of the free enzyme 56°C increased sharply with increase in pH. As a result of mechanical entrapment of the enzyme in a concentrated polymeric gel, made of 2-hydroxyethylmethacrylate copolymerized with methacrylic acid, the thermostability increased considerably. The pH dependence of the rate of thermoinactivation of this immobilized glucose oxidase was studied at a higher temperature, i.e., 65°C. The results showed that the rate of thermoinactivation of this enzyme became insensitive to pH changes as a result of immobilization (cf. the slopes of the lines in Fig. 13). Kulys et al. (84) explained this by the fact that the polymeric support forms hydrogen and electrostatic bonds, of the type indicated in Fig. 10, with the enzyme globule. The bound enzyme maintained its catalytically active

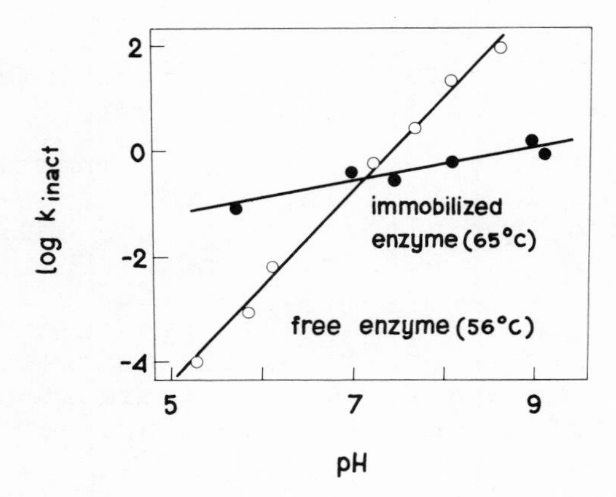

Fig. 13. pH dependence of the first-order rate constant (min^{-1}) characterizing the thermoinactivation of free glucose oxidase and glucose oxidase mechanically entrapped in polymeric gel (52% wt/wt 2-hydroxyethylmethacrylate copolymerized with 1% wt/wt of methacrylic acid). Conditions: 56°C and 65°C for the free and immobilized enzymes, respectively; 5x 10^{-3}M tris-HCl or phosphate (84).

conformation so strongly that the change in the ionic state of the globule (due to pH changes) was no longer important in the mechanism of denaturation.

B. Shift of the pH Profile of the Rate of Enzyme Inactivation

Since the classic studies of Katchalski et al. (85,86), the shift of the pH profile of the catalytic activity of enzymes on their being attached to polyelectrolytes has been an established fact (for review, see (87). On immobilization of enzymes on polyanions, the pH profile of the enzymatic activity shifts toward the alkaline region, because of the higher concentration of hydroxonium ions in the support phase as compared to that in the solution phase. On immobilization on polycations, the shift is toward the acidic region, because of the lower concentration of hydroxonium ions in the support phase as compared to the solution phase (e.g., see Fig. 14A). The shift of the pH profile of the catalytic activity of the enzymes on immobilization should also take place with a nonelectrolyte support because protons are distributed between the support and the medium in a certain specific way (87).

It is obvious that similar effects should take place during inactivation of the enzyme; i.e., the difference in the pH values of the solution and the support phases should induce a shift in the apparent pH profile of the inactivation rate. It was demonstrated by Katchalski et al. that, as was to be expected, enzymes attached to a polyanionic support have a higher stability in alkaline pH; and enzymes attached to a polycationic support become more stable in acidic solutions. This phenomenon illustrates the data of Streltsova et al. (88), in which the activity and stability of penicillin amidase immobilized on different supports were studied. One can see (Fig. 14B) that the pH profiles of the rate constants of enzyme inactivation shift depending on the nature

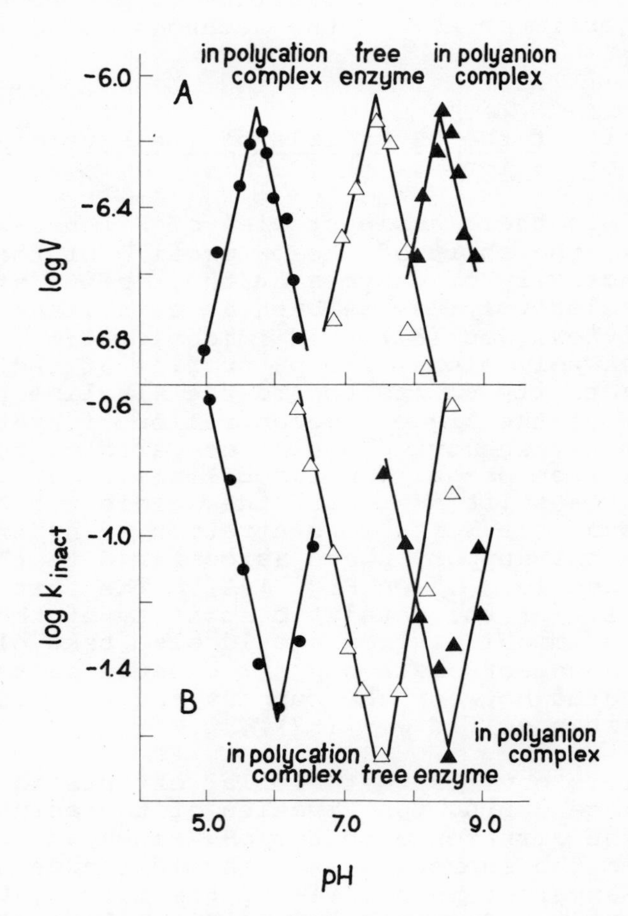

Fig. 14. Part A: pH dependence of the maximal rate of hydrolysis of benzylpenicillin V, catalyzed by penicillin amidase, in the native and immobilized states at 25°C.; Part B: pH dependence of the inactivation rate constant of penicillin amidase in the native and immobilized states at 50°C (88).

of the support, in complete agreement with the theory.

III. ENZYMATIC REACTIONS IN ORGANIC SOLVENTS WITH A LOW CONTENT (OR IN THE ABSENCE) OF WATER

Many enzymes have been adapted by nature for functioning primarily in aqueous solutions. This is a limitation for enzymes that are intended to be used in technological processes. Many chemical reactions are thermodynamically directed toward the desired product only in certain organic solvents. This is associated with specific solvation effects, with the solubility of the certain components of the reaction, and, last but not least, with the fact that sometimes, in addition to the required product, water is formed. Hence in aqueous solution the equilibrium is shifted toward the starting substances. This prevents biocatalysis from being introduced into some important processes, such as synthesis of esters or amides, polymerization of amino acids and sugars, and dehydration reactions.

To overcome this difficulty, water should be replaced by a nonaqueous solvent as a reaction medium. This idea has been developed in a great number of works and, there have been attempts at carrying out enzymatic reactions in organic solvents or aqueous-organic mixtures with a high concentration of the nonaqueous component (10,89-114). Many authors state that when water is replaced by an organic solvent as the reaction medium, the catalytic activity of enzymes drastically decreases and their substrate specificity disappears; however there are also reports that describe an increase in enzyme activity with the addition of small amounts of organic solvents to the medium.

A partial solution to the problem is to use an organic component which produces the lowest denaturing effect on the enzyme (114). Two other approaches seem, however, to be promising, i.e., the use of biphasic aqueous organic systems (mixtures of water with water-immiscible organic solvents) (11,115) and entrapment of enzymes in "reversed" surfactant micelles (116).

A. Enzymatic Reactions in Biphasic Systems Such as Water:Water Immiscible Organic Solvent

As immobilized enzyme research was developing, some works appeared describing how enzymes could be made to function in organic solvents, or in aqueous:organic mixtures with a high concentration of organic component. This functioning was enabled by the increase in stability due to attaching the enzymes to a support (98,101,102,106,107, 109,112,113,117-119). However even in the best systems, at a concentration of a nonaqueous component exceeding 90%, immobilized enzymes do inactivate. This indicates that, from the point of view of preparative enzymatic synthesis, the problem is not resolved. Quite possibly, it defies solution at all, because when water is replaced by a nonaqueous medium the conformation of enzymes will inevitably change significantly (10,120), with their catalytic function being thereby impaired.

We found (10,115) a basically different solution to the problem. In all the works known to us where the behavior of enzymes in water-organic media was studied, a solvent such as acetone, acetonitrile, dioxane, dimethylsulfoxide, dimethylformamide, methanol, or ethanol was used as a nonaqueous component. The common feature of all these solvents is their limitless ability to mix with water. This is why, in fact, the traditional problem arises of stabilization of enzymes against inactivation by an organic component. The gist of our approach (10,115) is that an organic solvent added to a water phase containing an enzyme should be *immiscible* with water, e.g., chloroform, ether, fatty aliphatic alcohols, or hydrocarbons. In other words, the system should contain two phases, aqueous and organic. The enzyme will persist only in the aqueous phase, because common proteins possessing a satisfactory solubility in water are almost insoluble in hydrophobic solvents (10). If necessary, the enzyme can be immobilized in an aqueous phase. The substrates, when dissolved in the organic phase, can freely (121,122) diffuse to the water phase, where they will undergo a chemical conversion; and the products will diffuse back to the organic phase.

Such a biphasic system possesses a number of basically important properties:

1. As the enzyme does not come into contact with the nonaqueous component of the reaction mixture, the problem of stabilization of the enzyme is eliminated (11,115).
2. The proportion of the organic phase volume can be made infinitely close to unity; hence the equilibrium in such a biphasic system can be infinitely close to the equilibrium in the purely organic medium (see the theoretical notes given in Refs. 11,115).
3. A high content of the nonaqueous component in the medium can also ensure solubility of hydrophobic reagents (e.g., steroids) that is sufficient for preparative synthesis (123,124). The enzymatic system water:water-immiscible organic solvent may be prepared as an emulsion of an aqueous solution of the enzyme in an organic medium. It can also be prepared by much more convenient methodology as a suspension in an organic medium of porous particles (e.g., porous glass or ceramics, hydrophilic gel) impregnated by an aqueous solution of the enzyme (11,115) (see Fig. 15). The aqueous solution of the enzyme can be entrapped in microcracks of a polymeric carrier by mechanical deformation of the carrier (125). The enzyme can be used in both a free and an immobilized state.

It may seem that preparative enzymatic synthesis in a biphasic system cannot be applied to the reactions where the reagents (and, most of all, reaction products) are ions, as they would not go from water to the organic phase. But this difficulty can be overcome (115) by selecting hydro-

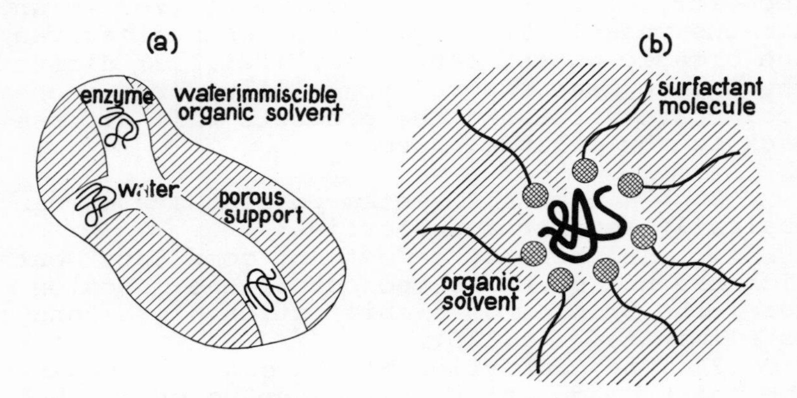

Fig. 15. Schematic diagram of enzyme molecules in water:water immercible biphasic systems.

phobic counterions for ionic reagents, as was the case in nucleotide chemistry (126) and as is the practice in phase-transfer catalysis (e.g., see Ref. 127).

B. Catalysis by Water-Soluble Enzymes Entrapped in "Reversed" Micelles of Surfactants in Organic Solvents

Molecules of many surfactants form in organic solvents associates of the "reversed micelle" type, where polar (ionic) groups of the surfactant molecule form the nucleus of the associate and the hydrocarbon fragments constitute the external layer (128-130). Such surfactant micelles are known to be able to solubilize ions, polar organic substances, and also several dozen water molecules per surfactant molecule. It has been demonstrated recently (116) that with the help of the micelle-forming surfactants, relatively high concentrations of biopolymers can be solubilized in organic solvents. Amounts to 1 mg/ml, which corresponds to 10^{-5}M of active centers with the molecular mass of the enzyme not exceeding 100,000, have been demonstrated. This by far exceeds the necessary level of the catalytic concentrations for the majority of enzymes (131,132). The enzyme almost completely retains its catalytic activity and substrate specificity. A molecule of the enzyme, when entrapped in a reversed micelle (see Fig. 15B), is protected against denaturation by the fact that the interface between the protein globule and the organic solvent phase is stabilized by the surfactant molecules. As a result, the biocatalyst cannot come into direct contact with the unfavorable organic medium. The enzyme is in a microreactor containing only several hundred water molecules per molecule of the enzyme, which corresponds to less than 1% vol/vol of the total amount of water in the organic solvent surfactant system. For example, the relative catalytic activity of chymotrypsin solubilized in octane can be maintained at room temperature for months (116).

All the above concepts have been verified (116) with the chymotrypsin-catalyzed hydrolysis of the nitroanilide of N-glutaryl-L-phenylalanine and with the peroxidase oxidation of ferrocyanide ions or pyrogallol. The reactions of these enzymes with their specific substrates were studied in hydrophobic (hydrocarbon) solvents, i.e., octane and benzene, with a surfactant commercially manufactured in thousands of tons, i.e., di(2-ethylhexyl) ester of sulfosuccinic acid sodium salt (Aerosol OT).

A typical experiment was the following: 2 ml of 0.1 M surfactant in hydrocarbon was supplemented with 0.01 ml or less of 10^{-3}M or less enzyme in an aqueous buffer. Then 0.01 ml or less of the substrate solution in water or acetonitrile was added; and the rate of the enzymatic reaction was measured spectrophotometrically in the resulting homogeneous (optically transparent) system. Figure 16 presents the data on peroxidase oxidation of pyrogallol. When water is replaced by a medium consisting of reversed micelles in octane, the

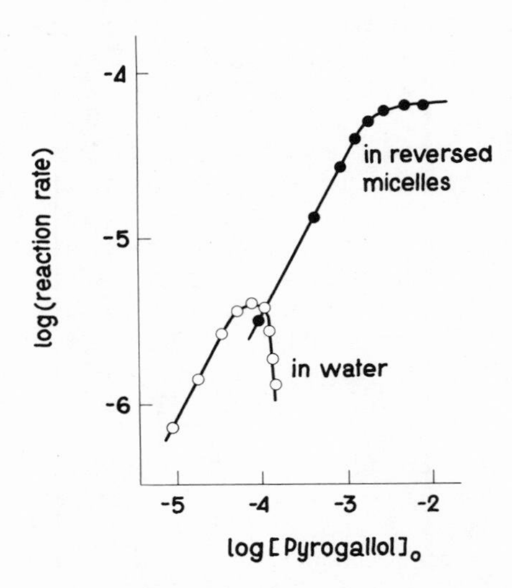

Fig. 16. Initial rate (moles/min) of peroxidase oxidation of pyrogallol (M) in water and in octane, with water content 2% wt/wt and 0.1 M Aerosol OT (micelle-forming component). Conditions: 26°C, pH 7, 2.5×10^{-3}M H_2O_2, 3×10^{-9}M peroxidase (116).

kinetic mechanism of catalysis (as shown in Fig. 16) changes, because substrate inhibition, characteristic of reactions in water, is absent.

IV. REACTIVATION OF IRREVERSIBLY DENATURED ENZYMES

Along with the stabilization of enzymes against denaturation, attention should be paid to the problem of reactivation of already denatured proteins. The unfolding of the protein globule leads, as a rule, first to reversible inactivation, and later, to irreversible inactivation of the enzyme. In other words, after the source of the denaturing action is removed, the protein sometimes remains catalytically inactive; and the reasons for such irreversibility can be most diverse.

A. Mechanisms of Irreversible Denaturation of Enzymes

As a rule, denaturation of enzymes induced by elevated temperature, extreme pH values, or organic solvents can only be irreversible as a result of secondary processes, e.g., aggregation, chemical modification of functional groups, autolysis or degradation induced by admixtures of other enzymes, or breakage of the native disulphide bonds (2,4, 133).

As early as forty years ago, Anson found when studying hemoglobin and other proteins that on thermoinactivation there occurs aggregation of macromolecules that can even result in the formation of a precipitate. Anson was also the first to reactivate irreversibly denatured proteins by dissolving a precipitate in a concentrated solution of urea or by varying the pH (see ref. for (133)). A similar reactivation procedure was successfully used with denatured β-lactamase, which formed a precipitate in a solution at 60°C; as a result of incubation of the precipitate in 5 M guanidine chloride or 8 M urea followed by the removal of denaturants, the enzyme was renatured (134).

Sometimes the SH-groups of proteins are involved in irreversible inactivation. For instance phosphorylase, a subunit enzyme with a large number of SH-groups, can be inactivated by blocking the SH-groups by p-mercuribenzoate. This causes the dissociation of the protein into monomers. Madsen and Cori (135) could reactivate the enzyme by adding cysteine to mercaptidated phosphorylase; as a result, the initial tetrameric structure of the enzyme was regained. Another example is renaturation of urease by an excess of iodide ions (136). Inactivation of this enzyme is due to the poisoning of its SH-groups by heavy metal cations. The mechanism of reactivation is that of binding metal ions into a complex with iodide.

However, there can be different reasons for the irreversibility of the denaturation process. It is known, for example, that irreversible denaturation is often accompanied by considerable conformational changes. This was demonstrated with the help of physico-chemical methods, e.g. for dihydropholatreductase inactivated at 45°C (137) and ATPase inactivated at 46°C (138). Proceeding from this, we have suggested (139) the following mechanism: at elevated temperature correct noncovalent interactions that maintain the structure at room temperature are violated and thermodynamically governed non-native bonds (Fig. 17) are formed. When the temperature is lowered, these incorrect noncovalent interactions, although thermodynamically unstable, can persist for purely kinetic reasons, as the molecular mobility of the polypeptide chains must decrease with the lowering of the temperature. In other words, on a sharp temperature decrease in spite of the native conformation being thermodynamically favorable, the protein keeps its kinetically determined metastable conformation.

However, there is a way to reactivate such an irreversibly denatured enzyme (140). One can attempt to destroy all non-native interactions in the irreversibly denatured state and to unfold the polypeptide chains e.g. in a concentrated solution of urea; simultaneously, the S-S bonds in the protein should be split. If the irreversibly denatured protein could be transformed into a random coil, the results of Anfinsen (141) should apply for reactivation. He showed that a polypeptide chain, existing in the stage of a random coil, is capable under optimal conditions of folding into the native conformation with high yield. In other words, if our arguments about the so-called irreversible denaturation are correct, the

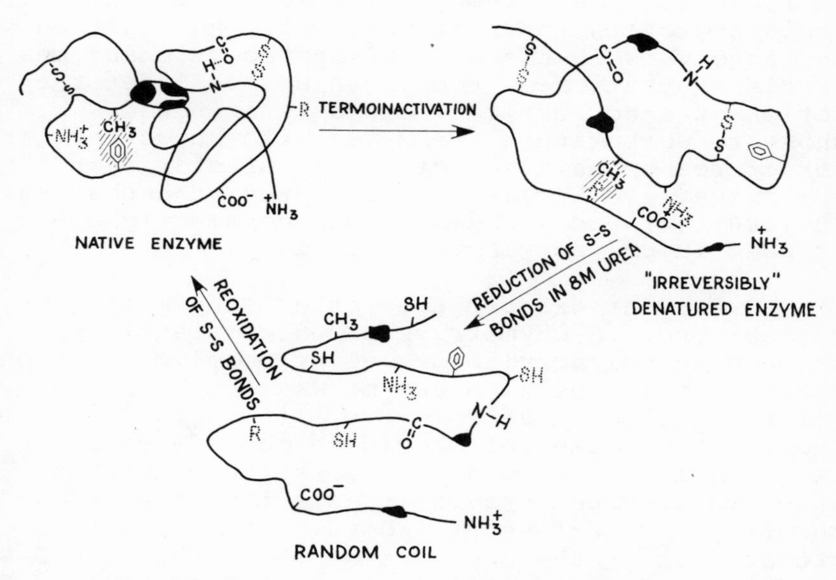

Fig. 17. Schemes for conformational changes and denaturation due to sulfur oxidation-reduction.

irreversibly denatured enzyme should reactivate if it is first made to acquire a random coil conformation, by splitting the S-S bonds in a concentrated solution of urea, and to refold into the native conformation, by removing the denaturating agent and reoxidizing the S-S bonds. An experiment of this nature was indeed realized (139,140, 142,143).

B. Reactivation of an Enzyme Irreversibly Inactivated as a Result of Monomolecular Thermal Denaturation.

The experiment was performed with immobilized enzymes covalently attached to a support via several linkages. As a result of immobilization, some protein-protein interactions, e.g., aggregation and autolysis in the case of proteolytic enzymes or disintegration of the given enzyme under the action of other enzymes, can be eliminated. We chose to work with the enzymes with known tertiary structures containing neither a cofactor, nor SH-groups, i.e. trypsin immobilized on Sepharose 4B (140,142) and α-chymotrypsin covalently entrapped in polyacrylamide gel (143).

By way of example the latter scheme is discussed here. α-chymotrypsin was covalently entrapped in polyacrylamide gel by copolymerization (56). The immobilized enzyme was incubated in a gel suspension at 95°C (curve 1 in Fig. 18). The inactivation observed should be considered irreversible, in terms of the accepted views, as after cooling the enzymatic activity remained at the low level for several days. Then with this preparation in the presence of 10 M urea, the S-S bonds were reduced at 25°, pH 8, 0.01 M EDTA, 0.01 M dithiotreithol, and with nitrogen blanketing the reaction mixture (144), with the catalytic activity being totally lost (curve 2 in Fig. 18). Finally, when the denaturing agent was removed, the S-S bonds were reoxidized (curve 3 in Fig. 18) at an optimal ratio of the catalysts for thiol-disulphide exchange; this consists of the oxidized and reduced forms of glutathione at 4×10^{-4}M and 4 10^{-3}M, respectively, at 25° and pH 8 essentially as described in (144). It is obvious from Fig. 18 that the method allows one to almost totally regenerate the catalytic activity of the irreversibly inactivated enzyme. Moreover, as seen in the same figure, the irreversible monomolecular thermoinactivation-reactivation cycle can be repeated many times.

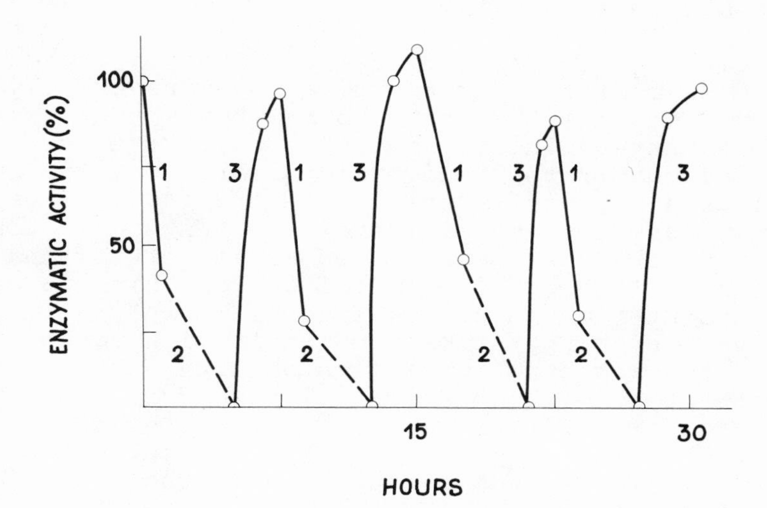

Fig. 18. Multiple regeneration of the catalytic activity of α-chymotrypsin, covalently entrapped in polyacrylamide gel. Curve 1 is after irreversible thermoinactivation at 95°, pH 8, 0.005 M tris - HCl. Curve 2 is after reduction of S-S bonds in the presence of 8 M urea and 10^{-2}M dithiotreithol, pH 8.5, 0,01 M tris - HCl. Curve 3 is after reoxidation of thioldisulphide exchange (reduced and oxidized glutathiones 4×10^{-3}M and 4×10^{-4}, respectively) pH 8,0, 0.05 M tris - HCl.

C. Perspectives

We believe that our approach can be extended to subunit enzymes. To this end, we must learn how to make proteins fold from a random coil into a native conformation. This has been accomplished for some proteins (141), e.g., for subunit alkaline phosphatase (145). Both immobilization of enzymes and the use of effectors can prove feasible here (146,147).

It is not excluded that the above approach could be modified for reactivating irreversibly denatured enzymes that have no S-S bonds. An example is intracellular proteases, which are very stable; but if they undergo denaturation, the result as a rule is irreversible (148-149). These enzymes usually contain several metal ions, such as Ca^{2+}, that can be regarded as structural analogs of S-S bonds. The metal cations bind in a multipoint fashion with the bends of polypeptide chains, endowing the macromolecule with added rigidity, and hence greater stability (148). On denaturation of such enzymes, instead of the cleavage of S-S linkages, the metal ions can be removed by dialysis or chelating agents. When the protein refolds from a random coil, the metal ions are added back.

V. CONCLUSION

In the present communication, only sufficiently general physico-chemical approaches that also have a certain technological value and allow the enzyme to be protected from the denaturing factors of the medium (elevated temperature, extreme values of pH, additions of organic solvents) have been considered. All of the various approaches discussed herein fall into two groups that correspond to the following two general principles:

1. Rigidification (fixation) of the protein globule prevents it from unfolding during denaturation and hence protects it from inactivation. This can be achieved either by attachment of brackets to the protein to form a water-soluble preparation or by multipoint covalent and physical attachment of the protein with a complementary surface on the support to give an insoluble heterogeneous catalyst.

2. Spatial isolation of the enzyme from the environmental factors unfavorable for maintenance of catalytic activity and specificity also can be achieved. Realization of this principle depends on the nature of the inactivating action. For example, to prevent pH inactivation, one can immobilize the enzyme on a support capable of providing a local pH shift. The effect of the organic medium can be coped with in two ways: either the enzymatic reaction can be carried out in a biphasic water:water immiscible organic solvent system or the enzyme can be entrapped in reversed micelles of a surfactant.

Our method for reactivation of irreversibly denatured enzymes has been described for an immobilized preparation. Immobilization plays a significant role here. During the refolding of the protein from a random coil the number of incorrect or non-native conformations is diminished; and most important, the formation of intermolecular S-S bonds is eliminated. For a more detailed discussion of this issue see (146,147). If immobilization is involved, there arises the question of the support. It should be taken into consideration that the intermediate state of the reactivation, i.e. an immobilized random coil (146), can effectively interact with the surface of a support. In other words, sorption of polypeptide chains on a support can appear more pronounced in an unfolded, compared to a folded, protein. In principle, sorption of polypeptide chains on a support can hinder the subsequent refolding of the protein into the native conformation (146).

Nevertheless, systems have already been devised that give 100% reactivation. Moreover, the system described in this review (Fig. 18) has another very important advantage. As a result of multipoint binding of the enzyme to a complimentary support, the enzyme thermostability has increased dramatically. For example the thermoinactivation of α-chymotrypsin immobilized in polyacrylamide gel (56) proceeds at an appreciable rate only above 80°C, whereas the native enzyme undergoes denaturation at 45-50°C. Thus, we have prepared an immobilized enzyme having the two technologically most essential properties: higher thermostability and full reactivation after irreversible thermal denaturation.

REFERENCES

1. JOLY, M. "Physicochemical Approach to Denaturation of Proteins," Academic Press, New York, 1965.
2. TANFORD, C. *Adv. Protein. Chem. 23:* 121, 1968.
3. ALEKSANDROV, V. Ya. "Cells, Macromolecules, Temperature", Nauka Press, Leningrad, and Springer-Verlag, Berlin, 1975.
4. KUSHNER, V. P. "Conformation Flexibility and Denaturation of Biopolymers" (Russ.), Nauka Press, Leningrad, 1977.
5. CHIBATA, I., TOSA, T. & SATO, T. *Meth. Enzymol. 44:* 739, 1976.
6. WEETALL, H. H., VANN, W. P., PITCHER, W. H., LEE, D. D., LEE, Y. Y., & TSAO, G. T. *Meth. Enzymol. 44:* 776, 1976.
7. LILLY, M. D. & DUNNILL, P. *Meth. Enzymol. 44:* 717, 1976.
8. PASTORE, M., MORISI, F. & TACCARDELLI, D. In "Insolubilized Enzymes" (M. Salmona, C. Saronio and S. Garattini, eds.) Raven Press, New York, 1974, p. 211.
9. WOODWARD, J. & WISEMAN, A. *J. Appl. Chem. Biotechnol 26:* 580, 1976.
10. SINGER, S. J. *Adv. Protein. Chem. 17:* 1, 1962.
11. KLIBANOV, A. M., SAMOKHIN, G. P., MARTINEK, K. & BEREZIN, I. V. *Biotechnol. Bioeng. 19:* 1351, 1977.
12. BEREZIN, I. V., KLYOSOV, A. A., MARGOLIN, A. L., NYS, P. S., SAVITSKAYA, E. M. & SVEDAS, V.-J. K. *Antibiotiki* (Russ.) *21:* 519, 1976.
13. WEETALL, H. H. *Biochim. Biophys. Acta 212:* 1, 1970.

14. COONEY, D. L., DEMAIN, A. L., DUNNILL, P., HUMPHREY, A. E., LILLY, M. D. & WANG, D. I. C. "Fermentation and Enzyme Technology," John Wiley, New York, 1977.
15. BEREZIN, I. V., ANTONOV, V. K. & MARTINEK, K., eds. "Immobilized Enzymes," Moscow University Press, Moscow, 1976.
16. GOLDMAN, R., GOLDSTEIN, L. & KATCHALSKI, E. In: "Biochemical Aspects of Reactions on Solid Supports" (G. R. Stark, ed.) Academic Press, New York, p. 1, 1971.
17. ZABORSKY, O. R. "Immobilized Enzymes", CRC Press, Cleveland, 1973.
18. WEETALL, H. H. & SUZUKI, S., eds. "Immobilized Enzyme Technology", Plenum Press, New York, 1975.
19. MOSBACH, K. *Meth. Enzymol. 44:* 1976.
20. BELL, G., BLAIN, J. A., PATTERSON, J. D. E. & TODD, R. *J. Appl. Chem. Biotechnol. 26:* 583, 1976.
21. Anon., "Enzymes and Proteins from Thermophilic Micro-organisms", Experientia, Suppl. 26, 1976.
22. VIETH, W. R. & VENKATASUBRAMANIAN, K. *Chem. Technol. 4:* 309, 1974.
23. MELROSE, G. J. H. *Rev. Pure Appl. Chem. 21:* 83, 1971.
24. GABEL, D., STEINBERG, I. Z. & KATCHALSKI, E. *Biochemistry 10:* 4661, 1971.
25. GIACOMETTI, G. M., COLOSIMO, A., STEFANINI, S., BRUNORI, M. & ANTONINI, E. *Biochim. Biophys. Acta 285:* 320, 1972.
26. BERLINER, L. J., MILLER, S. T., UY, R. & ROYER, G. P. *Biochim. Biophys. Acta 315:* 195, 1973.
27. MOORE, T. A. & GREENWOOD, C. *Biochem. J. 149:* 169, 1975.
28. LASCH, J. *Acta Biol. Med. Germ. 34:* 549, 1975.
29. BAREL, A. O. & PRIEELS, J.-P. *Eur. J. Biochem. 50:* 463, 1975.
30. BLUMENFELD, L. A. "Problems of Biophysics," Nauka Press, Moscow, 1974, chap. 5.
31. MARTINEK, K., TORCHILIN, V. P., MAKSIMENKO, A. V., SMIRNOV, V. N. & BEREZIN, I. V.

Dokl. Akad. Nauk SSSR (in press)

32. FEENEY, R. E., BLANKENHORN, G. & DIXON, H. B. F. *Adv. Protein Chem.* *29*: 135, 1975.
33. ZABORSKY, O. R. In "Enzyme Engineering", Vol. 2 (E. K. Pye and L. B. Wingard, Jr. eds.), Plenum Press, New York.
34. WOLD, F. *Meth. Enzymol.* *25*: 623, 1972.
35. KENNEDY, J. H., KRICKA, L. J. & WILDING, P. *Clin. Chim. Acta.* *70*: 1, 1976.
36. ZABORSKY, O. R. In "Enzyme Engineering", Vol. 1, (L. B. Wingard, Jr. ed.) John Wiley, New York, 1972, p. 211.
37. SAIDEL, L. J., LEITZES, S. & ELFRING, W. H. *Biochem. Biophys. Res. Commun.* *15*: 409, 1964.
38. HERZIG, D. J., REES, A. W. & DAY, R. A. *Biopolymers* *2*: 349, 1964.
39. WANG, J. H. & TU, J. *Biochemistry* *8*: 4403, 1969.
40. JOSEPHS, R., EIZENBERG, H. & REISLER, E. *Biochemistry* *12*: 4060, 1973.
41. BEAVEN, G. H. & GRATZER, W. B. *Int. J. Peptide Protein Res.* 5: 215, 1973.
42. SNYDER, P. D., JR., WOLD, J. F., BERNLOHR, R. W., DULLUM, C., DESNICH, R. J., KRIVIT, W. & CONDIE, R. M. *Biochim. Biophys. Acta* *350*: 432, 1974.
43. MAKSIMENKO, A. V., TORCHILIN, V. P., KLIBANOV, A. M. & MARTINEK, K. *Vestnik MGU* (*Bull. Moscow Univ.*) *19*: No. 5, 1978.
44. REINER, R., SIEBENEICK, H.-U., CHRISTENSEN, I. & DORING, H. *J. Mol. Catal.* *2*: 119, 1977.
45. WOODWARD, J. & WISEMAN, A. *J. Appl. Chem. Biotechnol.* *26*: 580, 1976.
46. TORCHILIN, V. P., MAKSIMENKO, A. V., SMIRNOV, V. N., BEREZIN, I. V., KLIBANOV, A. M. & MARTINEK, K. *Biochim. Biophys. Acta* *522*: 277, 1977.
47. KHORANA, H. G. *Chem. Rev.* *53*: 145, 1953.
48. REINER, R., SIEBENEICK, H.-V., CHRISTENSEN, I. & LUKAS, H. *J. Mol. Catal.* *1*: 3, 1975.
49. GOLDSTEIN, L. (1972) *Biochemistry* *11*: 4072, 1972.
50. CHANG, T. M. S., ed. "Biomedical Applications of Immobilized Enzymes and Proteins",

Plenum Press, New York, 1977.
51. GLASSMEYER, C. K. & OGLE, J. *Biochemistry* *10*: 786, 1971.
52. ROYER, G. P. & UY, R. *J. Biol. Chem.* *248*: 2627, 1973.
53. KLIBANOV, A. M., SAMOKHIN, G. P., MARTINEK, K. & BEREZIN, I. V. *Biochim. Biophys. Acta* *438*: 1, 1976.
54. GABEL, D. *Eur. J. Biochem.* *33*: 348, 1973.
55. DELAAGE, M. & LAZDUNSKI, M. *Eur. J: Biochem.* *4*: 378, 1968.
56. MARTINEK, K., KLIBANOV, A. M., GOLDMACHER, V. S. & BEREZIN, I. V. *Biochim. Biophys. Acta* *485*: 1, 1977.
57. GABEL, D. & KASCHE, V. *Biochem. Biophys. Res. Commun.* *48*: 1011, 1972.
58. MOSBACH, K. & GESTRELIUS, S. *FEBS Lett* *42*: 200, 1974.
59. MARTINEK, K., KLIBANOV, A. M. & BEREZIN, I. V. *USSR-USA Seminar on Immobilized Enzymes*, Moscow, 1975.
60. MARTINEK, K., GOLDMACHER, V. S., KLIBANOV, A. M., TORCHILIN, V. P., SMIRNOV, V. N., CHAZOV, E. I. & BEREZIN, I. V. *Dokl. Akad. Nauk SSSR* *228*: 1468, 1976.
61. JAWOREK, D., BOTSCH, H. & MAIER, J. *Meth. Enzymol.* *44*: 195, 1976.
62. GOLDMACHER, V. S., KLIBANOV, A. M. & MARTINEK, K. *Vestnik MGU (Bull. Moscow Univ.)* *19*: No. 4, 1978.
63. MARTINEK, K., KLIBANOV, A. M., TCHERNYSHEVA, A. V., MOZHAEV, V. V., BEREZIN, I. V. & GLOTOV, B. O. *Biochim. Biophys. Acta* *485*: 13, 1977.
64. KULYS, J. J. & KURTINAITIENE, B. S. *Biokhimiya* (Russ). *43*: 453, 1978.
65. KLYOSOV, A. A. & SVEDAS, V. YU.-K. In: "Itogi Nauki i Techniki", (V. L. Kretovich, ed.) Viniti Press, Moscow, 1978, p. 209.
66. O'DRISCOLL, K. F. *Adv. Biochem. Eng.* *4*: 155, 1976.
67. CHANG, T. M. S. "Artificial Cells", Thomas, Springfield, Illinois, 1972.
68. MESSING, R. A. *Meth. Enzymol.* *44*: 148, 1976.

69. POLTORAK, O. M. & CHIUKHRAI, E. S. "Physico-Chemical Fundamentals of Enzymatic Catalysis", High School Press, Moscow, 1971, Chap. 14.
70. CHAPELLE, E. W., RICH, E., JR. & MACLEOD, N. H. *Science 155*: 1287, 1967.
71. HORIGOME, T., KASAI, H. & OKUYAMA, T. *J. Biochem. 75*: 299, 1974.
72. SCHNEIDER, Z., STROINSKI, A. & PAWELKIEWICZ, J. *Bull. Acad. Polon. Sci. Ser. Sci. Biol. 16*: 203, 1968.
73. NORRIS, R. D. & PAWELKIEWICZ, J. *Phytochemistry 14*: 1701, 1975.
74. GOLDMACHER, V. S. Ph.D. thesis, Lomonosov State University, Moscow, 1977.
75. GOLDMACHER, V. S., KLIBANOV, A. M., MISHIN, A. A., MOZHAEV, V. V. & MARTINEK, K. *Vestnik MGU (Bull. Moscow Univ.) 19*: No. 3, 1978.
76. MARTINEK, K., GOLDMACHER, V. S., MISHIN, A. A., TORCHILIN, V. P., SMIRNOV, V. N. & BEREZIN, I. V. *Dokl. Akad. Nauk SSSR 239*: 227, 1978.
77. ALLEN, R. C. & MAURER, H. R., eds. "Electrophoresis and Isoelectric Focusing in Polyacrylamide Gel", DeGruyter, New York, 1974.
78. WHITE, M. L. *J. Phys. Chem. 64*: 1563, 1960.
79. GRESSEL, J. & ROBARDS, A. W. *J. Chromatogr. 144*: 455, 1975.
80. BEREZIN, I. V., KLIBANOV, A. M. & MARTINEK, K. *Biochim. Biophys. Acta 364*: 193, 1974 & *Meth. Enzymol. 44*: 571, 1976.
81. MARTINEK, K., GOLDMACHER, V. S., KLIBANOV, A. M. & BEREZIN, I. V. *FEBS Lett. 51*: 152, 1975.
82. OBRINK, B. & LAURENT, T. V. *Eur. J. Biochem. 41*: 83, 1974.
83. GRYSZIEWICZ, J. *Folia Biol. 19*: 119, 1971.
84. KULYS, J. J., KURTINAITIENE, B. S. & AKULOVA, V. F. *J. Solid-Phase Biochem.* (in press)
85. LEVIN, Y., PECHT, M., GOLDSTEIN, L. & KATCHALSKI, E. *Biochemistry 3*: 1905, 1964.
86. GOLDSTEIN, L., LEVIN, Y. & KATCHALSKI, E. *Biochemistry 3*: 1913, 1964.

87. BEREZIN, I. V., KLIBANOV, A. M. & MARTINEK, K. *Uspekhi Khim. (Russ. Chem. Rev.) 44:* 17, 1975.
88. STRELTSOVA, Z. A., SVEDAS, V. K., MAKSIMENKO, A. V., KLYOSOV, A. A., BRAUDO, E. E., TOBSTOGUZOV, V. B. & BEREZIN, I. V. *Bioorg. Khim.* (Russ.) *1:* 1464, 1975.
89. BARNARD, M. L. & LAIDLER, K. J. *Arch. Biochem. Biophys. 44:* 338, 1953.
90. KHURQIN, YR. I., ROSLYAKOV, V. YA., AZIZOV, YU. M. & KAVERZNEVA, E. D. *Izv. Akad. Nauk SSSR Ser. Khim. 12:* 2840, 1968.
91. BETTELHEIM, F. A. & LUKTON, A. *Nature 198:* 357, 1963.
92. BETTELHEIM, F. A. & SENATORE, P. *J. Chim. Phys. (Phys. Chim. Biol.) 61:* 105, 1964.
93. RAMMLER, D. H. In: "Dimethyl Sulfoxide", Vol. 1 (S. W. Jacob, E. E. Rosenbaum, and D. C. Wood, eds.) Marcel Dekker, New York, 1971, p. 189.
94. AZIZOV, YU. M., ZVERINSKAYA, I. B., NIKITINA, A. N., ROSLYAKOV, V. YA. & KHURGIN, YU. I. *Izv. Akad. Nauk SSSR Ser. Khim. 12:* 2843, 1968.
95. KLYOSOV, A. A., VIET, N. V. & BEREZIN, I. V. *Eur. J. Biochem. 59:* 3, 1975.
96. HUTTON, J. P. & WETMUR, J. G. *Biochem. Biophys, Res. Commun. 66:* 942, 1975.
97. LACHMAN, L. B. & HANDSCHUMACHER, R. E. *Biochem. Biophys. Res. Commun. 73:* 1094, 1974.
98. WEETALL, H. H. & VANN, W. P. *Biotechnol. Bioeng. 18:* 105, 1976.
99. CASE, S. I. & BAKER, R. F. *Anal. Biochem. 64:* 477, 1975.
100. ELODI, P. *Acta Physiol. Acad. Sci. Hung. 20:* 311, 1961.
101. BARTH, T., JOST, K. & RYCHLIK, I. *Coll. Czechoslov. Chem. Commun. 38:* 2011, 1973.
102. HORVATH, C. *Biochim. Biophys. Acta 358:* 164, 1974.
103. INAGAMI, T. & STURTEVANT, J. M. *Biochim. Biophys. Acta 38:* 64, 1960.
104. MOSOLOV, V. V., AFANASEV, P. V., DOLGICH, M. S. & LUSHNIKOVA, E. V. *Biokhimiya* (Russ.) *33:* 1030, 1968.

105. TAN, K. H. & LOVRIEN, R. *J. Biol. Chem. 247:* 3278, 1972.
106. TANIZAWA, K. & BENDER, M. L. *J. Biol. Chem. 249:* 2130, 1974.
107. WAN, H. & HORVATH, C. C. *Biochim. Biophys. Acta 410:* 135, 1975.
108. MYERS, J. S., & JAKOBY, W. B. *Biochem. Biophys. Res. Commun. 51:* 631, 1973.
109. HUSSAIN, Q. Z. & NEWCOMB, T. F. *Proc. Soc. Exp. Biol. Med. 115:* 301, 1963.
110. CASTANEDA-AGULLO, M. & DEL CASTILLO, L. M. *J. Gen. Physiol. 42:* 617, 1959.
111. BIELSKI, B. H. J. & FREED, S. *Biochim. Biophys. Acta 89:* 314, 1964.
112. TOSA, T., MORI, T. & CHIBATA, I. *Enzymologia 40:* 49, 1971.
113. INGALIS, R. G., SQUIRES, R. G. & BUTLER, L. G. *Biotechnol. Bioeng. 17:* 1627, 1975.
114. KLIBANOV, A. M., SEMENOV, A. N., SAMOKHIN, G. P. & MARTINEK, K. *Bioorg. Khim.* (Russ.) *3:* No. 11, 1977.
115. MARTINEK, K., KLIBANOV, A. M., SAMOKHIN, G. P., SEMENOV, A. N. & BEREZIN, I. V. *Bioorg. Khim.* (Russ.) *3:* 696, 1977.
116. MARTINEK, K., LEVASHOV, A. V. & BEREZIN, I. V. *Dokl. Akad. Nauk SSSR* (Russ.) *236:* 920, 1977.
117. BUTLER, L. & SQUIRES, R. *Enzyme Technol. Digest 4* (3): 108, 1975.
118. SCHWABE, C. *Biochemistry 8:* 795, 1969.
119. KELLY, S. J., BUTLER, L. G. & SQUIRES, R. G. *Enzyme Technol. Digest 5 (2):* 107, 1976.
120. BRANDTS, J. F. In "Structure and Stability of Biological Macromolecules" (S. N. Timasheff and G. D. Fasman, eds.) Marcel Dekker, New York, 1969, p. 213.
121. DAVIES, J. T. & RIDEAL, E. K. "Interfacial Phenomena", Academic Press, New York, 1961.
122. SCHOLTENS, J. R. & BIJSTERBOSCH, B. H. *FEBS Lett 62:* 233, 1976.
123. CREMONESI, P., CARREA, G., FERRARA, L. & ANTONINI, E. *Biotechnol. Bioeng. 17:* 1101, 1975.
124. BUCKLAND, B. C., DUNNILL, P. & LILLY, M. D. *Biotechnol. Bioeng. 17:* 815, 1975.

125. KLIBANOV, A. M., SAMOKHIN, G. P., MARTINEK, K. & BEREZIN, I. V. *Biotechnol. Bioeng. 19:* 211, 1977.
126. RAJBHANDARY, V. Z., YOUNG, R. J. & KHORANA, H. G. *J. Biol. Chem. 239:* 3875, 1964.
127. DEHMLOW, E. V. *Angew. Chem. Int. Ed. 13:* 170, 1974.
128. SHINODA, K., NAKAGAWA, T., TAMAMUSHI, B. I. & ISEMURA, T. "Colloidal Surfactants", Academic Press, New York, 1963.
129. FENDLER, J. H. & FENDLER, E. J. "Catalysis in Micellar and Macromolecular Systems", Academic Press, New York, 1975.
130. MITTAL, K., ed. "Micellization, Solubilization and Microemulsions", Plenum Press, New York, 1977.
131. LAIDLER, K. J. & BUNTING, P. S. "The Chemical Kinetics of Enzyme Action", Clarendon Press, Oxford, 1973.
132. BEREZIN, I. V. & MARTINEK, K. "The Physicochemical Fundamentals of Enzyme Catalysis", High-School Press, Moscow, 1977.
133. ANSON, M. L. *Adv. Protein Chem. 2:* 361, 1945.
134. DAVIES, R. B., ABRAHAM, E. P. & DALGLEISH, D. G. *Biochem. J. 143:* 137, 1974.
135. MADSEN, N. B. & CORI, C. F. *J. Biol. Chem. 223:* 1055, 1956.
136. MATTIASSON, B., DANIELSON, B., HERMANSSON, B. & MOSBACH, K. *FEBS Lett. 85:* 203, 1978.
137. KITCHELL, B. B. & HENKENS, R. W. *Biochim. Biophys. Acta 534:* 89, 1978.
138. AYALA, J. A. & NIETO, M. *Biochem. J. 169:* 371, 1978.
139. MARTINEK, K., MOZHAEV, V. V. & BEREZIN, I. V. *Dokl. Acad. Nauk SSSR 239:* 483, 1978.
140. MOZHAEV, V. V. & MARTINEK, K. "2nd All-Union Symposium on Immobilized Enzymes", October 3-7, Abovyan, Armenian SSR, 1977.
141. ANFINSEN, C. B. & SCHERAGA, H. A. *Adv. Protein. Chem. 29:* 205, 1975.
142. MARTINEK, K., MOZHAEV, V. V., SMIRNOV, M. D. & BEREZIN, I. V. *Biotechnol. Bioeng.* (in press).

143. MOZHAEV, V. V., SMIRNOV, M. D., MARTINEK, K. & BEREZIN, I. V. *Bioorg. Khim.* (Russ.) (in press).
144. SINHA, N. K. & LIGHT, A. *J. Biol. Chem. 250:* 8624, 1975.
145. LEVINTHAL, C., SINGER, E. R. & FETHEROLF, K. (*Proc. Natl. Acad. Sci. USA 48:* 1230, 1962.
146. MOZHAEV, V. V., MARTINEK, K. & BEREZIN, I. V. *Molekul. Biol.* (Russ.) (in press).
147. MOZHAEV, V. V., MARTINEK, K. & BEREZIN, I. V. *Molekul. Biol.* (Russ.) (in press).
148. MATTHEWS, B. W., WEAVER, L. H. & KESTER, W. R. *J. Biol. Chem. 249:* 8030, 1974.
149. IKAI, A. *Biochim. Biophys. Acta 445:* 182, 1976.
150. TORCHILIN, V. P., MAKSIMENKO, A. V., SMIRNOV, V. N., BEREZIN, I. V., KLIBANOV, A. M. & MARTINEK, K. *Biochim. Biophys. Acta 567:*1,1979.

THEORY OF RELIABILITY AND KINETICS OF INACTIVATION OF BIOCATALYTICAL SYSTEMS

S. D. Varfolomeev

Department of Chemical Enzymology
Moscow State University
Moscow, USSR

The phenomenon of aging, which is characteristic of biocatalysts and biostructures, is caused by the physical chemical processes of the denaturation of proteins and biomembranes. The phenomena accompanying denaturation have been described by Joly (1); and the physical chemical aspects of the problem also have been discussed elsewhere (2).

We have been interested in the kinetics of inactivation of enzyme and multienzyme systems. In this paper we treat these kinetics in terms of the theory of reliability of physical systems. The principal reliability characteristics being the reliability function, the failure intensity, and the mean time to failure. The approach is applied 1) to a system having an exponential reliability function which undergoes all or nothing inactivation, 2) to consecutive and parallel reactions in multienzyme systems, 3) to a system involving a change in the rate limiting step during inactivation, and 4) to the inactivation of organelles.

Usually, the process of inactivation is described in terms of a monomolecular reaction proceeding at a certain frequency:

$$E \xrightarrow{k} E_i \qquad \text{(Eq. 1)}$$

where E and E_i are catalytically active and

inactive forms of an enzyme, respectively, and k is the inactivation rate constant. The rate of inactivation is proportional to the active enzyme concentration;

$$-\frac{dE}{dt} = k\,E \qquad \text{(Eq. 2)}$$

and the enzyme concentration as a function of time can be given by

$$E = E_o e^{-kt} \qquad \text{(Eq. 3)}$$

where E_o is the initial enzyme concentration.

Inactivation processes described by Eqs. 1 to 3 are found frequently with enzyme catalysts; and a great deal of experimental data fit the exponential description. The kinetics of inactivation of glucose oxidase immobilized on various carriers provides an example (3).

However, Eqs. 2 and 3 rest on the assumptions that the inactivation frequency, characterized by the rate constant k, remains constant during the inactivation and that the inactivation of a given enzyme molecule does not affect the behavior of neighbor enzyme molecules. One may doubt whether these assumptions are always justified. For example, the inactivation frequency may depend on the history of the system and thus be time dependent. The kinetics of inactivation of formate dehydrogenase from *Arthrobacter* SP1, shown in Fig. 1, follows a complicated pattern which cannot be described by Eqs. 2 and 3.

The inactivation of multienzyme systems incorporated into complexes or entrapped in biological membranes poses other specific problems. The physical or chemical transformation of one of the enzymes involved may to a certain extent affect the stability of other species; and biomembrane

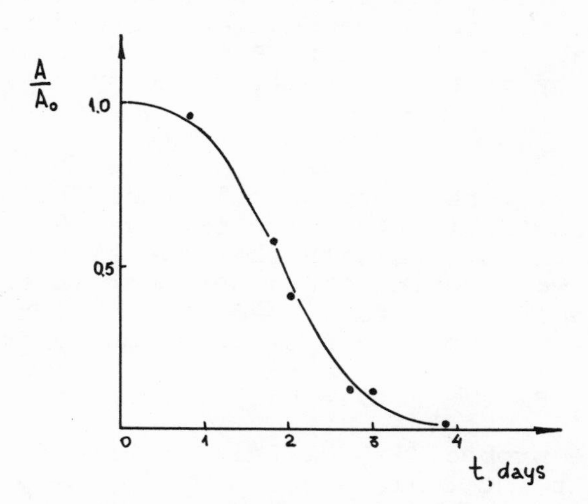

Fig. 1. Kinetics of inactivation of formate dehydrogenase from *Arthrobacter* SP 1 (Rodionov Yu. V., unpublished).

systems may undergo gradual modifications that cause alterations of the kinetic characteristics of the inactivation processes. It is clear from these considerations that a more general model for the description of enzyme inactivation is needed to account for a wider range of complications. The application of the reliability theory for describing general patterns of aging of physical systems seems very promising. Its utility for the study of enzymes, multienzyme systems, and more complex biological systems will be demonstrated by analysis of experimental data collected mainly in this laboratory. The mathematics of the theory was elaborated by Gnedenko, Belyaev, and Solov'ev (4).

I. PRINCIPAL RELIABILITY CHARACTERISTICS

The probability of failure free functioning

of a system for a given period of time is the quantitative characteristic of major concern for the reliability theory. Generally, this probability value is time dependent and is described by the reliability function, $P(t)$. Accordingly, the probability of a failure is given by

$$Q(t) = 1 - P(t) \qquad \text{(Eq. 4)}$$

Let $P(t)$ be the probability of failure free functioning during the time interval 0 to t and $P(\Delta t)$ be the probability that the element will continue to work during the subsequent interval Δt). The probability of failure free work for the interval from 0 to $(t + \Delta t)$ is given by the product

$$P(t + \Delta t) = P(t)\, P(\Delta t) \qquad \text{(Eq. 5)}$$

Hence the probability of failure free functioning for the Δt interval is

$$P(\Delta t) = \frac{P(t + \Delta t)}{P(t)} \qquad \text{(Eq. 6)}$$

The probability of failure during the same interval is given by

$$Q(\Delta t) = 1 - P(\Delta t) = \frac{P(t) - P(t + \Delta t)}{P(t)} \qquad \text{(Eq. 7)}$$

As Δt approaches zero, Eq. 7 becomes

$$Q(\Delta t) \approx - \frac{d\,P(t)}{dt} \quad \frac{\Delta t}{P(t)} \qquad \text{(Eq. 8)}$$

Using notation

$$- \frac{dP(t)}{dt} \quad \frac{1}{P(t)} = \lambda(t) \qquad \text{(Eq. 9)}$$

we can rewrite Eq. 8 in the form

$$Q(\Delta t) = \lambda(t)\, \Delta t \qquad \text{(Eq. 10)}$$

Thus, the parameter λ (t) characterizes the failure probability at each given time interval. $\lambda(t)$, with the dimension of reciprocal time, is referred to as the failure intensity and represents the frequency of breakdowns in a physical sense.

For most physical systems the failure intensity function has the form shown in Fig. 2. The time of functioning may be conventionally subdivided into three periods. The first period is characterized by a decreased failure intensity during which the most defective elements get out of order. Then the failure intensity reaches some constant value, characteristic of the operating period when the system elements get out of order at a constant frequency. The accumulation of irreversible defects and the aging of the system elements caused by certain physical or chemical processes leads to an increase in the failure intensity with time (period 3).

Eq. 9 relates the reliability function to the failure intensity. The solution of this differential equation gives the reliability function of Eq. 11 (Fig. 3). Accordingly, the probability of failure free functioning of a system over a period from t_1 to t_2 is given by Eq. 12 (Fig. 3). Eqs. 11 and 12 are the key equations of the theory of reliability, which makes it possible to calculate the probability of failure free functioning of any system during a given time interval provided the failure intensity function is known. These equations are kinetic ones since they describe the behavior of a system in time.

The experimental reliability function can be determined rather easily by monitoring the variation in the number, n, of active elements with time. Suppose the function n(t) is known; then the empirical reliability function is given by

$$P_N(t) = \frac{n(t)}{N_o} \qquad \text{(Eq. 13)}$$

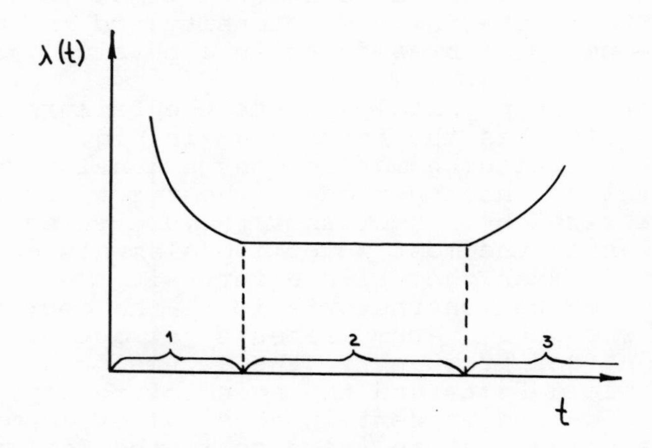

Fig. 2. The most frequent time dependence of failure intensity (4).

where N_o is the initial number of active elements. The function $P_N(t)$ provides a good enough approximation to P(t) if the system comprises a large number of working elements and n(t) is determined from a sufficient number of experiments.

Another important reliability parameter, the mean time to failure, is given by

$$T_o = \int_o^{\infty} P(t)dt \qquad \text{(Eq. 14)}$$

This parameter provides a convenient quantitative characteristic of an element or a system. Graphically, it is given by the area below the P(t) curve. The dispersion of the lifetimes of elements

$$\sigma^2 = \frac{1}{N} \sum_i (\tau_i - T_o)^2 \qquad \text{(Eq. 15)}$$

$$P(t) = e^{-\int_{o}^{t} \lambda (t)\, dt} \qquad \text{(Eq. 11)}$$

$$P(t_1, t_2) = e^{-\int_{t_1}^{t_2} \lambda (t)\, dt} \qquad \text{(Eq. 12)}$$

Fig. 3. Key reliability function equations.

provides yet another important characteristic of the reliability in cases where it has a finite value. In Eq. 15 τ_i is the lifetime of a given element, T_o is the mean lifetime of the elements, and N is the number of elements in a sample.

II. EXPONENTIAL RELIABILITY LAW AND THE KINETICS OF INACTIVATION OF ENZYMES

The purpose of this work is to describe the stability of biocatalysts and biocatalytic systems in terms of the reliability theory. We will concentrate on determination of an adequate reliability function, elucidation of a kinetic failure intensity function, and calculation of the mean time to failure.

The reliability approach is directly applicable to problems of enzymology. Determination of empirical reliability functions is based on the determination of the number of active elements present in the system at a given time. As applied to enzymology, $n(t)$ represents the number of enzyme molecules that retain their catalytic activity. If the reaction rate (or assayed activity) is proportional to the number of active biocatalyst molecules, then Eq. 16 holds;

$$A(t) = \alpha\, n(t) \qquad \text{(Eq. 16)}$$

and the empirical reliability function in the form of Eq. 13 is given by

$$P(t) = \frac{A(t)}{A_o} \qquad \text{(Eq. 17)}$$

where A_o is the initial activity of an enzyme or enzyme system. Thus, the determination of enzyme activity as a function of time yields the empirical reliability function.

Fig. 2 shows that the functioning of most systems is characterized by periods of time when the failure intensity remains constant, so that

$$\lambda(t) = \lambda = \text{constant} \qquad \text{(Eq. 18)}$$

The corresponding reliability functions have the form (cf. Eq. 11)

$$P(t) = e^{-\lambda t} \qquad \text{(Eq. 19)}$$

This exponential law is widely applied in the reliability theory.

The exponential law of inactivation is very typical of catalytic reactions involving enzymes. Most enzyme systems undergo exponential inactivation (1). The kinetics of inactivation of hydrogenase from the phototrophic bacteria *Thiocapsa roseopersicina*, shown in Fig. 4, provides an example. This enzyme accelerates the formation of molecular hydrogen from reduced organic electron carriers (5); and the kinetics of inactivation in air follows the exponential law (Eq. 19) quite well. In this case the failure frequency λ

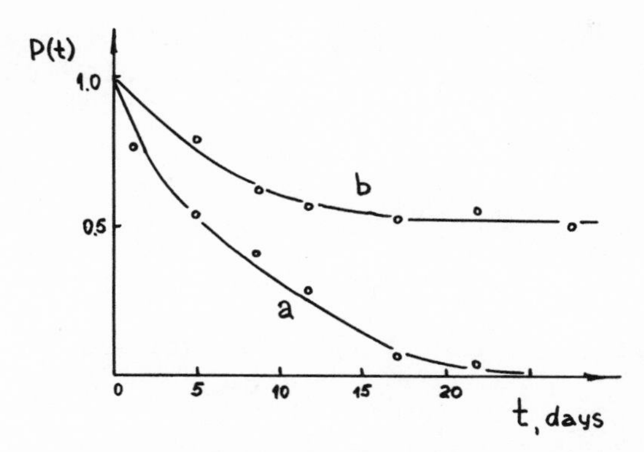

Fig. 4. Kinetics of inactivation of hydrogenase from *T. roseopersicina*. Curve A: in air; Curve B: under argon (5).

represents the enzyme inactivation rate constant k_i. It is related to a certain physical or chemical process which occurs at the active center of the enzyme and leads to a loss of catalytic activity. Thus, the exponent in Eq. 20 can be interpreted in two ways. Kinetically, it is the rate constant of a monomolecular reaction for conversion of the enzyme to the inactive form. In terms of the reliability theory, it is the failure intensity. Evidently, both characteristics have the same physical sense and represent the constant frequency of the enzyme inactivation events.

$$P(t) = \frac{A}{A_o} = e^{-k_i t} \qquad \text{(Eq. 20)}$$

Eq. 20 describes the variation of the concentration of the active form of the enzyme with time. On the other hand, Eq. 19 stresses the probabilistic aspects of the inactivation. This equation gives the probability that a given enzyme molecule will retain its catalytic activity for the period of time denoted t.

According to Eq. 14, the statistical mean for the time to failure over all molecules of an enzyme undergoing exponential inactivation is given by

$$T_o = \int_o^\infty e^{-k_i t} = \frac{1}{k_i} \qquad \text{(Eq. 21)}$$

Linearization of the data represented in Fig. 4A on semilogarithmic coordinates gave a k_i of 5.1 x 10^{-3} hr^{-1} for Hydrogenase from *T. roseopersicina*. Accordingly, the mean time of functioning of the enzyme in the air was about 8.15 days.

An important feature of exponential inactivation is that the probability of functioning of a system during a period Δt is independent of t and represents a function of the interval Δt only.

Thus, we have obtained an equation describing enzyme inactivation from considerations of the reliability theory, without recourse to chemical kinetics.

Next we consider a system of enzymes operating in series and capable of undergoing exponential inactivated; then we assume the loss of catalytic activity of the system as a whole. If the enzyme inactivation events can be considered to be independent of one another, then the probability of functional activity of a system is given by the product of the respective probabilities for the individual enzymes. The resulting reliability function has the form of Eq. 22 in Fig. 5. If the intensity of inactivation of the ith enzyme is expressed as $\lambda_i = k_i$, then Eq. 23 (Fig. 5) follows. Substitution of Eq. 23 into 22 gives Eq. 24 (Fig. 5). According to Eq. 14, the mean time to failure can be expressed as in Eq. 25 of Fig. 5. The largest term in the denominator of Eq. 25 corresponds to the fastest inactivation reaction. This means that the mean time to failure is determined by the stability of the most labile protein. If the enzyme chain contains a protein which undergoes inactivation at a far higher rate than do the other proteins involved, then the inactivation of the system is governed by the kinetics of inactivation of the protein in question.

Consider now a series of enzymes with parallel reactions. If two enzymes catalyze one and the same reaction and undergo inactivation at different rates, the experimental activity varies according to Eq. 26 (Fig. 5), where λ_1 and λ_2 are the failure intensities and A_1 and A_2 are the initial activities of the respective enzymes. The sum of the initial activities gives the experimental initial activity of the system A_o. According to Eqs. 14 and 26, the mean time to failure is given by Eq. 27 (Fig. 5).

The inactivation of hydrogenase from *T. roseopersicina* during incubation of the enzyme under argon (5) (Fig. 4B) may serve as an example. Variation of the catalytic activity with aging of the enzyme follows a pattern that can be decomposed into at least two exponential curves. Rapid inactivation, characterized by a half-life of 6.4 days, is followed by a period where the activity remains practically constant. Thus, the mean time to failure for this hydrogenase under argon is very large.

In the general case of n enzyme fractions of various activities and stabilities, analysis of the mean time to failure is shown in Eq. 28 (Fig. 5). Here A_j and K_j are the initial activities and rate constants, respectively, for the individual enzyme fractions; and A_o is the sum of the individual initial activities for the system. Note that the mean time to failure of enzyme systems in the case of parallel reactions depends on the characteristics of the most active (the largest A_j) or the most stable (the smallest k_j) elements. This feature distinguishes such systems from those catalyzing consecutive reactions, the reliability

$$P(t) = P_1(t) \ldots P_i(t) \ldots P_n(t) \qquad \text{(Eq. 22)}$$

$$P_i(t) = e^{-\lambda_i t} \qquad \text{(Eq. 23)}$$

$$P(t) = \exp\left(-\sum_{i=1}^{n} k_i t\right) \qquad \text{(Eq. 24)}$$

$$T_o = 1/\sum_{i=1}^{n} k_i \qquad \text{(Eq. 25)}$$

$$A(t) = A_1 e^{-\lambda_1 t} + A_2 e^{-\lambda_2 t} \qquad \text{(Eq. 26)}$$

$$T_o = \frac{1}{A_o}\left(\frac{A_1}{\lambda_1} + \frac{A_2}{\lambda_2}\right) \qquad \text{(Eq. 27)}$$

$$T_o = \frac{1}{A_o}\sum_{j=1}^{n}\frac{A_j}{k_j} \qquad \text{(Eq. 28)}$$

Fig. 5. Reliability functions (P(t) and mean time to failure T_o for multienzyme systems.

of which depends on the reliability of the most labile protein involved.

III. CHANGE IN THE RATE LIMITING STEP DURING INACTIVATION OF MULTIENZYME SYSTEMS

Multienzyme systems responsible for sequential transformations of substrates usually contain enzymes of various stabilities towards inactivation. This implies the possibility of a change in the rate limiting step during inactivation. Suppose that the slowest step of the sequence involves an enzyme of considerable stability, while some other enzyme involved in another step of the same sequence undergoes rapid inactivation. The development of the process may lead to a situation in which the activity of the most labile enzyme will determine the rate of the process. Thus, a change in the rate limiting step may occur as the inactivation proceeds. Such a change should affect markedly the kinetics of the process. The corresponding kinetic patterns are described below.

Consider a linear chain of n enzymes with the rate of the stationary process involved described by Eq. 29 in Fig. 6 (6, 7). S_o is the initial concentration of the first substrate; and V_i and K_i are the maximum velocity and Michaelis constant, respectively, for the ith chain fragment. Consider the case when two enzymes have comparable K_i/V_i values which by far exceed those characteristic of the other enzymes of the chain. The stationary reaction rate is then given by Eq. 30 (Fig. 6), where the parameters with subscripts i or j characterize the two enzymes in question.

Let inactivation of both enzymes follow the exponential law, which means that the failure intensities λ_i and λ_j are time independent. Variation of parameters V_i and V_j with time is then given by Eqs. 31 and 32 of Fig. 6. Substitution of these two equations into Eq. 30 yields Eqs. 33 (Fig. 6), which at zero time gives Eq. 34. The reliability function then takes the form of Eq. 35.

$$P(t) = \frac{V_t}{V_o} = \frac{\left(1 + \frac{A_i/K_i}{A_j/K_j}\right) e^{-\lambda_i t}}{1 + \frac{A_i K_i}{A_j K_j} e^{-(\lambda i+\lambda j)t}} \quad \text{(Eq. 35)}$$

The kinetic behavior of the system during inactivation will depend on the ratio of the kinetic parameters.

$$\frac{A_i/K_i}{A_j/K_j} = \alpha \quad \text{(Eq. 36)}$$

and on the difference between the respective failure intensities.

Three particular cases are of importance.

Case 1: Failure frequencies of the two enzymes are identical, so that $\lambda_i = \lambda_j$. This is the simplest case. The kinetics of the process are determined by the failure intensity λ_i, irrespective of which of the two enzymes limits the overall rate of reaction.

Case 2: Rate determining enzyme undergoes inactivation more rapidly than do any of the other enzymes in the chain. Then we have $\alpha >> 1$ and $\lambda_j >> \lambda_i$; and the reliability function depends only on the exponent $-\lambda_j t$. The kinetics of inactivation still follow the exponential law (Fig. 7A); and the reliability and mean time to failure are determined by the failure intensity of the rate determining enzyme.

Case 3: Rate of inactivation of rate limiting enzyme is well below that of other enzymes. This condition can be written as $\alpha >> 1$ and $\lambda_i >> \lambda j$. The value for P(t) becomes $\alpha \exp(-\lambda_i t)/(1 + \alpha \exp(-\lambda_i t))$, with the kinetic curve of inactivation having a rather complicated shape. The failure

$$v_{st} = S_o \Big/ \left(\sum_{i=1}^{n} \frac{K_i}{V_i} \right) \qquad \text{(Eq. 29)}$$

$$v_{st} = \frac{S_o \; V_i \; V_j}{K_i \; V_j + K_j \; V_i} \qquad \text{(Eq. 30)}$$

$$V_i = k_i \, E_{io} \, e^{-\lambda_i t} = A_i \, e^{-\lambda_i t} \qquad \text{(Eq. 31)}$$

$$V_j = k_j \, E_{jo} \, e^{-\lambda_j t} = A_j \, e^{-\lambda_j t} \qquad \text{(Eq. 32)}$$

$$V_t = \frac{S_o A_i A_j e^{-(\lambda_i + \lambda_o) t}}{K_i A_j e^{-\lambda_j t} + K_j A_i e^{\lambda_i t}} \qquad \text{(Eq. 33)}$$

$$V_o = \frac{S_o A_i A_j}{K_i A_j + K_j A_i} \qquad \text{(Eq. 34)}$$

Fig. 6. Change in rate limiting step during inactivation.

intensity of the non rate limiting enzyme determines the kinetics of inactivation (Fig. 7B). The mean time to failure for the system as a whole is equal to ln α/λ_i, which corresponds to a two fold

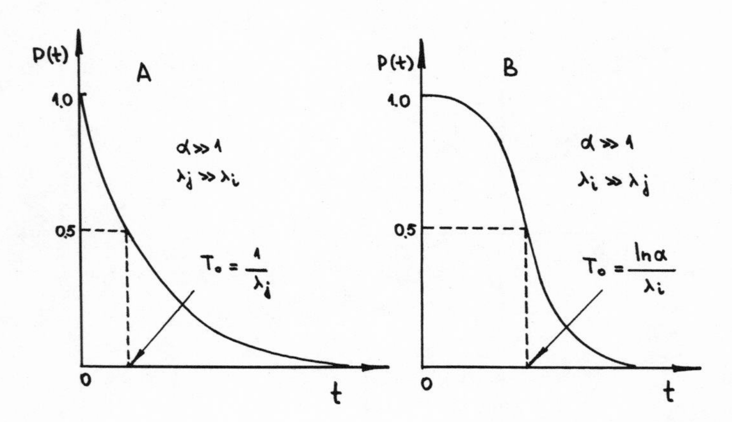

Fig. 7. Kinetics of inactivation of multienzyme systems: Part A: rate limiting enzyme has the highest failure intensity. Part B: a non rate limiting enzyme has the fastest inactivation.

decrease in the initial activity. Thus, a change in the rate limiting step occurs in this case during inactivation of the multienzyme system.

Similar kinetic patterns to Case 3 have been described in the literature (8); an example is shown in Fig. 8 depicting the kinetics of inactivation of the mitochondria multienzyme complex which catalyzes the oxidation of NADH with oxygen; the data suggest that a change in the rate limiting step may have occurred. The curve shown in Fig. 8 was calculated using the expression for P(t) given for Case 3. It should be noted that agreement between the experimental and calculated values in Fig. 8 cannot be considered unambiguous proof of a change in the rate limiting step. The observed pattern also can be explained from basically different considerations.

A change in the rate limiting step may also occur in systems functioning under diffusional

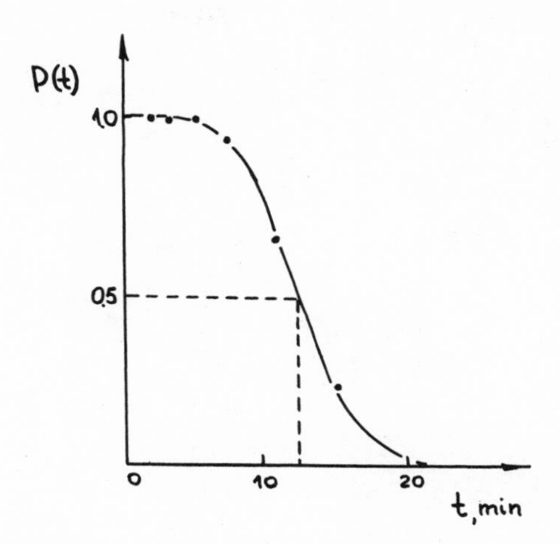

Fig. 8. Kinetics of inactivation of a multienzyme system for the oxidation of NADH, complex I+II+III+IV, of the respiratory chains of mitochondria (8).

limitations. Consider an enzyme immobilized on a substrate impermeable membrane. Accounting for diffusion layer effects gives a relation for the steady state velocity (9) (Eq. 37, Fig. 9) that includes the membrane surface A, the system volume V, and the surface concentration of enzyme E_s. S^{ℓ} is the volume concentration of the substrate, k is a surface analogue of the parameter k_{cat}/K_m, ℓ is the thickness of the diffusion layer, and D is the diffusivity of the substrate. Assuming that inactivation of the enzyme follows the exponential law, we have the expression for P(t) given by Eq. 38 (Fig. 9). Here λ is the failure intensity for the immobilized enzyme. If the initial stage of the process goes under strictly diffusional control, then the rate of reaction rate is by far higher than the rate of diffusion of the substrate, so that $k\ E_s\ \ell >> D$. Under this condition, the 1 in the numerator of Eq. 38 disappears to give Eq. 39 (Fig. 9) for P(t). This is identical to that

$$v_{st} = \frac{A}{V}\left(\frac{k\ E_s\ D\ S^{\ell}}{k\ E_s\ \ell + D}\right) \quad \text{(Eq. 37)}$$

$$P\ (t) = \frac{\left(1 + \frac{k\ E_s\ \ell}{D}\right) e^{-\lambda t}}{1 + \left(\frac{k\ E_s\ \ell}{D}\ e^{-\lambda t}\right)} \quad \text{(Eq. 38)}$$

$$P\ (t) = \frac{\frac{k\ E_s\ \ell}{D}\ e^{-\lambda t}}{1 + \left(\frac{k\ E_s\ \ell}{D}\ e^{-\lambda t}\right)} \quad \text{(Eq. 39)}$$

Fig. 9. Equations with diffusional limitations included.

for Case 3 above except that α now is equal to $(k\ E_s\ \ell/D)$. Fig. 10 shows a family of curves calculated for various α values using Eq. 38. When α far exceeds unity, the curves fit Eq. 39.

It is worth mentioning that kinetic studies of the inactivation of immobilized enzymes provides a convenient means of experimentally determining the diffusion modulus $k\ E_s\ \ell/D$, if inactivation follows the exponential law.

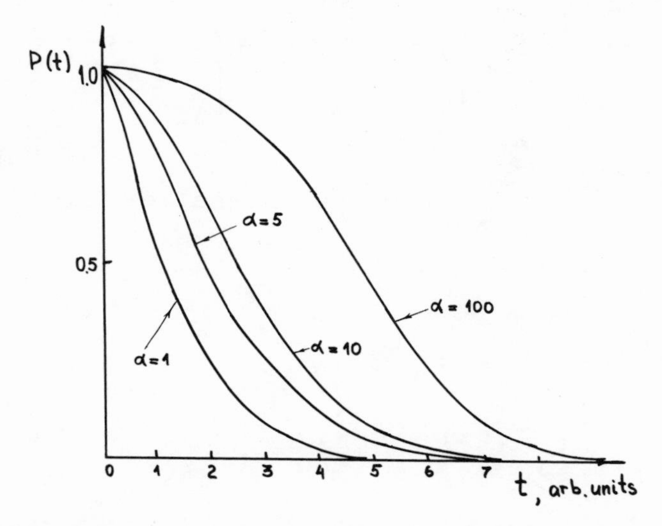

Fig. 10. Kinetics of inactivation of systems with change of the rate limiting step, as calculated using Eq. 38. The failure intensity was assumed to be 1.

IV. EXPONENTIAL RELIABILITY LAW IN THE INACTIVATION OF ORGANELLES

One can expect the exponential reliability law to be applicable to complex biological systems as well as to single or multiple enzymes. Using this approach, accounting should be made of the ability of biological systems to undergo conversion to forms that differ from the parent ones. The kinetic model of inactivation of organelles, based on the exponential law, was developed (10,11) to describe the inactivation and subsequent activation of chloroplasts during aging. The aging of complex structures like organelles may involve consecutive transformations leading to species of somewhat modified structures which to a certain degree retain the functional activity of the

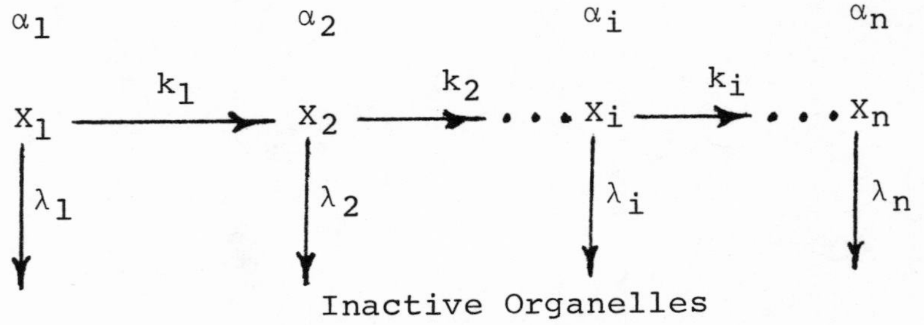

$$\begin{array}{ccccccc} \alpha_1 & & \alpha_2 & & \alpha_i & & \alpha_n \\ X_1 & \xrightarrow{k_1} & X_2 & \xrightarrow{k_2} \cdots & X_i & \xrightarrow{k_i} \cdots & X_n \\ \downarrow \lambda_1 & & \downarrow \lambda_2 & & \downarrow \lambda_i & & \downarrow \lambda_n \end{array}$$

Inactive Organelles

$$\frac{dX_1}{dt} = -(k_1 + \lambda_1)\ X_1$$

$$\frac{dX_i}{dt} = k_{i-1}\ X_{i-1} - (k_i + \lambda_i)\ X_i$$

$$\frac{dX_n}{dt} = k_{n-1}\ X_{n-1} - \lambda_n\ X_n$$

Fig. 11. Kinetic scheme for inactivation of organelles.

starting material. Structural transformations may generate a variety of forms (X_1, $X_2 \ldots X_n$) including species of higher as well as lower activities, compared with the parent organelles. Structural transformations develop with time as a result of processes which can be described by individual rate constants (k_1, $k_2 \ldots k_{n-1}$). Evidently, modified organelles not only possess activities of their own, but are also characterized by specific stabilities towards denaturation. Assuming that each form undergoes inactivation independently from all other forms, and that inactivation of individual forms follows the exponential law, each modification can be characterized by a certain value of the failure intensity, $\lambda_1, \ldots \lambda n$.

The functional activities of various modifications, which can be determined by measuring their catalytic activity in some reaction of functional importance, are proportional to the concentrations of the respective forms, with proportionality factors $\alpha_1 \ldots \alpha_n$. The experimental activity is the sum of the activities of the individual forms. The mechanism of aging or organelles can be described by the kinetic scheme shown in Fig. 11. The symbols X_1, $X_2 \ldots X_n$ are structurally and catalytically different forms, $k_1 \ldots k_{n-1}$ are rate constants of structure transformations caused by aging, and $\lambda_1 \ldots \lambda_n$ are failure intensities for the individual modifications. Simple rate expressions can be written for the concentrations of the individual forms to give the system of differential equations of Fig. 11. The experimental activity is given by

$$A = \sum_{i=1}^{n} \alpha_i X_i \qquad \text{(Eq. 40)}$$

Thus, in order to determine the kinetics of the inactivation of organelles during their aging, one must find solutions to the Fig. 11 system of linear differential equations with constant factors (12,13). This can be done relatively easily using the method of determinants, as described in many texts on differential equations. The solution gives all the real single valued roots, which can lie incorporated into Eq. 40 to give the experimental activity of the organelle suspension as follows:

$$A = \sum_{i=1}^{n} \alpha_i C_i e^{Bit} = \sum_{i=1}^{n} \alpha_i C_i e^{-\frac{t}{\tau_i}} \qquad \text{(Eq. 41)}$$

where

$$\tau_i^{-1} = k_i + \lambda i, \; i = 1,\ldots n-1$$

$$\tau_n^{-1} = \lambda n$$

The analysis carried out in this section leads to the conclusions that 1) the aging of organelles may be accompanied by activity variation which, at least in theory, can be described by the sum of exponential terms and 2) the kinetics of aging, in theory, can be analyzed in terms of certain elementary steps.

The kinetic model described above was applied to the aging of isolated chloroplasts (10,11). Suspensions of isolated chloroplasts were incubated at constant temperature and assayed for their activity in the photodecomposition of water and photoreduction of potassium ferricyanide. The photoreduction activity of the chloroplasts increased with time during incubation and aging, reached a maximum value, and then fell to zero. It has been shown that the kinetics of inactivation can be described by a sum of at least two exponential curves

$$A = C_i e^{-\frac{t}{\tau_1}} + C_2 e^{-\frac{t}{\tau_2}} \qquad \text{(Eq. 42)}$$

Fig. 12 shows the data for the aging of the chloroplasts (Part A), and exemplifies the treatment of the kinetic data using the appropriate coordinate systems in order to determine the characteristic times τ_1 and τ_2. From the point of view of the mechanism, the dependence of Eq. 42 implies that inactivation involved at least two forms of

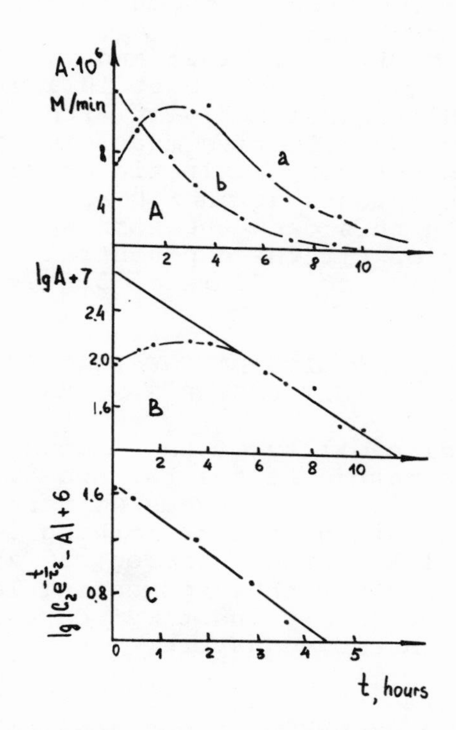

Fig. 12. Analysis of the kinetics of inactivation of isolated chloroplasts. Part A: activity data versus time for photoreduction of ferricyanide, inactivation at pH 7.8 in Curve a and at pH 6.0 in Curve B; Part B: determination of τ_2 and C_2 from semilogarithmic plot of data at long times; Part C: determination of τ_1 and C_1 using semilogarithmic coordinates (11).

chloroplasts with different activities and failure intensities. Such a scheme was developed based on using only X_1 and X_2 of Fig. 11 and assuming only pure X, called X_o, at time zero. This was used to develop a modified bioexponential expression for chloroplast reducing activity, as discussed elsewhere (11). The results shown in Fig. 12 give the following parameter values: A_o, 8.8 x 10^{-6} M/min; C, 4 x 10^{-5} M/min; τ_1, 1.42; and τ_2, 3.3 hr. The mean time to failure was about 10.1 hr. The increase of chloroplast activity with aging markedly increased the effective mean time of functioning. For comparison, a kinetic curve of inactivation of chloroplasts at pH 6.0 also is shown in Fig. 12A. In this case the corresponding reliability function fit the exponential law equation, with a mean time to failure of 3.3 hr.

V. NORMAL RELIABILITY LAW OF INACTIVATION OF ENZYMES AND MULTIENZYME SYSTEMS

The applicability of the exponential law depends on the nature of the failure events. As it has been shown above, systems with randomly occurring failures of the all or nothing type, which lead to complete loss of biological activity, at least in theory fit the exponential law. The corresponding elementary inactivation events are of a sudden and decisive nature.

There are other cases in which the simple exponential law probably is not applicable to the kinetics of inactivation of biological catalysts and biological systems. For example, situations may produce lag times to changes in activity. We believe that such aging mechanisms are rather frequent in biological systems.

Defects resulting from irreversible physical or chemical structural transformations that do not lead to immediate losses of activity and occur as latent errors are treated as gradual failures in the reliability theory. The corresponding reliability function has been developed (4), with the parameters that determine the type of time dependence again being the mean time to failure and the dispersion of the times to failure. Solution of these equations requires numerical integration.

It is likely that the inactivation of formate dehydrogenase from *Arthrobacter* SP1 involves accumulation of latent defects (Fig. 1), since the sigmoid shape of the inactivation curve may point to either a change in the rate limiting step or a transition from diffusion to kinetic control. Such curves may also be indicative of the accumulation of latent defects, leading to irreversible aging of a system. At present we are not able experimentally to distinguish between the two mechanisms.

Physical or chemical processes responsible for the inactivation of enzymes and multienzyme systems provide a variety of features that need to be studied thoroughly. It should be noted that mechanistic studies of inactivation furnish valuable information about these processes. The reliability theory provides a general quantitative approach to the aging of systems of various types. The use of reliability models may significantly improve our basic mechanistic concepts and also may allow a kinetic description of the aging of complex biological systems. We emphasize that the reliability theory at least in principle allows one to predict the reliability parameters of rather complex multienzyme systems from the data on individual enzymes. However, considerably more definitive experimental studies are required before the practical usefulness of this approach can be determined.

ACKNOWLEDGMENT

The author is indebted to I.V. Berezin for valuable discussions.

REFERENCES

1. JOLY, M., "A Physico-Chemical Approach to the Denaturation of Proteins," Academic Press, London, 1965.

2. TIMASHEFF, S. N. & FASMAN, G. D. "Structure and Stability of Biological Molecules," Marcel Dekker, New York, 1969.
3. HERRING, W. M., LAURENCE, B. L. & KITTRELL, J. R. *Biotechnol. Bioeng.* *14*: 975, 1973.
4. GNEDENKO, B. V., BELYAEV, Y. K. & SOLOVIEV, A. D. "Mathematical Methods in Theory of Reliability", Nauka, Moscow, 1965.
5. TOAJ, CH.-D., VARFOLOMEEV, S. D., GOGOTOV, I. N. & BEREZIN, I. V. *Mol. Biol.* (Russ.) *10*: 452, 1976.
6. VARFOLOMEEV, S. D., "Kinetics of Reactions in Multienzyme systems", Moscow State University, 1976.
7. VARFOLOMEEV, S. D. & BEREZIN, I. V. *Mol. Biol.* (Russ.) *10*: 818, 1976.
8. LUSIKOV, V. N., D. Sc. Dissertation, Moscow State University, 1973.
9. GOLDMAN, R. & KATCHALSKI, E. *J. Theor. Biol.* *32*: 243, 1971.
10. BEREZIN, I. V., VARFOLOMEEV, S. D., ZAITZEV, S. V. & ILYINA, M.D. *Dokl. Akad. Nauk. SSSR* *219*: 213, 1974.
11. VARFOLOMEEV, S. D., ZAITZEV, S. V., ILYINA, M. D. & BEREZIN, I. V. *Mol. Biol.* (Russ.) *9*: 893, 1975.
12. BEREZIN, I. V., VARFOLOMEEV, S. D. & MARTINEK, K. *Chem. Revs.* (Russ.) *43*: 835, 1974.
13. VARFOLOMEEV, S. D. & BEREZIN, I. V. "Kinetics of Enzyme Reactions" Moscow State University, 1975.

SECTION II
CARBOHYDRATE CONVERSIONS WITH ENZYMES

ENZYMATIC CONVERSION OF CELLULOSE TO GLUCOSE: PRESENT STATE OF THE ART AND POTENTIAL

A. A. Klyosov and M. L. Rabinowitch

Department of Chemistry
Moscow State University, Moscow, USSR

The production of glucose from cellulosic materials seems to have a very promising future. Food, chemical, and microbiological industries need an inexpensive source of glucose; and there exists a continually renewable supply of cellulose in the world. Cellulose processing also is closely connected with the utilization of domestic, industrial, and agricultural wastes, in which cellulose containing materials constitute a considerable proportion.

Among the problems that must be resolved before a successful industrial process for glucose production from cellulose is realized are the following:

1. Selection of a readily available cellulosic raw material, the processing of which could be feasible.
2. Development of an effective pretreatment process which could significantly increase the subsequent rate or enzymatic hydrolysis and final product yield.
3. Development of cellulase complexes for glucose production at an optimum conversion.
4. Optimization of the glucose production process from cellulosic substances.
5. Development of an optimal enzyme reactor for the most efficient conversion of cellulose into glucose in terms of continuity of operation and insoluble residues recycling.

6. Design and operation of a pilot plant for the enzymatic production of glucose.

These problems are being studied at the Laboratory of Chemical Enzymology of Moscow State University in cooperation with other research groups in the USSR and elsewhere.

The purpose of this paper is to present some of the results and ideas on a number of these problem areas and to outline the future potential of our approach to studying the enzymatic hydrolysis of cellulose.

I. FUNDAMENTAL ASPECTS OF THE CELLULASE PROBLEM

In the context of contemporary understanding, a cellulase complex contains four groups of carbohydrases (1-4): endo-1,4-β-glucanase (1,4-β-D-glucan 4-glucanohydrolase, EC 3.2.1.4), exo-1,4-β-glucanase (exo-cellobiohydrolase, or 1,4-β-D-glucan cellobiohydrolase, EC 3.2.1.91), exo-1,4-β-glucosidase (1,4-β-D-glucan glucohydrolase, EC 3.2.1.74), and cellobiase (β-glucosidase, or β-D-glucoside glucohydrolase, EC 3.2.1.21). The overall flow chart for enzymatic hydrolysis of cellulose is shown in Fig. 1. Depending on the physical state of the initial substrate, G_n may be either a partially degraded insoluble cellulose with a relatively low degree of polymerization or a set of substituted cellodextrines, produced as a result of the hydrolysis of soluble cellulose derivatives such as carboxymethyl cellulose.

The available literature contains a number of detailed descriptions of the enzymatic hydrolysis mechanisms for cellulose (5-10). In our opinion, those descriptions are either particular cases of a more general mechanism or are derived from insufficient or *a priori* assumptions, such as the idea that there is a prehydrolytic C_1-factor or an endoglucanase specific only for crystalline parts of the insoluble cellulose (6,10). In our understanding, the absence of a more or less generally

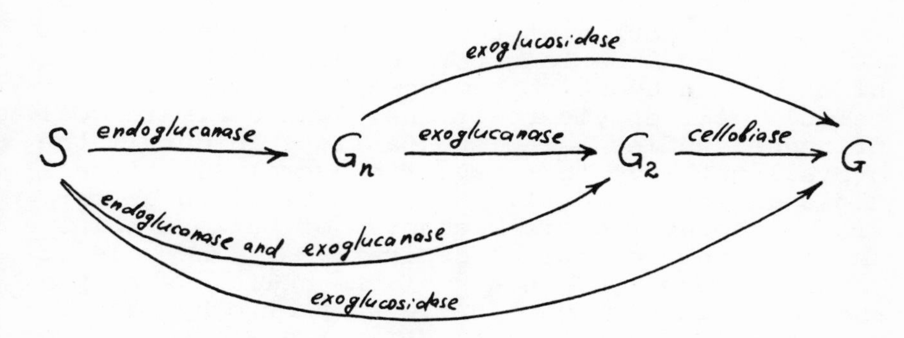

Fig. 1. Enzymatic hydrolysis of cellulose. S is the starting cellulose; G_n represents cello-oligosaccharides; G_2 is cellobiose; and G is glucose (2).

exceptable mechanism for the enzymatic hydrolysis of cellulose can be explained by at least two basic reasons. First, different experimenters have been using different composition cellulase complexes, thereby causing confusion in defining the rate determining factors for the enzymatic hydrolysis. As a result, the conclusions of the studies in this field seem to be incompatible, especially when different cellulosic substrates are used. The situation becomes even more aggravated in the absence of appropriate methods to determine the activities of different components of the cellulolytic enzyme complexes. Second, the kinetics of enzymatic cellulose hydrolysis have not received adequate study; and in almost all studies the multienzyme cellulase complex has been treated as a single enzyme for simplicity. In reality, this simplification is unacceptable in many instances, because the rate limiting step of the hydrolysis sequence may vary for cellulase complexes of different composition. Moreover, the rate limiting step may change in the course of an enzymatic reaction in the transition period from the beginning of the reaction to the point of achieving a certain degree of substrate conversion.

In the course of the last few years we have been engaged in a long term program to define the kinetics and hydrolysis mechanisms involved in cellulolytic enzyme reactions. The ultimate goal of these studies is to substantiate in detail the overall pathways for enzymatic hydrolysis of cellulose shown in Fig. 1. Also, it is aimed at defining the hydrolysis mechanism of various soluble and insoluble cellulose derivatives. The role of the individual components of the cellulase complex and their involvement in the enzymatic hydrolysis of specific substrates at various stages of conversion is also to be studied. A part of the problem also must include the development of new analytical methods to determine more specifically the properties and compositions of cellulase complexes of various origins. Other topics that need to be considered include the adsorption of cellulolytic enzymes on the surface of insoluble substrates and the resulting kinetics, the stability of individual components of the cellulase complex at different stages of the enzymatic cellulose hydrolysis, the function and regulatory effects of multienzyme cellulase systems in the overall reaction sequence, and the effects of the physical characteristics of the insoluble cellulose on the kinetics of hydrolysis. For the purpose of developing a quantitative description of enzymatic cellulose hydrolysis, we have launched a program to develop a kinetic theory to explain the action of cellulase multienzyme complexes during both nonsteady state and steady state phases of the reaction.

II. ACTIVITY AND COMPOSITION OF THE CELLULASE COMPLEXES OF DIFFERENT ORIGINS

This section of the paper is devoted to specific methods for determination of the activity of individual components of cellulase complexes. Also, we discuss the results of a comparative study of various cellulase complexes, that we carried out to establish a common criterion upon which to characterize different cellulolytic enzymes. We expressed the activity of the individual components in International Units, with one unit defined as the amount of enzyme that converts one micromole of substrate per minute. A summation of the data is given in Tables 1 and 2.

A. The Activity of Endo-1,4-β-Glucanases

A modified version of the viscometric technique (11,12) was used to determine the activity of endoglucanases of cellulase complexes. Contrary to other similar methods (13-17), the modified method can be used to determine the initial rate of enzymatic hydrolysis of soluble high polymer substrates, in our case carboxymethyl cellulose (CM cellulose). This approach appears to be more justifiable because of several reasons. First, the assumptions that are normally used when molar concentrations are calculated on the basis of the viscometric characteristics of the solution are most rigorously satisfied only at the initial stage of the reaction. Second, the exoenzymes which are often present in cellulase complexes produce a less pronounced effect, particularly during the initial period of polymer substrate hydrolysis. Finally, the information obtained from the viscometric data about the initial stage of the reaction covers the hydrolysis of the parts of the substrate which are the most sensitive to endoglucanase action and are of the most interest for the investigation.

A definition of the initial rate of decrease of the carboxymethyl cellulose viscosity effected by the enzyme is a cornerstone of this modified viscometric technique (11,12,18). If the weight concentration of polymer substrate in solution is denoted by C, and M stands for the number-average molecular weight of the substrate undergoing reaction, then the following expression can be obtained for the initial velocity of the reaction:

TABLE 1

ACTIVITIES OF CELLULASE COMPLEXES OF VARIOUS ORIGIN

Preparation (Source)	Total Activity (a,b)	Activities of Individual Components (b)			
		Endoglucanase	Exoglucosidase	Cellobiase	Aryl-β-glucosidase
I. Crude Technical Preparations (c)					
Cellolignorine PX (*T. lignorum*)	9.2	18	0.8	6	3.7
Cellobronine G3X (*T. longibrachiatum*)	11	26	10	5	0.2
Cellocandine G3X (*G. candidum*)	11	41	7	3.5	7.5
Celloviridine G3X (*T. viride*)	6.5	70	12	0.5	1.6
Cellolignorine PIOX (*T. lignorum*)	16	36	4	12	15
Cellokoningine PIOX (*T. koningii*)	24	120	35	13	12
Cellocandine GIOX (*G. candidum*)	-	130	-	45	30
Pectofoetidine G3X (*Asp. foetidus*)	5.2	22	-	28	3.7
Pectofoetidine PIOX (*Asp. foetidus*)	7.9	50	-	120	34

Table 1 (cont'd)

TABLE 1 (Cont'd.)

Preparation (Source)	Total Activity[a,b]	Activities of Individual Components[b]			
		Endoglucanase	Exoglucosidase	Cellobiase	Aryl-β-glucosidase
II. Purified Technical Preparations[d]					
(*T. lignorum*)	83	730	110	25	150
(*T. lignorum*)	88	830	130	100	160
(*T. lignorum*)	270	830	200	270	250
(*T. longibrachiatum*)	170	1000	175	70	37
(*T. viride*)	93	500	120	33	15
(*G. candidum*)	-	5700	250	90	-
(*G. candidum*)	430	4000	600	800	420
(*T. koningii*)	120	1300	180	57	150
(*Asp. foetidus*)	37	170	<10	3400	43

(a) Mandels Weber method; 1 I.U. = 1 micromole of reducing sugars produced as glucose from filter paper per minute.
(b) Specific enzyme activities may differ between batches of cellulase preparation
(c) Manufactured in USSR
(d) Purified by gel filtration.

TABLE 2

ACTIVITIES OF CELLULASE COMPLEXES OF VARIOUS ORIGIN

Preparation (Source)	Total Activity[a,b]	Activities of Individual Components[b]			
		Endoglucanase	Exoglucosidase	Cellobiase	Aryl-β-glucosidase
I. Commercial Preparations					
Nagase (--)	14	60	20	17	<0.5
Rapidase (--)	74	700	5	<1	1.5
Serva (*Asp. niger*)	37	120	20	425	50
Sigma (*Asp. niger*)	42	110	<7	735	53
Koch-Light (*Asp. niger*)	35	290	5	50	25
Sigma (*T. viride*)	10	20	3	11	37
Novo (*T. reesei*)	200	3000	350	20	28

(a) Mandels Weber method 1 I.U. = 1 micromole of reducing sugars produced as glucose from filter paper per minute

(b) Specific enzyme activities may differ between batches of cellulase prepa-

TABLE 2 (Cont'd)

Preparation (Source)	Total Activity(a,b)	Activities of Individual Components(b)			
		Endoglu-canase	Exoglu-cosidase	Cello-biase	Aryl-β-gluco-sidase
II. Laboratory Preparations					
(Thermoactinomyces species)(c)	7.4	43	<0.3	<0.5	3
(Bacterial)(d)	10	70	<0.3	<0.5	0.8
(*M. verrucaris*)(e)	10	66	5	1	0.9
(*Asp. terreus*)(f)	74	340	70	35	9.5
(*T. reesei*)(g)	290	2000	650	35	40

(c) From A.E. Humphrey, University of Pennsylvania USA

(d) Associate of *Bacterium alcaligenes*, *Pseudomonas desmolyticum*, *Bacterium album*, *Chromobacterium rheni*, *Bacillus megaterium*, *Cytophaga hutchinsoni*; from A.B. Pauluconis, National Research Institute for Applied Enzymology, Vilnus, USSR.

(e) From N.A. Rodionova, Institute for Biochemistry, Academia of Sciences of the USSR, Moscow.

(f) From L.G. Loginova, Institut for Microbiology, Academia of Sciences of the USSR, Moscow.

(g) From E.T. Reese, U.S. Army Nitick Research and Development Command, USA.

$$v_o = \left[\frac{d}{dt}\left(\frac{c}{M}\right)\right]_{t=0} \qquad \text{(Eq. 1)}$$

$$v_o = -\frac{C}{M_o^2}\left(\frac{dM}{dt}\right)_o \qquad \text{(Eq. 2)}$$

where M_o is a characteristic parameter of the initial polymer. The number-average molecular weight, M, and the viscosity-average molecular weight M_V, of the polymer are related as

$$M_V = k\ M \qquad \text{(Eq. 3)}$$

where k is a constant greater than or equal to one and dependent on the form of the polymer molecular weight distribution curve. On the other hand, the viscosity-average molecular weight of the polymer is related to the intrinsic velocity η, according to the Mark-Houwink equation:

$$[\eta] = H\ M_V{}^x \qquad \text{(Eq. 4)}$$

In Eq. 4 H and x are constants dependent on the characteristics of the polymer solvent interaction. Eqs. 3 and 4 permit one to find a relation between the number-average molecular weight of the polymer and the intrinsic viscosity of its solution. By combining Eqs. 1 to 4, we can see how the initial rate of the enzymatic reaction depends on the initial rate of the reduction of the solution intrinsic viscosity:

$$v_o = -\frac{C}{X\ M_o\ [\eta]_o}\left(\frac{d[\eta]}{dt}\right)_o \qquad \text{(Eq. 5)}$$

In practice, simple viscometers of the Ostwald or Ubbelohde type are commonly used to measure relative, but not intrinsic, viscosity. The link between the two viscosity measurements for numerous polymers over a fairly wide range of concentrations is described by the Hess-Philipoff equation:

$$\eta_{rel} = \left(1 + \frac{[\eta]C}{8}\right)^8 \qquad \text{(Eq. 6)}$$

Differentiating Eq. 6 for the initial moment of time gives Eq. 7.

$$\left(\frac{d[\eta]}{dt}\right)_o = \frac{1}{C\left(\eta_{rel,o}\right)^{7/8}} \left(\frac{d\eta_{rel}}{dt}\right)_o \qquad \text{(Eq. 7)}$$

Then, by combining Eqs. 5 and 7, we develop a formula which relates the initial rate of the enzymatic reaction to the initial rate of the relative viscosity reduction of the polymer substrate solution under the action of endoglucanase:

$$v_o = - \frac{1}{X \;\; M_o \;\; [\eta]_o \;\; \left(\eta_{rel}\right)^{7/8}} \left(\frac{d\eta_{rel}}{dt}\right)_o \qquad \text{(Eq. 8)}$$

The x in Eq. 4 can be determined either from tables or experimentally, using polymer fractions of known viscosity-average molecular weights. If CM cellulose, for example, of medium viscosity from Sigma Chemical Co. (No. C-4888, Lot 67C-0441) is used, then Eq. 2 can be simplified to give:

$$A = \frac{14\ (t_o - t_1)}{t_2\left(t + \frac{t_1}{2}\right)\ (t_o/t_2)^{7/8}\ \ [E]_o} \qquad \text{(Eq. 9)}$$

In Eq. 9 A is the activity of endoglucanase per g of dry cellulase preparation; t_o, t_1, and t_2 are the efflux times in sec of the pure buffer solution, the initial 0.2 to 0.4% CM cellulose solution, and the reaction mixture at approximately 1 min after enzyme addition, respectively; t is the elapsed time in sec from the start of the reaction up to the beginning of the measurements; and $[E]_o$ is the cellulase preparation concentration in g/ℓ in the reaction mixture. A unit of endoglucanase activity is the amount of activity that cleaves 1 micromole of glucoside bonds of soluble CM cellulose per minute during the initial period of reaction at pH 4.5, 0.05 M acetate buffer, 0.1 M NaCl, and 40°C.

Satisfactory performance of the technique in determining the endoglucanase activity is well illustrated in Table 3. This Table shows the performance of the viscometric technique in comparison with non glucose reducing sugar determinations by the Somogyi-Nelson method. As follows from (19), under the conditions of our experiment the reducing end groups formed belong mainly to oligosaccharides produced by endoglucanase. As follows from Table 3, both methods produce similar results; although the cellulase complexes came from different sources.

B. The Activity of Exo-1,4-β-Glucosidases

The only quantitative method to determine the activity of exoglucosidases of the cellulase complex was described ten years ago by Reese (20). Since that time this method has not been widely used on a practical basis. The reasons behind this are that the method requires the use of a rather expensive substrate, i.e. cellotetraose, and the method itself is limited by questionable assumptions. For example, it is assumed that for all cellulase complexes the ratio of the rates of glucose formation from cellotetraose and from cellobiose is 100 for exoglucosidase and 0.2 for cellobiase. This assumption seems to be unjustified; in the first place, differences in the substrate specificity of the enzymes from different sources are most likely to exist; and in the second place, the assumption does not take into account a probable contribution of endoglucanases to the rate of cellotetraose hydrolysis. Nevertheless, this work (20) was very important because it proved the exis-

TABLE 3

COMPARISON OF THE INITIAL VELOCITIES OF HYDROLYTIC CLEAVAGE OF SOLUBLE CM CELLULOSE BY THE CELLULASES OF DIFFERENT ORIGINS AS MEASURED BY VISCOMETRIC TECHNIQUE AND NON GLUCOSE REDUCING SUGAR METHOD

Preparation	v_o (μmoles/min/g)	
	Viscometric Method	Somogyi-Nelson Method[(a)]
Cellobronine G3X	26	17
Cellocandine G3X	41	45
Celloviridine G3X	70	95
Cellolignorine P10X	36	43
Cellokoningine P10X	120	85
Cellocandine G10X	130	100
Pectofoetidine G3X	22	27
Thermoactynomyces	43	47
T. reesei	2000	1800
T. longibrachiatum	1000	720
Rapidase	700	760
Asp. niger, Serva	120	110
Asp. niger, Koch-Light	290	300
Asp. niger, Sigma	110	120

(a) Reducing sugars minus glucose

tence of exoglucosidases in a great number of different cellulase complexes.

We have developed a quantitative method to determine the activity of exoglucosidases in cellulase complexes (18), which lacks for the most part the disadvantages of the Reese method. In particular, our method envisages the use of CM cellulose as a substrate. This new method does not require any assumption whatsoever on the similarity of the reactivity of exoglucosidases from different origins. Figs. 2 and 3 show some results using this method to determine the activity of exoglucosidases in cellulase complexes from *T. reesei* and *G. candidum*. As can be seen from Fig. 2, the same experiment also can help to determine the activity of another enzyme, cellobiohydrolase. The method is based on the fact that glucose may be produced from CM cellulose under the effect of two enzymes of the cellulase complex, exoglucosidase by itself or cellobiase in the presence of endoglucanase. After determining the rates of D-glucose formation in the presence of the cellulase plus various concentrations of added cellobiase, the data can be extrapolated to the intersection with the ordinate (Figs. 2 and 3). This intercept represents the exoglucosidase activity in the cellulase.

Since both V_m and the Michaelis constant for the cellobiase proper and the added cellobiase may differ, we increased the accuracy of the experiment by measuring the V_m/K_m ratio for both cellobiase enzymes in separate experiments; the results are plotted on the abscissa in Figs. 2 and 3. A detailed kinetic description of the proposed method is given in (18). As shown earlier (19), the contribution of endoglucanase to glucose formation from CM cellulose was negligible. The values of exoglucosidase activities for a number of cellulase complexes of different origins are shown in Table 2.

C. The Activity of Exo-1,4-β-Glucanases (Cellobiohydrolases)

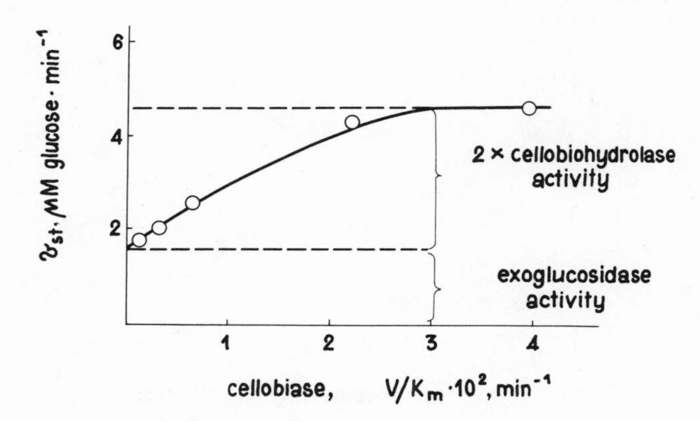

Fig. 2. Determination of the steady state activity of exoglucosidase and cellobiohydrolase in cellulase complex from *T. reesei*. Cellobiase preparation added was isolated from *Asp. foetidus*. Conditions: CM cellulose 0.1%, cellulase 0.01 g/l, pH 4.5, 40°C. Horizontal axis V_m/K_m.

To the best of our knowledge, no data have been published on the activity of cellobiohydrolases from cellulase complexes; and there has been no method to determine the activity of this enzyme up to now. A graphical method to determine the activity of cellobiohydrolase in a mixture with other cellulolytic components has been described by us (18). A practical example of cellobiohydrolase activity determination is illustrated in Fig. 2 for the case of cellulase complex from *T. reesei*. As follows from this figure, the activity of the enzyme is determined concurrently with the activity of the second exoenzyme of the cellulase complex, exoglucosidase. The method works when the added cellobiase increases the corresponding rate of D-glucose formation from CM cellulose under the action of the cellobiase enriched cellulase complex. It can be shown easily that the steady state rate of glucose formation reaches the limit v.

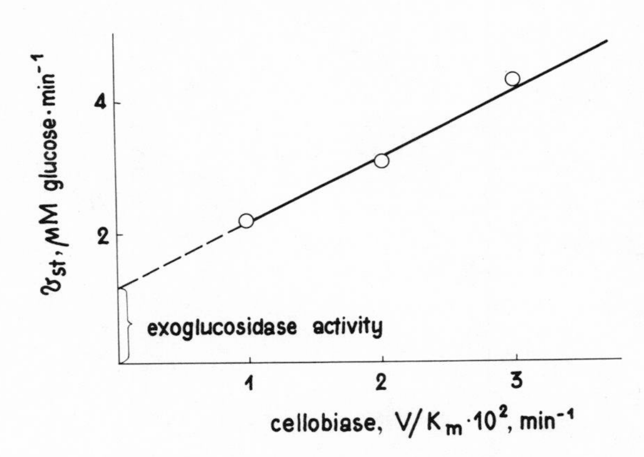

Fig. 3. Determination of the steady state activity of exoglucosidase in a cellulase complex from *G. candidum*. The cellobiase preparation added was isolated from *Asp. foetidus*. Conditions: CM cellulose 0.1%, cellulase 0.01 g/l, pH 4.5, 40°C. Horizontal axis V_m/K_m.

$$v = v_1 + 2\ v_2 \qquad \text{(Eq. 10)}$$

Here v_1 is the rate of glucose formation from higher oligosaccharides as effected by exoglucosidase (Fig. 1), and v_2 is the steady state rate of cellobiose formation from the oligosaccharides under the effect of cellobiohydrolase and probably also endoglucanase. The latter is a demonstration of cellobiohydrolase activity; although the actions of both enzymes in this case are equivalent kinetically. The number 2 in Eq. 10 arises because two glucose molecules are formed as a result of the hydrolytic cleavage of cellobiose. Thus, by determining the rate of glucose formation in the presence of added cellobiase, we can determine the steady state rate of cellobiose formation under the effect of the cellulase complex. This approach makes it possible to estimate the activity of cellobiohydrolase in a mixture with other cellulolytic enzymes. Fig. 2 shows that this activity is determined as half of the distance along the ordinate axis between the two dashed lines. The verification and generalization of this method is underway at the present time in our laboratory.

D. The Activity of Cellobiases

In the literature the activity of cellobiases is usually determined by the rate of accumulation of reducing groups in the solution where enzymatic cellobiose hydrolysis takes place. Our experience has shown that the determination of the D-glucose concentration in such solutions by means of a coupled glucose oxidase peroxidase reaction is more accurate (18,21). In this case, a unit of cellobiase activity is the amount of enzyme that converts 1 micromole of cellobiose, or produces 2 micromoles of glucose, per minute with a cellobiose concentration of 2×10^{-3}M and pH 4.5, 0.05 M acetate buffer, 0.1 M NaCl, and 40°C. Such a concentration of substrate is close to the value of the Michaelis constant for major cellobiases of different origins. Also, at this concentration of cellobiose, the transferase activity of the enzymes is negligible (12).

E. Effect of Ultrasound for Studying the Composition and Properties of the Cellulase Multienzyme Complex

The problem of studying the composition of multienzyme systems and in particular the cellulase complexes, is difficult. The approach normally used is to resolve the enzymatic system into various components by means of biochemical procedures followed by identification of each individual component. Recently, we have developed a new technique (22) to determine the composition and properties of the individual components of such multienzyme systems without the need for resolution into the individual components; and we have tested the technique on the cellulase complex from *G.*

candidum. Our method uses ultrasonics to study the pH induced conformational transition in the active centers of the enzymes (23-25). The method utilizes the observation that ultrasonic cavitation produces hydroxyl and hydroperoxyl free radicals in water solution (23). Penetration of these radicals into the enzyme active center results in enzyme inactivation due to destruction of certain functional groups important for the catalytic activity, typically tryptophan residues. It has been shown elsewhere (23-25) that the first order rate constant for ultrasonic enzyme inactivation is sensitive to the conformational state of the enzyme active center. This dependence is a consequence of the change in accessability of the active center functional groups to the free radicals following a modification in the enzyme conformation. Thus, a kinetic study of the enzyme ultrasonic inactivation allows us to determine the parameters associated with the conformational changes in the enzyme active center.

For example, a study of the pH dependence of the first order rate constant of inactivation, as effected by ultrasound, must have as an outcome certain activity versus pH-profiles. The characteristics of these profiles should be influenced by the pK values of the ionizable groups which control the native conformation of the active center and also by the rate constant of inactivation of the active center in the most or least stable conformation. Since the combination of these values very likely is highly individualized for many enzymes, a certain set of substrates selected for a particular multienzyme system becomes a specific tool for investigating the number of enzymes in the system, their properties with relation to the pH dependencies of conformational transitions, and the specificity of individual components of these multienzyme systems. It should be pointed out that the advantage of the ultrasonic method over a well known alternative reporter group method is that the former does not demand the presence of any reporters, such as chromophores or fluorescent probes which might themselves cause perturbation in the enzyme active center.

When studying the enzymes of the cellulase complexes, we used CM cellulose, cellobiose, and p-nitrophenyl-β-D-glucoside as a particular set of substrates. The enzymatic hydrolysis of these substrates normally gives a clue about the presence in the cellulase complexes of at least two enzymes, endoglucanase and cellobiase. The endoglucanase rapidly reduces the viscosity of the CM cellulose solutions, and the cellobiase cleaves cellobiose with the formation of glucose. The study of other enzymatic activities of the complexes, such as glucose formation from CM cellulose and hydrolysis of p-nitrophenyl glucoside, does not produce unambiguous conclusions about the presence of other enzymes in the cellulase complexes. For example, the formation of glucose from CM cellulose under the action of a cellulase complex, which may contain endoglucanase, may be an indication of the presence of both exoglucosidase and (or) cellobiase in the complex. Furthermore, the hydrolysis of p-nitrophenyl glucoside together with cellobiose may be preconditioned by the presence in the complex of either β-glucosidase or cellobiase and aryl-β-glucosidase. However, the ultrasonic method very definitely identifies at least four enzymes: endoglucanase, exoglucosidase, cellobiase, and aryl-β-glucosidase in the cellulase complex from *G. candidum*.

Fig. 4 shows the kinetic curves for glucose formation from CM cellulose in the presence of a cellulase complex from *G. candidum* before and after sonication. As seen from Fig. 4, the rate of glucose formation after sonication remained significant. Under the same conditions cellobiase was inactivated completely and did not cleave cellobiose. These observations were taken as a basis for the conclusion that the glucose formation from CM cellulose under the effect of the cellulase complex could not be controlled only by cellobiase. Over one third of the glucose yield was produced by exoglucosidase (Fig. 4).

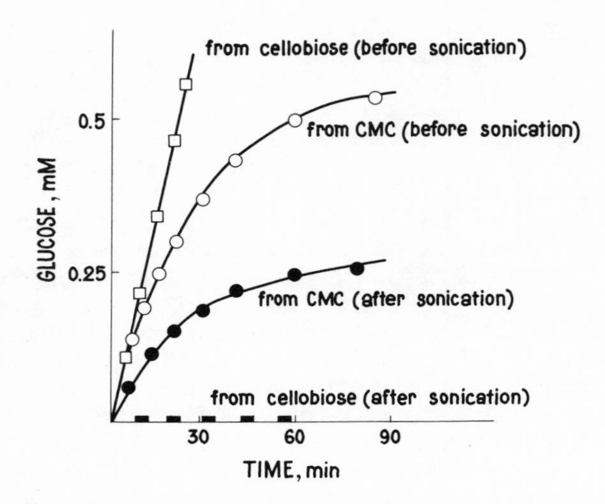

Fig. 4. The effect of sonication of the cellulase complex from *G. candidum* on the rate of glucose formation from CM cellulose (CMC) and cellobiose. The intensity of the ultrasound was 2 watts/cm^2 at 880 kHz. Sonication was carried out for 10 min at pH 9.2 and 40°C with 0.5 g/l cellulase preparation. Conditions for the enzymatic hydrolysis were: CM cellulose 10 g/l, cellobiose 2 mM, pH 4.5, 25°C.

A study of the pH dependence of ultrasonic inactivation of the cellulase complex from *G. candidum* produced additional evidence for the presence of exoglucosidase in the complex. A kinetic interpretation of the results, in terms of the pH dependency of the first order inactivation rate constant, as shown in Fig. 5, led us to conclude that the cellulase complex from *G. candidum* contained at least four cellulolytic enzymes. These enzymes had different ionization group constants, shown in Table 4, which controlled the pH profiles of the ultrasonic inactivation. The most stable

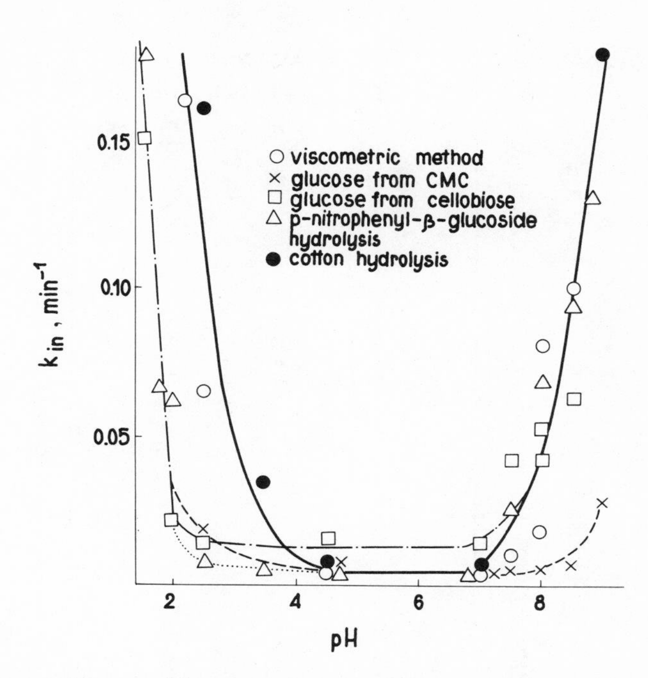

Fig. 5. The effect of pH on the ultrasonic inactivation rate constants of the components of the cellulase complex from *G. candidum*. The viscometric method (O) tests for endoglucanase; the glucose from CM cellulose (x) tests for exoglucosidase and cellobiase; the glucose from cellobiose (□) tests for cellobiase; and the nitrophenyl hydrolysis tests for aryl-β-glucosidase.

pH independent forms of the enzymes had somewhat different inactivation rate constants, also shown in Table 4. As follows from Fig. 5, the ultrasonic inactivation rate constants of the enzymes changed sharply at low or high pH values. It is probable that changed parts of the enzymes formed

TABLE 4

IONIZATION GROUPS OF ENZYMES FROM *G. candidum* CELLULASE COMPLEX

Enzyme	pK_1	PK_2	k_{in} (min^{-1})
endoglucanase	2.2	8.8	0.005 ± 0.001
exoglucosidase	~1.5	~10	0.005 ± 0.001
cellobiase	~1.5	~ 9	0.015 ± 0.001
aryl-β-glucosidase	~1.5	8.8	0.004 ± 0.002

salt bridges, which stabilized the native conformation at the active center and kept it closed with relation to the environment. As soon as the pH became higher or lower than the pK values of the ionic groups, the salt bridge was destroyed; and the active center became accessible to free radicals. As a result, an irreversible enzyme inactivation was accelerated.

Fig. 5 also shows that the most stable component, with respect to ultrasonic cavitation at alkaline pH values, was exoglucosidase in comparison with other enzymes. Thus, the ultrasonic treatment technique may be useful for the preservation of active exoglucosidase in the complex mixture of cellulolytic enzymes after partial inactivation.

III. KINETICS OF MULTIENZYME SYSTEMS AND THE MECHANISM OF CELLULOSE HYDROLYSIS

As was mentioned earlier, the available literature contains no generally accepted concepts about the mechanism of the enzymatic hydrolysis of cellulose. Some authors believe that the differences in the enzymatic hydrolysis of soluble and insoluble cellulose are indicative of different mechanisms of hydrolysis. The authors speculate that special enzymes may participate in native cellulose hydrolysis but not with soluble substrates. In our opinion, this dichotomy is the result of different kinetics for the same mechanism which controls the enzymatic hydrolysis of both soluble and insoluble cellulose. If this is the case, then one should be able to incorporate these different kinetic expressions into an integrated kinetic theory for multienzyme cellulase systems.

We approach the development of an integrated kinetic theory for the cellulase system by ascribing a specific rate expression to most of the pathways of Fig. 1, and shown here in Fig. 6. The four enzymes are indicated by E_1 for endoglucanase, E_2

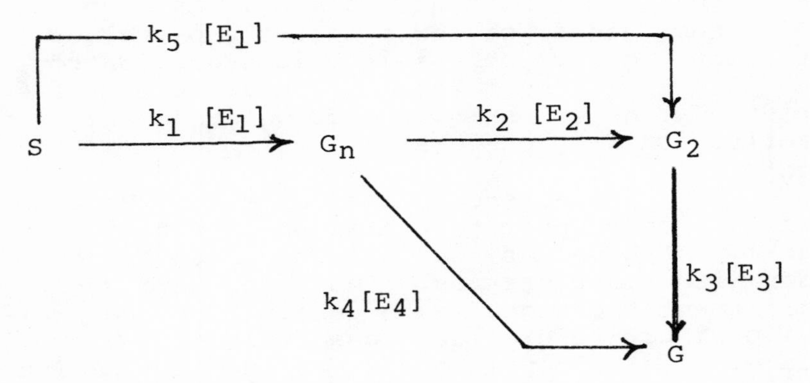

Fig. 6. Kinetic pathways for integrated kinetic theory of the cellulase system.

for cellobiohydrolase, E_3 for cellobiase, and E_4 for exoglucosidase. The rate constants for the individual enzymatic reactions in Fig. 6 may be complicated functions of the concentrations of initial substrate and intermediate metabolites. However, in this paper we will discuss only the simple case in which the initial substrate concentrations and the concentrations of transient products are considerably lower than the values of the Michaelis constants for the corresponding stages of the reaction. These simple kinetics can be observed easily in experiments with the hydrolysis of both soluble and insoluble cellulosics. In this case, the constants k_1 to k_5 can be determined for the soluble substrates as the ratio $k_{cat}/K_{m(app)}$ for each stage. For the insoluble substrates this ratio also is influenced by the adsorption and diffusional characteristics of the reacting materials. In comparing Figs. 6 and 1, the direct glucose formation from the primary polymer substrate and the direct cellobiose formation due to cellobiohydrolase have been omitted in the kinetic description. This simplification is justified since at typical concentrations of the cellulase complex the glucose formation from both soluble and insoluble substrates is characterized by a significant lag time; this is an indication that direct formation of glucose from the initial substrate does not contribute significantly to the overall process. The negligible effect of exoglucosidase on the primary high polymer substrate, observed at relatively low concentrations of substrate, is obvious because the concentration of the end parts of the polymer molecule would be extremely low. That is why the effect of cellobiohydrolase on the initial cellulosic substrate also can be neglected.

Let us now discuss the basic kinetic regularities observed in the process of glucose formation under the action of the cellulase multienzyme complex. The enzymatic hydrolysis of cellulose can take place under different kinetic conditions, which are determined by the composition of the cellulase complex and by the conditions of the experiment. An analysis of all possible conditions goes beyond the scope of this paper; so that we use for our discussion here only the following conditions:

1. The conversion of substrate into glucose is assumed to be small.
2. The consumption of the primary substrates into the intermediates G_n and G_2, may be small or large, depending on the composition of the cellulase complex and the activities of the components. If the consumption is small, then the substrate concentration [S] is essentially the same as the initial substrate concentration $[S]_o$. However, if the consumption is large, the $[S]_o$ equals the sum of the S, G_n, and G_2 concentrations.
3. The steady state concentration of intermediate oligosaccharides, G_n, is attained very rapidly and remains practically constant throughout the experiment.
4. The concentration of intermediate cellobiose, G_2, in one case may be constant with time or in another case may be continuously increasing during the reaction period under study.

$$\frac{d[G]}{dt} = k_3\ [E_3][G_2] + k_4\ [E_4][G_n] \qquad \text{(Eq. 11)}$$

$$\frac{d[G_2]}{dt} = k_5\ [E_1][S]_o + k_2\ [E_2][G_n] - k_3\ [E_3]\ [G_2] \qquad \text{(Eq. 12)}$$

$$[G_n] = \frac{k_1\ [E_1][S]_o}{k_2\ [E_2] + k_4\ [E_4]} \qquad \text{(Eq. 13)}$$

$$[G_2] = \left[\frac{k_1[E_1]k_2[E_2] + k_5[E_1](k_2[E_2] + k_4[E_4])}{k_3[E_3](k_2[E_2] + k_4\ [E_4])}\right]\left[(1-e^{-k_3[E_3]t})\right][S]_o \qquad \text{(Eq. 14)}$$

$$\frac{d[G]}{dt} = (k_1+k_5)[E_1][S]_o - \left[\frac{k_1[E_1]k_2[E_2] + k_5\ [E_1](k_2[E_2]+k_4[E_4])}{k_2[E_2] + k_4\ [E_4]}\right][S]_o e^{-k_3[E_3]t} \qquad \text{(Eq. 15)}$$

Fig. 7. Quantitative expressions based on the kinetic pathways of Fig. 6.

A. Kinetics of Enzymatic Hydrolysis of Cellulose Under Nonsteady State Conditions

A kinetic analysis of the reactions in Fig. 6, when the initial substrate is converted only to a small degree, provides the relationships shown in Fig. 7 for the rates of formation of glucose (Eq. 11) and cellobiose (Eq. 12) as well as for the steady state concentration of intermediate oligosaccharides G_n (Eq. 13). By substituting Eq. 13 into Eq. 12, followed by integration with $[G_2]$ equal to 0 at time zero, we obtain Eq. 14, which describes the concentration of intermediate cellobiose G_2 as a function of time. Further substitution of Eq. 13 and Eq. 14 into the equation for the rate of glucose formation (Eq. 11) gives Eq. 15 for the presteady state rate of glucose formation. This rate depends on the concentrations and activities of all four components of the cellulase complex.

B. Kinetics of Enzymatic Hydrolysis of Cellulose Under Steady State Conditions

Eq. 15, the expression for the rate of end product formation, when only a small amount of the initial substrate has been reacted, contains an exponential term, which disappears as the steady state is approached $(t \quad \infty)$. Eq. 15 then is reduced to

$$\left(\frac{d[G]}{dt}\right)_{st} = k_{app}\,[E_1][S]_o \qquad \text{(Eq. 16)}$$

where $k_{app} = k_1 + k_5$. An important conclusion can be drawn from this equation, in that the rate of formation of the end product, glucose, under steady state conditions is determined only by the kinetic parameters of the first enzyme in a multienzyme system; for a general case see (26). Using the overall kinetic scheme shown in Fig. 6, one can conclude that the steady state rate of glucose formation must be proportional to the activity of the endoglucanase in the cellulase complex, provided that the cellulose has been transformed only to a very limited extent. Consequently, the important question of which component of the cellulase complex attacks native cellulose can be answered by finding a linear correlation between the steady state rate of glucose formation, for a number of cellulase complexes, and the activity of the individual components of the cellulase complex. As will be shown further, this kind of linear correlation can be observed for insoluble substrates in relation to the activity of only one cellulolytic component, endoglucanase.

The above considerations hold true for the case when the concentration of the primary substrate is significantly higher than that of the intermediate metabolites. This condition is fulfilled, in particular, when the first enzyme in the multienzyme sequence is rate limiting, i.e. and $k_1[E_1] << k_2[E] + k_4[E]$. However, these relations are often not observed. Then, for kinetic analysis of such a system, one must take into account the fact that some of the initial substrate has reacted so that $[S]_o$ is made up of the concentrations of S, G_n, and G_2. With this expression, the steady state concentrations of G_2 and G_n can be obtained in a manner similar to obtaining Eq. 13. These can then be incorporated as before (Eqs. 11, 14, and 15) to produces Eq. 17 (Fig. 8) for the steady state rate of glucose formation. As before $k_{app} = k_1 + k_5$. It is obvious that at $k_1[E_1] <<$ $k_2[E_2] + k_4[E_4]$, $k_1[E_1] << k_3[E_3]$, and $k_5[E_1] <<$ $k_3[E_3]$ that Eq. 17 reduces to Eq. 16. The denominator of Eq. 17 (Y) is directly related to the degree of conversion of the primary substrate into the intermediate metabolites G_n and G_2 before the system attains a steady state. Thus, the rate of the end product formation, glucose, becomes Y times lower than the rate of the first enzymatic reaction in the multienzyme sequence. From a practical standpoint this means that as soon as the value of Y becomes greater than unity, the rate limiting action of the endoglucanase E_1, is replaced by the rate limiting action of E_2, E_3, or E_4.

$$\left[\frac{d[G]}{dt}\right]_{st} = \frac{k_{app}[E_1][S]_o}{1 + \frac{k_1[E_1]}{k_2[E_2]+k_4[E_4]} + \frac{k_1[E_1]k_2[E_2]+k_5[E_1](k_2(E_2]+k_4[E_4])}{k_3[E_3](k_2[E_2]+k_4[E_4])}}$$

(Eq. 17)

Fig. 8. Steady state rate of glucose formation when a considerable amount of the initial cellulose substrate has been converted to intermediate products.

From Eqs. 14 and 15 one can draw the important conclusion that the times required to attain a steady state concentration of intermediate cellobiose and a constant rate of glucose formation may not be the same. From Eq. 14 it follows that the cellobiose concentration approaches within 10% of its steady state value as soon as the conditions given in Fig. 9 are satisfied. On the other hand, from Eq. 15 it may be concluded that the rate of glucose production should be within 10% of being constant as soon as the conditions also shown in Fig. 9 are satisfied. By comparing Eqs. 18 and 19 we may conclude that in general the rate of glucose formation approaches a stationary level before the steady state concentration of cellobiose is established. Furthermore, the difference between the time required for these steady state conditions to be attained will be greater for low activity cellobiase, E_3, and a high activity ratio of exoglucosidase, E_4, to cellobiohydrolase, E_2. These time periods will be equal only in the case when exoglucosidase activity becomes negligibly small in comparison with the activity of cellobiohydrolase.

cellobiose:

$$e^{-k_3[E_3]t} \leqslant 0.1$$

or

$$t_{st} \geqslant \frac{2.3}{k_3[E_3]} \qquad \text{(Eq. 18)}$$

Glucose:

$$\left[\frac{k_1[E_1]k_2[E_2]}{k_2[E_2]+k_4[E_4]} + k_5[E_1]\right] e^{-k_3[E_3]t} \leqslant 0.1(k_1+k_5)[E_1]$$

or

$$t_{st} \geqslant \frac{2.3}{k_3[E_3]} - \frac{1}{k_3[E_3]} \ln\left[\frac{k_1 + k_5}{k_1 \dfrac{k_2[E_2]}{k_2[E_2]+k_4[E_4]} + k_5}\right]$$

(Eq. 19)

Fig. 9. Criteria for attainment of steady state cellobiose concentration and constant glucose production rate.

And finally, it follows from Eqs. 14 and 15 that the steady state enzymatic hydrolysis of cellulose can be reached not only by stretching out the reaction period but also by increasing the concentration of cellobiase, E_3, in the reaction system.

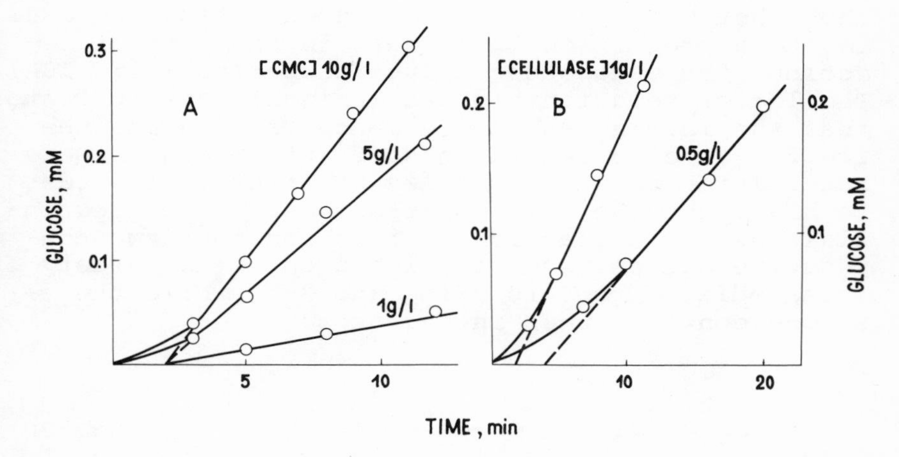

Fig. 10. The kinetics of glucose formation during the initial period of CM cellulose hydrolysis by a cellulase complex from *G. candidum*. Part A: variation of CM cellulose (CMC) concentration, cellulase at 1 g/l. Part B: variation of cellulase complex concentration, CM cellulose at 5 g/l. Conditions: pH 4.5, 40°C.

The latter development is expected to be accompanied by a decrease in the steady state concentration of intermediate cellobiose and an additional increase in the stationary rate of glucose formation. Even if the first enzyme is still rate limiting; this effect takes place at the expense of a decrease of the inhibitory action of cellobiose on endoglucanase.

C. Hydrolysis of Soluble CM Cellulose

The kinetics for the production of glucose by a cellulase complex acting on soluble CM cellulose were typical of a sequential enzymatic reaction (Fig. 10). With an increase in enzyme concentration, the stationary rate of glucose formation also increased proportionally, while the induction period for the reaction decreased (Fig. 10B). On the other hand, a substrate concentration increase led to a proportional increase in the rate of reaction, for a constant induction period (Fig. 10A). The latter relationship was a significant indicator that the initial substrate concentration and the steady state concentrations of intermediates were not related to the Michaelis constants. In order to highlight the most important aspects of sequential enzymatic reactions let us consider two consecutive stages, one of which produces a variety of intermediate products P (G_n and G_2), while the second converts them into glucose:

$$S \xrightarrow[E_1,\ E_2]{k_{1,2}} P \xrightarrow[E_3,\ E_4]{K_{3,4}} G \qquad \text{(Eq. 20)}$$

In this case, $k_{1,2}$ and $k_{3,4}$ are apparent first order rate constants. The first constant depends only on the activities of endoglucanase and cellobiohydrolase, while the second is related only to the activities of cellobiase and exoglucosidase. This kind of transformation is a simplification; but as will be shown later, this simplification agrees well with experimental data.

An analysis of Eq. 20 with due account for substrate consumption and the extent of its transformation into intermediate products brings us to a series of equations:

$$\frac{d[P]}{dt} = k_{1,2}\,[S] - k_{3,4}[P] \qquad \text{(Eq. 21)}$$

$$\frac{d[G]}{dt} = k_{3,4}\,[P] \qquad \text{(Eq. 22)}$$

$$[G] = \frac{k_{1,2}k_{3,4}}{k_{1,2}+k_{3,4}} \left[t - \frac{1}{k_{1,2}+k_{3,4}} \right] [S]_o + \qquad \text{(Eq. 24)}$$

$$\frac{k_{1,2}k_{3,4}}{(k_{1,2}+k_{3,4})^2} [S]_o \, e^{-(k_{1,2}+k_{3,4})t}$$

$$[G] = \frac{k_{1,2}k_{3,4}}{k_{1,2}+k_{3,4}} \left(t - \frac{1}{k_{1,2}+k_{3,4}} \right) [S]_o \qquad \text{(Eq. 25)}$$

Fig. 11. Simplified kinetics of glucose formation from CM cellulose.

$$[S]_o = [S] + [P] \qquad \text{(Eq. 23)}$$

The solution of these expressions leads to Eq. 24 (Fig. 11), which describes the kinetics of glucose formation for a limited degree of conversion. One can see that at $t > 1/(k_{1,2} + k_{3,4})$ this expression is reduced to that of Eq. 25. Now Eq. 24 can be rewritten as

$$[G] = v_{st}\,(t - \tau) + v_{st}\,t e^{-t/\tau} \qquad \text{(Eq. 26)}$$

where τ is the duration of the induction period and v_{st} is the steady state rate of reaction for Eq. 20, as given by

$$v_{st} = \frac{k_{1,2}\,k_{3,4}}{k_{1,2} + k_{3,4}} [S]_o \qquad \text{(Eq. 27)}$$

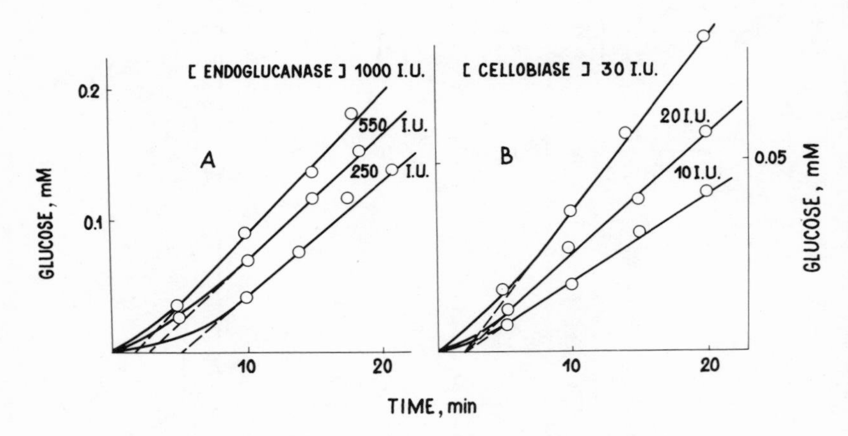

Fig. 12. The effect of added endoglucanase or cellobiase on the kinetics of glucose formation from CM cellulose under the action of a cellulase complex from *G. candidum*. Part A: crude cellulase complex at 0.5 g/l with endoglucanase activity of 250 I.U.); added endoglucanase was the Rapidase form; conditions: CM cellulose 0.5%, pH 4.5, 40°C. Part B: purified cellulase complex 0.01 g/l with cellobiase activity of 10 I.U.; added cellobiase was from *Asp. foetidus*; conditions: CM cellulose 0.1%, pH 4.5, 40°C.

If we now assume that the concentrations of the intermediate products do not change with time, the induction period of the reaction may be obtained by extrapolating to the time axis the linear portion of the glucose production. The extrapolated intercept, τ, is equal to $1/(k_{1,2} + k_{3,4})$. Eq. 28 and the induction period expression confirm the obvious conclusion that the stationary rate of glucose formation is always controlled by the slowest step, while the induction period is controlled by the fastest step in the sequence. This means that if a change in the concentration of one of the enzymes results in a different lag period but the same stationary rate of glucose formation, then this enzyme is not rate limiting. However, if the change in enzyme concentration effects the rate of glucose formation under steady state conditions, then this enzyme may be considered as the rate limiting factor. This conclusion is substantiated by the data in Fig. 12, which shows that endoglucanase is not the enzyme which limits the rate of glucose formation in the hydrolysis of soluble CM cellulose. Actually, during the initial period of CM cellulose hydrolysis, the rate of glucose formation is considerably lower than that of other reducing oligosaccharides (Fig. 13). Moreover, as can be seen from Tables 1 and 2, in many cellulase preparations the activities of cellobiase and exoglucosidase are substantially lower than those of endoglucosidase. These

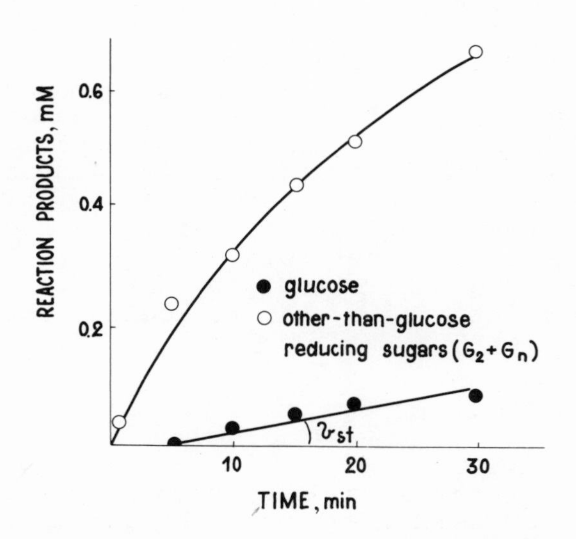

Fig. 13. The kinetics of the formation of glucose and other reducing sugars during the initial period of the hydrolysis of CM cellulose by a cellulase complex from *G. candidum* at 1 g/l. Conditions: CM cellulose 0.5%, pH 4.5, 40°C.

considerations prompt us to conclude that the role of endoglucanase in the process of glucose formation from CM cellulose resides apparantly in a rapid conversion of substrate into intermediate oligosaccharides, which are then slowly cleaved by other enzymes to produce the end product.

Fig. 12B shows that the addition of cellobiase to the cellulase complex increases significantly the rate of glucose formation with the time lag of the reaction intact. Thus, when the glucose production process goes through a stage of cellobiose formation, the cleavage of the latter is a rate limiting step in the sequence.

Let us see now if cellobiase is the only component which is really responsible for the rates of glucose formation during the hydrolysis of CM cellulose. If glucose production is actually controlled by cellobiase, then the concentration of cellobiose in the reaction mixture must relate quantitatively to the observed rate of glucose formation, according to the Michaelis Menten equation. In reality, however, this is not the case. Theoretical estimates of the cellobiose concentrations that should be present in the reaction mixture if all of the cellobiose was used for glucose formation are considerably greater than experimentally observed concentrations of cellobiose and oligosaccharides. In other words $[G_2]_{theor.} > ([G_2] + [G_n])_{exp.}$ The theoretical concentration of cellobiose is calculated by the following expression:

$$[G_2]_{theor.} = \frac{K_3}{V_3/v_{st} - 1} \qquad \text{(Eq. 28)}$$

where v_{st} is the stationary rate of glucose formation from CM cellulose under the action of the cellulase complex and K_3 and V_3 are the Michaelis constant and maximum velocity respectively, for cellobiase in the cellulase complex, as determined in a separate experiment. It should be emphasized that even the cellulase preparations of high cellobiase activity, for example the purified preparations from *T. lignorum* and *G. candidum* listed in Table 1 and for which $[G_2]_{theor.} < ([G_2] + [G_n])_{exp.}$, showed a significant role for exoglucosidase in the process of glucose formation from CM cellulose. This experiment was carried out by establishing that with moderate concentrations of cellobiase added to the cellulase complex, the rate of glucose formation from CM cellulose was linearly dependent on the total activity of cellobiase in the complex (Fig. 3). The extrapolation of this straight line dependence to zero concentration of cellobiase in the cellulase complex showed, that even in the virtual complete absence of cellobiase, the rate of glucose formation remained high. These data point to the important and even decisive role of exoglucosidase in the cellulase complexes in regard to glucose formation from CM cellulose. This enzyme, as was discussed above, produces glucose from oligosaccharides, which are the products of endoglucanase action. Since endoglucanase does not limit glucose formation in the multienzyme sequence (see above), the exoglucosidase becomes a rate limiting enzyme in this case.

This kinetic analysis suggests that in the sequence described by Eq. 20, the factor which controls the glucose formation is the stage of cleavage of the intermediate products, cellobiose and oligosaccharides, under the action of cellobiase and exoglucosidase ($k_{1,2} >> k_{3,4}$). So having in mind the fact that for many cellulase complexes the exoglucosidase appears to be a decisive factor in the control of glucose formation, the enzymatic hydrolysis of CM cellulose can be represented by the following pathway:

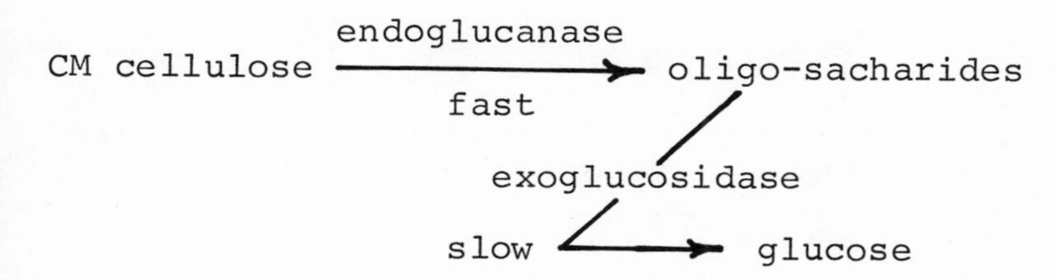

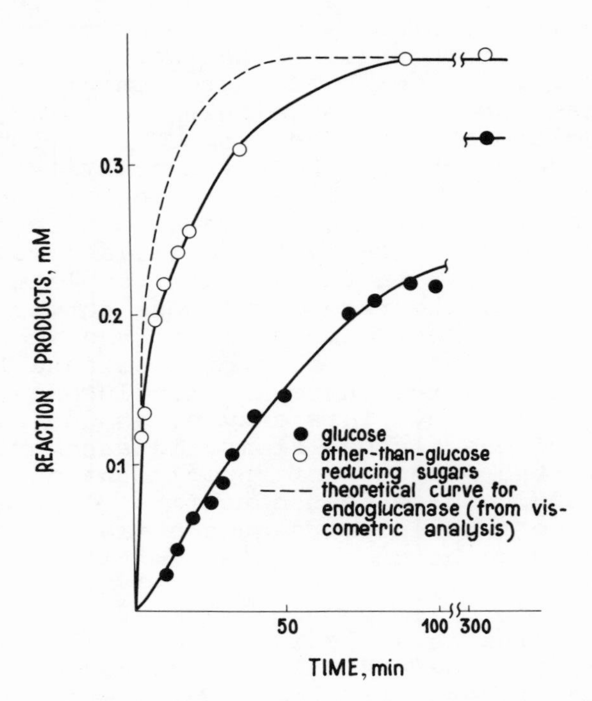

Fig. 14. The kinetics of the extensive hydrolysis of CM cellulose by the cellulase complex from *G. candidum* at 0.5 g/l. Conditions: CM cellulose 0.2%, pH 4.5, 40°C.

Let us now discuss the kinetics of the formation of glucose and other reducing sugars in the case of extensive enzymatic hydrolysis of CM cellulose. Fig. 14 shows that the concentration of other reducing sugars reaches a nearly constant level in a relatively short period of time and remains at this level almost to the end of the reaction. As is known, the incomplete enzymatic conversion of CM cellulose to glucose is determined by the presence of substituted glucoside residues, which hinder the enzyme action (27,28). Let us see which factor determines the extent of CM cellulose conversion under the action of the cellulase complex. In accordance with the contemporary understanding of the phenomenon, a random reaction between endoglucanase and CM cellulose takes place only at the portions of the substrate which are composed of some minimal number of unsubstituted glucoside residues (28). On the other hand, depolymerization of CM cellulose under the effect of exoenzymes, which cleave glucose or cellobiose units from the nonreducing end of the polymer molecule, stops as soon as the successive cleavage reaches the substituted residue. That is why the individual exoenzymes are incapable of producing an extensive hydrolysis of CM cellulose (29,30).

Endoglucanases open up internal nonsubstituted regions of the CM cellulose molecule and makes them accessible for exoenzymes. This effect leads to synergism, where the added effect of the endoenzymes and exoenzymes acting together is greater than that which could be produced by the same enzymes acting independently (29,30).

If the cellulase complex contained all four components, the end products of CM cellulose hydrolysis would be only glucose and substituted oligosaccharides, resistant to the further action of endoenzymes and exoenzymes. The concentrations of the reducing groups which belong to these substituted oligosaccharides ceases to build up upon termination of the action of endoglucanases, since cleavage of oligosaccharides from the nonreducing end may take place only further along. It was noted above, that the rate of endoglucanase action is considerably greater than that of glucose formation. That is why the formation of other reducing groups, which belong mainly to oligosaccharides is completed faster than that of glucose (Fig. 14). Thus, the limiting concentration of other reducing sugars seems to be determined only by a substrate specificity of endoglucanase. If so, the minimal length of the nonsubstituted part of CM cellulose, which is effected by the action of endoglucanase, can be calculated. Based on the assumption that substituted glucose units are randomly distributed in the CM cellulose molecules, one can prove easily that the proportion of the parts of the molecule containing $i \geqslant \ell$ nonsubstituted residues is equal to $(1 - s)^{\ell}$. Here s is the portion of the substituted residues in the polymer, and ℓ is the minimal length of the nonsubstituted part of the molecule which is effected by endoglucanase. The total concentration of the parts with $i \geqslant \ell$ and which are effected by endoglucanase is

$$\frac{C}{M}(1-s)^{\ell}$$

where C is the weight concentration, and M is the average molecular weight of the monomeric radical of the polymer chain. If the concentration of nonsubstituted residues accessible to endoglucanase is considered to be equal to the final concentration of other reducing sugars, we can carry out some transformations and get the expression for ℓ:

$$\ell = \frac{\log([P]_{t\to\infty} \quad M/C)}{\log(1-s)} \qquad \text{(Eq. 29)}$$

Our estimates show that for most of the endoglucanases tested, this value applies to the nonsubstituted residues (discussed later and in Table 8).

The limiting concentration of glucose apparently may serve as an indicator of the number of nonsubstituted residues that can be cleaved by exoenzymes from oligosaccharides. On the average one molecule of oligosaccharide is equivalent to one molecule of glucose. This, however, does not mean that glucose is cleaved from every oligosaccharide molecule. The more probable case is when the action of endoglucanase produces both susceptible and inert oligosaccharides; the former are later cleaved and produce a few glucose molecules while the latter produce no molecule of glucose. We believe that the formation of substrates for the exoenzymes takes place at the beginning of the endoglucanase action, when the longest nonsubstituted parts of CM cellulose and those most vulnerable to hydrolysis are attacked; although their concentration in the polymer is low. The evidence suggests that a reduced rate of formation of glucose actually is present during the accumulation of oligosaccharides (Fig. 14).

A comparison of the results for the viscometric analysis and for the kinetics studies of the other reducing sugars shows that not all accessible parts of CM cellulose have similar vulnerability to hydrolysis. As follows from Fig. 14, a theoretical progress curve for the endoglucanase action, based on the kinetic parameters determined by viscometric measurements, differs substantially from the experimental curve for the other reducing sugars. The theoretical and experimental data agree well only for the initial reaction period. This is another illustration of the fact that endoglucanases first split the most accessible nonsubstituted randomly distributed parts of the CM cellulose molecule, and then act on the nonsubstituted parts having a minimal length acceptable for action by the enzyme. We are convinced that under such experimental conditions there is no significant product inhibition or inactivation of the enzyme, which could bring about similar kinetic consequences.

One of the very convincing facts that the hydrolysis of the most resistant parts of CM cellulose does not produce a substrate for exoenzymes was obtained when we observed a concurrent action of cellulase complexes from *G. candidum* and *Asp. niger*. As follows from Fig. 15, the hydrolysis of CM cellulase under the action of the *Asp. niger* cellulase complex produces a limiting concentration of other reducing sugars which is twice as low as that produced by the *G. candidum* enzyme complex. In accordance with the theoretical considerations discussed earlier, this means that the minimal length of the nonsubstituted part of the CM cellulose molecule sensitive to the hydrolytic action of endoglucanase from *Asp. niger* is greater than

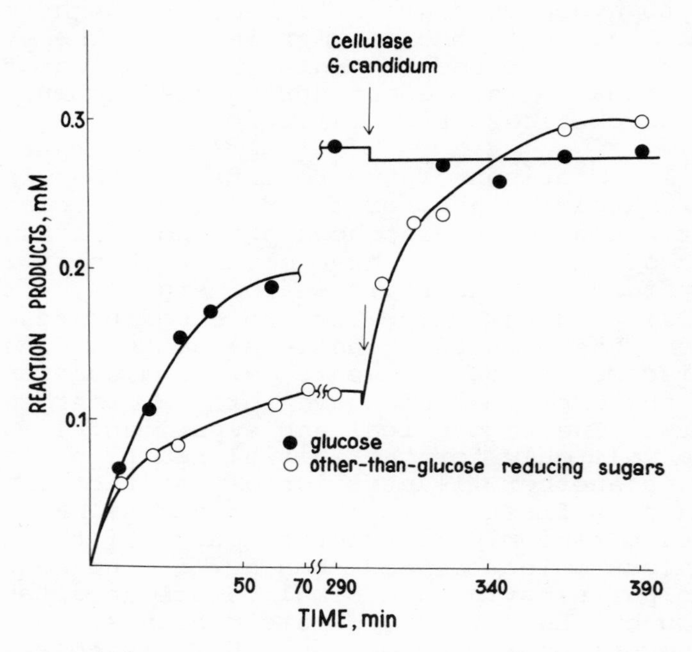

Fig. 15. The kinetics of the extensive hydrolysis of CM cellulose under the consecutive action of cellulase preparations from *Asp. niger* at 0.5 g/l and *G. candidum* at 0.4 g/l. Conditions: CM cellulose 0.2%, pH 4.5, 40°C.

that for the endoglucanase from *G. candidum* (also see Table 6). However, the limiting concentration of glucose which is produced by the preparation from *Asp. niger* is about the same as that produced by the preparation from *G. candidum*. Thus, every longer oligosaccharide gives in the former case two instead of one molecule of glucose. If the preparation from *G. candidum* is added to the products of the CM cellulose hydrolysis produced by the preparation from *Asp. niger*, a considerable increase of other reducing sugars is observed without concurrent glucose formation. It means that the substrate for glucose formation is really produced only during the initial period of endoglucanase reaction with CM cellulase.

In summary, the basic conclusion of this part of the study is that the CM cellulose hydrolysis under the effect of the cellulase complex goes as follows: the hydrolysis is started by rapid cleavage of a small number of the longest and the most accessible nonsubstituted parts of the initial substrate under the action of endoglucanase; this produces substrates for exoenzymes, mainly for exoglucosidase. The exoglucosidase at a later stage of the reaction cleaves glucose from these substrates at a comparatively lower rate. Concurrently, endoglucanase slowly splits the shorter nonsubstituted parts; but the oligosaccharides thus produced do not serve as substrates for glucose formation. The action of endoglucanase stops as soon as the remaining parts of the CM cellulose molecule contain no more than four anhydroglucose units in length. During the initial stages of the hydrolysis, endoglucanase produces the major portion of the reducing sugars, while exoglucosidase produces the major portion of the glucose.

D. Hydrolysis of Native Cellulose (Cotton)

An ultimate goal of our study of the enzymatic hydrolysis of native cellulose is to identify the component of the cellulase multienzyme complex which is the first to attack native cellulose. A kinetic analysis of the Fig. 6 pathways was selected as a first approximation of cotton hydrolysis. From this analysis we concluded that if this scheme is adequate and if the first component attacking cellulose is really endoglucanase, then when this first enzyme becomes rate limiting there must be a linear correlation between the stationary rate of glucose formation and the activities of endoglucanases in the cellulase complexes. Since the above considerations must be true only for low conversion of the primary substrate, we designed an experiment in such a way that the solubilization of ball milled cotton did not exceed 10%, and its conversion to glucose was no more than 3-7%.

In order to identify the component which limited the rate of glucose formation from the insoluble substrate, we determined the steady state rates of ball milled cotton hydrolysis under the effect of the cellulase preparations shown in Table 5. The preparations were substantially different as judged by the absolute activities of their components and by the activity ratios. For example, the endoglucanase and cellobiase activity ratio changed from 0.05 in the complex from *Asp. foetidus* to 60 for the complex from *G. candidum*, the latter being enriched by endoglucanase. As was predicted in an earlier section, the concentration of intermediate cellobiose was either constant over a certain period of time or increased gradually throughout the part of the reaction under study. The constant part was demonstrated by the preparations of *G. candidum* and *Asp. niger* (Fig. 16) and the changing part by preparations of *T. lignorum* and *G. candidum* enriched by endoglucosidase (Fig. 17). These variations are predetermined by the composition of the cellulase complexes and the activities of the components. It must be noted, however, that in all instances the time period required to attain a stationary rate of glucose formation was much shorter than that to attain a steady state concentration of intermediate cellobiose. A relatively short lag phase of the reaction made it possible to determine the stationary rates of glucose formation in all cases (Table 6) and to relate them to the activities of the individual components of the cellulase complexes. By doing this we found that a linear correlation existed between the stationary rate of glucose production and only the activity of endoglucanase of these cellulase complexes.

In all of the Table 6 cellulase complexes, except the *G. candidum* preparation which was enriched with added endoglucanase, we observed a practically constant ratio of 19 ± 3 for the

TABLE 5

SPECIFIC ACTIVITIES OF COMPONENTS OF
CELLULASE COMPLEXES FROM DIFFERENT SOURCES

Source	Specific Activity (I.U./g)		
	Endoglucanase	Exoglucosidase	Cellobiase
Asp. foetidus	170	-	3400
Asp. niger (Serva)	120	<20	420
T. lignorum	36	11	9
G. candidum	4000	1000	1000
Asp. niger (Koch-Light)	76	<5	27
T. koningii	64	35	13
T. longibrachiatum	37	<1	2.3
G. candidum (a)	5700	250	90

(a) preparation was enriched by endoglucanase

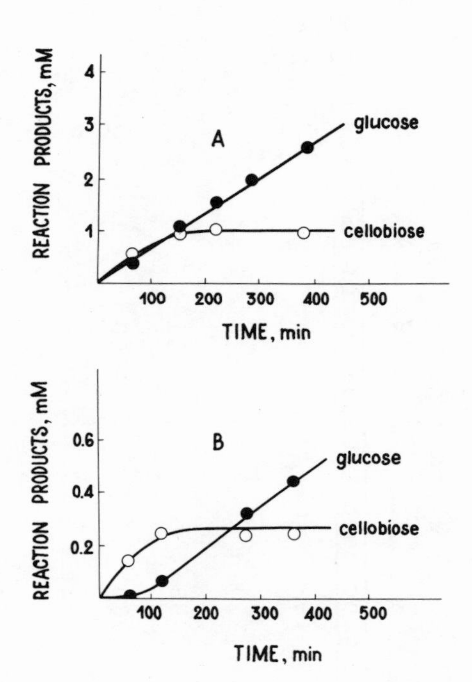

Fig. 16. The kinetics of the hydrolysis of milled cotton by cellulase preparations. Part A: cellulase from *G. candidum* at 0.03 g/l. Part B: cellulase from *Asp. niger* at 0.5 g/l. Conditions: cotton 0.5%, pH 4.5, 40°C.

stationary rate of glucose formation from cotton as compared to the endoglucanase activity of the corresponding cellulase complexes. The latter were determined using CM cellulose as the soluble substrate. In the case of the *G. candidum* preparations listed in Tables 5 and 6 we observed a 7 fold lower stationary rate of glucose formation from cotton as compared to our expected rate. However, the addition of some cellulase preparation from *Asp. foetidus* to this complex increased the endoglucanase concentration by 1.2 fold and that

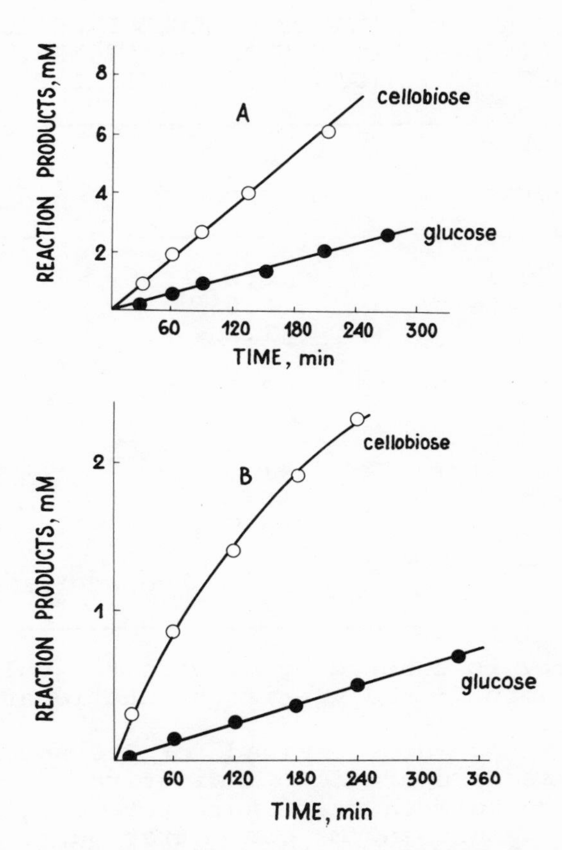

Fig. 17. The kinetics of the hydrolysis of milled cotton by cellulase preparations. Part A: cellulase from *T. lignorum* at 5.0 g/l. Part B: cellulase from *G. candidum* at 0.05 g/l and enriched by endoglucanase. Conditions: cotton 0.5%, pH 4.5, 40°C.

of cellobiase by 56 fold; this addition facilitated the complete hydrolysis of the intermediate cellobiose. As a result, the stationary rate of glucose formation increased and reached a theoretically

TABLE 6

ENDOGLUCANASE ACTIVITY OF CELLULASE COMPLEXES FROM DIFFERENT ORIGIN AND STATIONARY RATE OF GLUCOSE FORMATION FROM MILLED COTTON

Source	v_{st} (μmoles/min/g)	A/v_{st}.[a]
Asp. foetidus	8.8	19
Asp. niger (Serva)	8.0[b]	15
T. lignorum	1.8	20
G. candidum	220	18
Asp. niger (Koch-Light)	3.2	24
T. koningii	4.2	15
T. longibrachiatum	1.8	21
G. candidum	40	140[d]
	350[c]	19[c]
	Average:	19 ± 3

[a] Activity is denoted by A, with the values taken from Table 5; and v_{st} is the stationary rate of glucose formation.
[b] Same values were obtained for 2.0 and 5.0 g/l of cellulase preparation, indicative of linear de - pendence between stationary rate of glucose formation and enzyme concentration.
[c] With added preparation from *Asp. foetidus* to 0.3 g/l to 0.05 g/l of *G. candidum* cellulase complex.
[d] Omitted from average.

expected value proportional to the endoglucanase concentration in the other cellulase complexes (Table 6). Thus, all eight cellulase preparations demonstrated a linear correlation between the endo-glucanase activity and the stationary rate of glu-cose production from cotton such that

$$v_{st} = \frac{\text{activity}}{19 \pm 3} \qquad \text{(Eq. 30)}$$

or

$$v_{st} = 5.3 \times 10^{-2}(\text{activity}),$$

$$\mu\text{mole/min/g} \qquad \text{(Eq. 31)}$$

This correlation substantiates the fact that endoglucanase is a cellulolytic component which initiates the hydrolysis of native cellulose; this evidence also favors the above theoretical considerations. The existence of such a correlation is also an indication of the rate limiting action exerted by endoglucanase on glucose formation in most of the cellulase complexes we studied (Table 6).

The data so far suggest the rejection of the hypothesis on the presence in cellulase complexes of a prehydrolytic C_1-enzyme which was thought to be the first enzyme to attack the native cellulose. In reality, if such a factor did exist, then independent of the mechanism of its action it would have manifested itself in the kinetics of glucose accumulation. In such a case, the steady state rate of glucose formation would have been proportional to the activity of the C_1-enzyme. In actuality, however, the proportionality is observed only for endoglucanase. In the light of our data, the C_1-enzyme hypothesis can not be saved, even by a recently forwarded suggestion of its author (10) that the C_1-enzyme is an endoglucanase with unique specificity that produces only very limited conversion of the insoluble substrate, undetectable by modern physicochemical analytical techniques, but makes the substrate vulnerable with respect to other so called C_x enzymes of the cellulase complex. In terms of chemical kinetics this assumption means, that the C_1-enzyme would need to be a rate limiting factor in cellulose hydrolysis; this in turn would have to result in a linear dependence between the stationary rate of glucose formation and activity of the C_1-enzyme. This is not the case, however, based on our experiments which showed that such a dependence is really linear only for a common endoglucanase of random action (Table 6).

The only argument in favor of the prehydrolytic C_1-enzyme hypothesis would arise if the concentrations of this enzyme happened to be proportional to the concentrations of endoglucanases in all eight of the cellulase preparations we studied; but this seems like an extremely unlikely occurrence. Therefore, our conclusion is that a hypothetical C_1-enzyme, which supposedly first attacks native cellulose, is really the endoglucanase of random action. The question why in some instances cellulase complexes have low reactivity with respect to insoluble cellulose, despite relatively high endoglucanase activity, belongs to the field of regulatory kinetics and will be discussed elsewhere. It must be noted here only that such cellulase complexes do not contain, as a rule, other components, particularly cellobiase which facilitate conversion of intermediate metabolites into glucose, or contain them in insufficient quantities. The cellulase preparations Rapidase and *M. verrucaria*, as well as bacterial and some other cellulases, (Tables 1 and 2) are examples.

It is important to note that the efficiency of splitting glucoside bonds of soluble substrate, shown by endoglucanases of different origin, is proportional to the efficiency of splitting similar bonds in insoluble cellulose. This conclusion which is substantiated by the data in Table 6 is rather unexpected because it seems that endoglucanases of different origin would behave differently with respect to the substrate in these two different physical states. In other words, it might be expected that allegedly different sorption properties of the endoglucanases with respect to the soluble and insoluble substrates would result in different effects on their activities. However, the Table 6 ratios of activities of endoglucanases from different sources, with respect to the glucoside bonds in CM cellulose and milled cotton, are very similar and equal to 19 $\pm$ 3. This fact is evidence in favor of the similarity of endoglucanases of different origin in terms of their catalytic action with respect to native cellulose.

An important outcome of this study is a conclusion about priority of exoglucosidase, E_4, action when compared with the consecutive action of cellobiohydrolase, E_2, and cellobiase, E_3, during initial period of glucose formation from milled cotton. Actually, as follows from the earlier theoretical analysis of the kinetics of enzymatic cellulose hydrolysis, if the rate of glucose formation attains a stationary level faster than the concentration of intermediate cellobiose reaches its constant value, then this must be the result of a higher activity of exoglucosidase in comparison with the activities of cellobiohydrolase and cellobiase. The experimental data related to this question show that in the course of the action on cotton of all the cellulase complexes we have studied, a stationary rate of glucose formation is established within minutes, while the concentration of intermediate cellobiose approaches its steady state level in no less than dozens of minutes (Fig. 16). Sometimes, the steady state conditions for cellobiose formation did not develop within the time limits of our experiment, which normally continued for hours (Fig. 17). In other words, a significant portion of the glucose produced by hydrolysis of both native and soluble cellulose is formed in two consecutive stages, i.e. the action of endoglucanase followed by exoglucosidase.

It must be noted that in the hydrolysis of native cellulose, the relation between the rates of these two stages is inverted in comparison with that for the hydrolysis of a soluble substrate, CM cellulose. In the latter case the second step (Eq. 20) is a rate limiting factor, while in the hydrolysis of cotton a rate limiting role is assumed by the first stage. This changeover in the rate limiting stage of glucose formation associated with the transition from a soluble to an insoluble substrate obviously is affected by the lower rate of action of endoglucanases on the surface of the insoluble substrate. The reason behind this slowdown may be either a nonproductive adsorption of endoglucanases on the insoluble substrate or a diffusional limitation in the course of the degradation of native cellulose.

The role of exoglucosidase was underestimated by many authors (32,33) because they believed that the glucose formation was controlled only by cellobiase. The reason behind that, in our opinion, lays in the lack of appropriate methods to determine the exoglucosidase activity in the cellulase complexes. It is also worthwhile to note that in the hydrolysis of insoluble substrates, exoglucosidase apparently cleaves glucose from the surface of the substrate which already has been partially degraded by endoglucanase. The fact that higher oligosaccharides are not found in the reaction mixture is a result of such a mode of exoglucosidase action.

The kinetic analysis of the cellulase multienzyme systems described in this paper shows the integrity of the mechanism of enzymatic hydrolysis of native cellulose and its soluble derivatives. The integrity of the mechanism follows, in particular, from the observed linear correlation between the rate of native cellulose hydrolysis and that of its soluble derivative, determined under specific experimental conditions. At the same time, the literature contains a generally accepted idea that a correlation of this kind cannot be observed; and moreover, that soluble polymer derivatives of cellulose cannot be used to evaluate the ability of the cellulase complex to hydrolyze native cellulose. The kinetic analysis of the reactions makes clearer the grounds on which our viewpoint is based. Correlations similar to that obtained by us can be observed only under certain kinetic conditions, particularly when the rate of glucose formation reaches its stationary value so that the concentration of glucose varies linearly with time.

Possible reasons why such correlations were not observed before are the following: 1) measurement of the total amount of reducing sugars, including both glucose and other groups, during the hydrolysis of insoluble cellulose, 2) recording a nonstationary rate of glucose formation, 3) a high degree of conversion of cellulose, 4) inadequate methods to determine the endoglucanase activity, or 5) use of such cellulase complexes where a rate limiting stage with respect to insoluble cellulose was controlled by a component other than endoglucanase. The validity of these conclusions about the presence or absence of such correlations and the reasons behind this phenomenon can be evaluated only after an analysis of the kinetics of the cellulase complex as a multienzyme system. This is summarized in Table 7.

IV. ROLE AND PROPERTIES OF THE INDIVIDUAL COMPONENTS OF CELLULASE COMPLEXES OF DIFFERENT ORIGINS

A. Endoglucanases

As was shown in the previous Section, endoglucanases are the enzymes which initiate hydrolysis of both soluble and insoluble (native) forms of cellulose. Depending on the physical state of the substrate, endoglucanase can act as a rate limiting or non rate limiting enzyme for glucose formation. Products of the endoglucanase action, depending on the state of the substrate, can be either substituted cellodextrines or partially degraded insoluble cellulose (Table 7).

A very important property of endoglucanases is their ability to be regulated by the action of cellobiose. We found (36) that cellobiose may effect the activities of various endoglucanases differently. A number of endoglucanases are noticeably suppressed by cellobiose at concentrations above 3 mM. The endoglucanases of *T. reesei* and the cellulase preparation Rapidase are examples of such

TABLE 7

CHARACTERISTIC BEHAVIOR OF TYPICAL CELLULASE COMPLEX WITH RESPECT TO SOLUBLE AND INSOLUBLE CELLULOSE

	Characteristic	Soluble Cellulose	Insoluble Cellulose
1.	The enzyme which determine stationary rate of glucose formation	Exoglucosidase	Endoglucanase
2.	The enzyme which produces main portion of other than glucose reducing sugars	Endoglucanase	Mainly cellobiohydrolase, partially endoglucanase
3.	The composition of reducing sugars		
	a) at the beginning of reaction	Higher oligosaccharides	Cellobiose
	b) at the end of reaction	Substituted oligosaccharides plus glucose	Glucose
4.	Products of random action of endoglucanase	Oligosaccharides	Reducing end groups of the surface of insoluble cellulose

enzymes. An example of endoglucanase which is activated by cellobiose is a low molecular weight endoglucanase of the cellulase complex from *T. koningii* (37). We also found (36) that this enzyme is activated by cellobiose by a mechanism of transglycosilation, where cellobiose acts as an acceptor; and the activation reaches 6 fold at saturating concentrations of cellobiose (K_s is equal to 15 mM). Such a specific mechanism of activation manifests itself in an acceleration of the random cleavage of the polymeric molecules of cellulose under the effect of endoglucanase. However, its action does not accelerate the production of soluble reducing sugars. It is worthwhile to note that the total endoglucanase activity of the preparation from *T. koningii* is almost uneffected by cellobiose. The reason for this is the presence of other endoglucanases together with a low molecular weight enzyme in the preparation from *T koningii*. These other enzymes apparently are suppressed by cellobiose, similar to the endoglucanases from *T. reesei* or Rapidase. Thus, cellobiose appears as a regulator of endoglucanase activities in the cellulase complexes. Contrary to cellobiose action, glucose does not produce such regulatory effects (36).

In the previous Section we showed that the substrate specificity of the endoglucanases with relation to CM cellulose can be characterized by limiting concentration of other reducing sugars, which are produced in the reaction. We used this relation to determine the substrate specificity of a number of endoglucanases, as shown in Table 8.

Table 8 shows that endoglucanases of various origins are discriminated by a minimal length of the nonsubstituted part of CM cellulose. However, many of them are capable of cleaving the parts containing four or greater nonsubstituted glucose units. Published data show that some endoglucanases can attack the parts of CM cellulose which contain three (27) or two (28) nonsubstituted glucose units as well; and even those consisting of

TABLE 8

SUBSTRATE SPECIFICITY OF ENDOGLUCANASES OF VARIOUS ORIGINS WITH RELATION TO CM CELLULOSE

Source	Minimal Length of Nonsubstituted Glucose Units in CM Cellulose Sensitive to Endoglucanase Action[a]
G. candidum	4
T. lignorum	4
T. reesei[b]	4
Rapidase[b]	4
Asp. niger	5

[a] Calculated from Eq. 29

[b] Preparations contained insufficient amount of cellobiase; additional cellobiase from *Asp. foetidus* was added to facilitate determination of other reducing sugars of substituted oligosaccharides.

TABLE 9

ENDOGLUCANASE ADSORPTION ON INSOLUBLE CELLULOSE IN A COLUMN REACTOR

Source	Adsorbed endoglucanase (%)	
	Filter paper	Avicel
G. candidum	30	73
T. lignorum	64	97
Asp. niger	-	13
Rapidase	8	4

a pair of nonsubstituted and 6-0-substituted moieties (28). We do not discard this possibility. However, our data show that this kind of reaction would require a considerably longer time for enzyme incubation with the substrates, which is an indication of a nonspecific action of endoglucanase.

We found also, that endoglucanases of different origins have variable ability to be adsorbed on insoluble substrates (Table 9).

Despite the differences in substrate specificity and adsorption characteristics of endoglucanases from different sources, they all have the common feature that the ratio of their activities with respect to soluble (CM cellulose) and insoluble (cotton) substrates is almost the same, and equal to 19 $\pm$ 3 (see Table 6). This evidence points to their similarity in terms of their contribution to the hydrolysis of cellulose.

B. Exoglucosidases

Exoglucosidases belong to the least known components of the cellulase complex. Even since the importance of cellobiohydrolase was realized (38), many investigators came to the conclusion that glucose was formed by a consecutive action of cellobiohydrolase and cellobiase (32,33). However, our observations demonstrate that exoglucosidase plays a decisive role in glucose formation from both soluble and insoluble cellulosics. A comparative study of the exoglucosidase and cellobiase contributions to glucose formation from CM cellulose is shown in Table 10. The exoglucosidases from *Trichoderma* and *Geotrichum* are the main producers of the glucose formed from CM cellulose. Therefore, the quantity of glucose formed from CM cellulose by these preparations can be taken as a measure of the exoglucosidase activity. On the other hand, the preparations of *Aspergillus* produce glucose mainly under the effect of cellobiase. At the present time our laboratory is engaged in a comparative study of the roles played by cellobiase and exoglucosidase during the later stages of the conversion of insoluble cellulose under the action of cellulase complexes.

C. Cellobiases

A well accepted viewpoint of the role of cellobiase is that it produces a major portion of the glucose from cellulose by way of hydrolysis of the intermediate cellobiose; it also reduces the inhibitory effect of cellobiose on cellobiohydrolase and on endoglucanase. Our observations show that cellobiase in splitting cellobiose becomes a factor which can also regulate the endoglucanase behavior; the latter is the first enzyme that attacks cellulose and supplys the substrate for exoglucosidase and cellobiohydrolase. Thus, cellobiase can conceivably regulate the activities of all three other components of the cellulase complex. However, the role it exercises in the direct production of glucose seems to be not so important as was believed

TABLE 10

COMPARATIVE CONTRIBUTIONS OF EXOGLUCOSIDASE AND CELLOBIASE TO GLUCOSE FORMATION FROM CM CELLULOSE

Source	Cellobiase Activity (2 x I.U./g)	Cellobiase Contribution (%)	Exoglucosidase Contribution (%)
G. candidum	10 - 1600[a]	15 - 40	60 - 85
T. reesei	50	0	100
T. lignorum	40 - 550[a]	15 - 40	60 - 85
T. koningii	15	20	80
T. viride	20	⩽ 20	⩾ 80
T. longibrachiatum	10	⩽ 40	⩾ 60
Asp. niger	800 - 1500[a]	70 - 100	0 - 30
Asp. foetidus	4000	100	0

[a] Preparations differed in degrees of purification

earlier, because a major portion of the glucose is produced by exoglucosidase.

It must also be noted that a substantial role in the regulation of cellobiase activity and, as a result, of the activity of the whole cellulase complex, is played by the surface of the insoluble substrate. In the presence of an essentially insoluble substrate, the cellobiase appears to be a single component of the cellulose complex; however the cellobiose is in solution. Therefore, the adsorption of cellobiase on the surface of the insoluble substrate is unnecessary. At the same time, cellobiase can be adsorbed irreversibly by an insoluble substrate at a sufficiently high substrate concentration. And although the adsorbed cellobiase is able to split cellobiose, it does so at a considerably lower rate in comparison with soluble cellobiase. The cellobiases of different origins often have different sorption characteristics which depend on the type of insoluble substrate (Table 11).

TABLE 11

ADSORPTION OF CELLOBIASES ON INSOLUBLE CELLULOSICS IN A PLUG FLOW REACTOR

	Adsorbed cellobiase (%)	
Source	Avicel	Filter paper
G. candidum	85	43
Asp. niger	35	-

The literature contains a number of indications that the cellulase complex may contain various cellobiases which differ from each other by molecular weight, substrate specificity, and other properties. It is also noted from the literature that cellobiases of different origins may be substantially different in their properties (39,40). For the purpose of clarification of the identity of different cellobiases in terms of their kinetic properties, we determined the kinetic parameters for cellobiose hydrolysis under the effect of a number of different cellulase preparations. The results, shown in Table 12, indicate that in terms of kinetics, the cellobiases of different sources are almost identical; and the values of the apparent Michaelis constants for all of them are around 2 ± 0.5 mM. Also, the purificaiton of the cellulase complexes does not change appreciably the values of K_m for cellobiose hydrolysis.

V. ADSORPTION OF CELLULOLYTIC ENZYMES ON THE SURFACE OF CELLULOSE AND THE PROPERTIES OF THE BOUND ENZYMES

Adsorption of enzymes on the surface of the substrate is a very important feature of the enzymatic degradation of an insoluble substrate. This makes the process substantially different from that of the splitting of soluble compounds. This adsorption may be quite specific for a given enzyme and substrate, yet it must also be a reflection of nonspecific interaction of the enzyme as a surfactant with the interface. This interaction typically causes a reduction in the strength of the solid particles, in particular those having a crystalline structure, known as the adsorption induced strength lowering, or Rebinder effect (41, 42). As far as we know, this idea has not been introduced into enzymology as yet, at least not into the field of the enzymatic hydrolysis of native cellulose. At the same time, we can surmise that this effect may be one of the factors which accelerates the degradation of an insoluble substrate under the effect of the adsorbed enzyme.

TABLE 12

MICHAELIS CONSTANTS (K_m) AND MAXIMUM VELOCITIES (V_m) OF CELLULASE COMPLEXES OF VARIOUS ORIGINS WITH RESPECT TO CELLOBIOSE[(a)]

Cellulase preparations[(b)]		K_m (mM)	V_m $\left(\frac{\mu\text{moles of } G_2}{\text{min, g}}\right)$
T. lignorum	I	1.3	7
	II	1.3	12
	III	1.2	40
	IV	1.3	550
G. candidum	I	2.5	5
	II	2.0	45
	III	2.5	180
	IV	2.8	270
	V	2.4	550
T. koningii	I	1.0	6.5
	II	1.8	20
T. reesei	I	2.0	20
	II	1.8	70
Asp. niger	I	2.5	35
	II	2.7	1200

TABLE 12
(Cont'd)

Cellulase preparations[b]	K_m (mM)	Vm $\left(\frac{\mu\text{moles of } G_2}{\text{min, g}}\right)$
Asp. foetidus I	2.0	75
II	2.0	10000
III	1.8	9000
T. viride	2.0	23
T. longibrachiatum	1.5	10

(a) pH 4.5 and 40°C.

(b) Different preparations from the same microbial source had different degrees of purification or came from different manufacturers.

Therefore, in the context of the study of the mechanism of enzymatic hydrolysis of cellulose, it is important to examine the characteristics of the adsorption of the cellulase complex components onto the insoluble substrate and to examine the functioning of the adsorbed enzymes. Unfortunately, the literature on this very important subject contains very little data (43-46).

A. Reversible and Irreversible Adsorption of Cellulolytic Enzymes on Cellulose

The process of adsorption of the components of the cellulase complex onto an insoluble substrate can be described by the Langmuir isotherm, as in heterogeneous catalysis:

$$\Theta = \frac{E_a}{E_{a,max}} = \frac{K[E]}{1 + K[E]} \qquad \text{(Eq. 32)}$$

E_a is the amount of adsorbed enzyme; $E_{a,max}$ is the maximum possible amount of adsorbed enzyme, which is proportional to the area of the accessible substrate surface; [E] is the concentration of enzyme in solution; θ is the fraction of the surface of the substrate which is occupied by enzyme; and K is equilibrium adsorption constant. Fig. 18 shows the adsorption isotherms for different concentrations of insoluble substrate, which fit the description of Eq. 32. Generally speaking, the conformity of the isotherm behavior with Eq. 32 does not mean that there was a specificity or non-specificity in the interaction between the enzyme and the insoluble substrate. The Langmuir isotherm often fits the non-specific sorption of proteins on solid interfaces. However, when only a limited fraction of the substrate surface is engaged in an interaction with the enzyme, then the Langmuir isotherm can be transformed into a form similar to the Michaelis Menten equation, which corresponds to a specific interaction between the enzyme and the substrate in solution. By specific interaction,

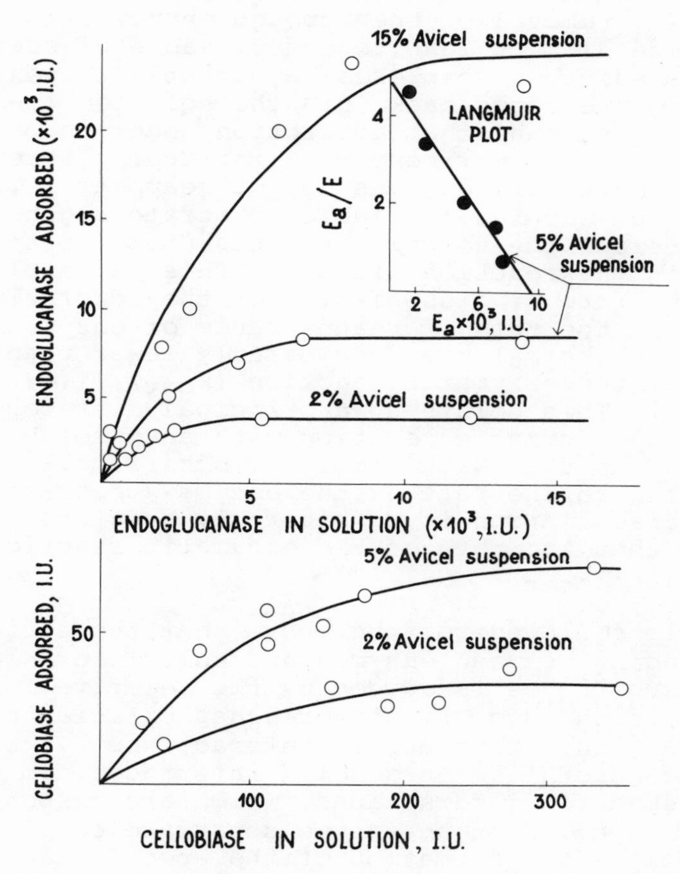

Fig. 18. The adsorption isotherms for components of the crude cellulase complex from *T. koningii* on Avicel. Conditions: pH 4.5, 40°C.

we mean the interaction between the substrate and the active center of the enzyme.

A study of the kinetics of interaction between the enzyme components of the cellulase complex and cellulose in a batch reactor (Fig. 19) shows that adsorption equilibrium occurred rapidly. However, removal of the unbound enzymes from the solution did not lead to a noticeable dissociation of the adsorbed complexes; apparently the enzymes were not released back into the solution (see Fig. 19). This means that adsorption under the conditions of the experiment was practically irreversible. Moreover, enzymes do not reappear in solution even when an insoluble substrate begins to degrade and the soluble products start to accumulate in the reaction mixture. This can explain why the rate of dissociation of the adsorptive complex and the rate of reappearance of enzymes in solution (k_{diss}) are considerably lower than the rate of the enzymatic reaction (k_{cat}), i.e. $k_{diss} < k_{cat}$. This points up a principal difference between an enzymatic reaction with an insoluble substrate and that with a soluble substrate. In the latter case the rate of the enzyme substrate complex dissociation is considerably greater, as a rule, then the rate of the enzymatic reaction ($k_{diss} > k_{cat}$).

It follows from the above that the efficiency of binding between enzymes and soluble or nonsoluble substrates is guided by the relative importance of the specificity or nonspecificity of the interaction. Nonspecific interactions, like enzyme adsorption on a solid interface, dominate over specific interactions, which are responsible for production of an enzyme substrate complex. As a result after formation of the specific enzyme substrate complex, the enzyme from this complex remains in the adsorption layer and retains its capacity to form new specific complexes at later stages of the reaction. This phenomenon can be called a delocalized adsorption of enzymes; which means that the enzymes may migrate on the surface of the solid particles and initiate hydrolytic reactions at different points on the substrate.

Our experiments showed that along with this kind of interaction between enzymes and insoluble substrates, there was another kind of adsorption

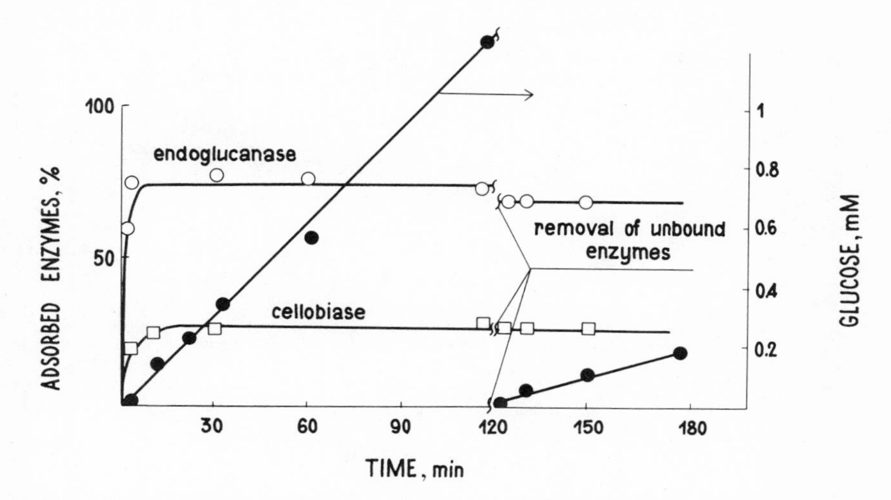

Fig. 19. The kinetics of interaction of the crude cellulase complex from *T. koningii* at 50 g/l with microcrystalline cellulose at 2% in a batch reactor. Conditions: pH 4.5, 40°C.

of cellulolytic enzymes on cellulose that was reversible. In this reversible interaction $k_{cat} < k_{diss}$, similar to the reactions with soluble substrates. Reversible adsorption was studied in a column reactor, charged with microcrystalline cellulose powder. The column also was loaded with a small volume of solution, containing a high concentration of a cellulase preparation, followed by elution with a buffer of pH 4.5 corresponding to the pH optimum for cellulase activity. We found that the maximum concentration of the products of the cellulolytic reaction at the outlet of the reactor was observed in the fractions which contained cellulolytic enzymes desorbed from the column. This concentration peak was a little behind the peak for inert proteins at the outlet of the column (Fig. 20). In the course of further elution,

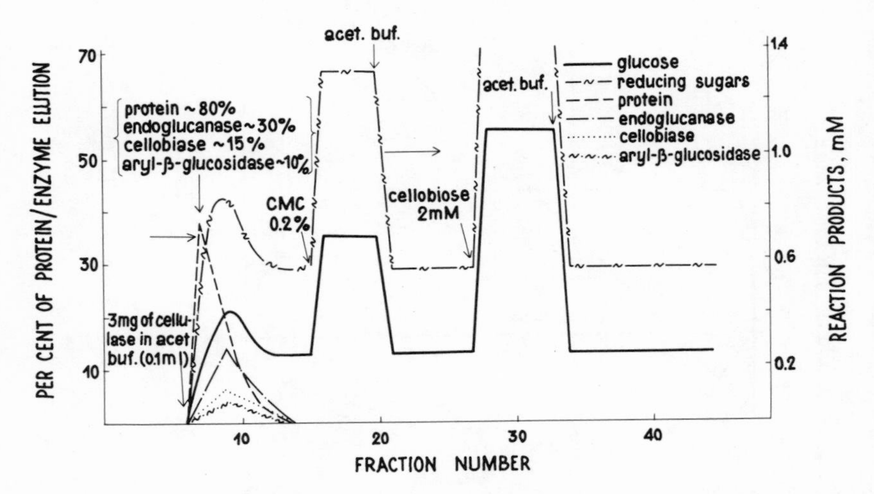

Fig. 20. The elution profile for a crude cellulase preparation from *G. candidum* as determined at the outlet of a column charged with 3g of microcrystalline cellulose. Conditions: pH 4.5, 25°C.

desorption of the enzymes from the column stopped; and the concentration of the products at the column outlet decreased to a constant level. This phenomenon, in our opinion, can be explained by assuming that the dissolved enzymes are adsorbed onto the cellolosic particles in both a reversible and an irreversible mode. Both types of adsorption can lead to enzyme substrate complexes that produce end products. However, the reversibly adsorbed enzymes are eluted from the column together with the end products shortly after the inert proteins, which do not interact effectively with the cellulose. Contrary to the behavior of the adsorbed enzymes, the irreversibly adsorbed enzymes are not eluted by the buffer and continue their catalytic action on the surface of the insoluble substrate to give a steady state output of end products (Fig. 20).

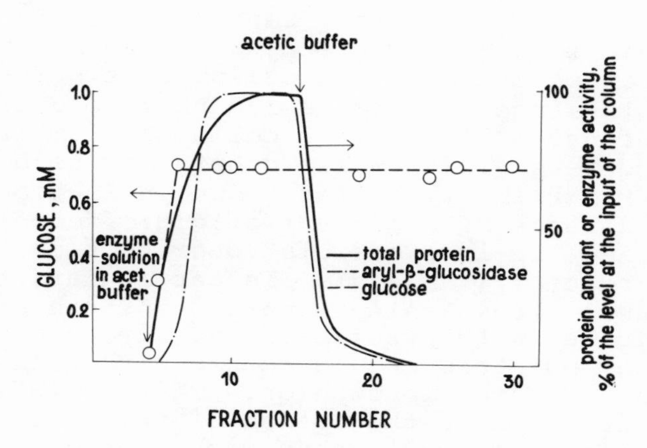

Fig. 21. The elution profile of a cellulase preparation from *G. candidum* at 10 g/l and the reaction products at the outlet of a column charged with microcrystalline cellulose and following an extended time of elution. Conditions: pH 4.5, 15°C.

The question arises as to why the presence of the reversibly adsorbed enzymes cound not be detected in the desorption experiments in a batch reactor (Fig. 19). We feel that during the batch reaction the enzymes which were first adsorbed reversibly gradually were converted into irreversibly adsorbed ones. This assumption becomes more valid if we look at the results of the experiment where we tried to pass the enzyme through a column charged with cellulose for a longer period of time. As seen in Fig. 21, the yield of glucose did not decrease after replacement of the enzyme solution with buffer. Apparently, during this long period of enzyme passage through the column, all of the adsorbed enzymes had sufficient time to be bound irreversibly yet remain in a catalytically active state on the surface of the substrate. This changeover from one type of binding to another may be explained by an evolution of the surface of the substrate during its enzymatic transformation.

As was noted in many reports (37,39), the cellulase complex often includes a number of different endoglucanases and cellobiases. Therefore, we are inclined to hypothesize that some of the endoglucanases of the same cellulase complex may be bound reversibly, while others may be bound only irreversibly. As was shown in Section IV, the endoglucanases of different origin offered variable capacities for irreversible adsorption in the column reactor (Table 9). In particular, the endoglucanase from Rapidase passed through a column of cellulose with practically no irreversible adsorption; but this enzyme reacts with cellulose only in a reversible fashion. This is indicated by the fraction at the outlet of the column which contained the endoglucanase peak plus a considerable amount of end products of the cellulolytic reaction.

We may conjecture that the hydrolysis of insoluble substrates under the effect of irreversibly adsorbed cellulases may be more efficient than hydrolysis produced by reversibly adsorbed enzymes, both at the level of biochemical utilization of the cellulose by microorganisms and at the level of the industrial application of this reaction sequence. The reasons behind such an efficiency may be based on the irreversible adsorption having the advantage of being almost insensitive to losses of enzymes or on the enzymes which fixed themselves tightly on the surface of the substrate acting as surfactants to reduce the surface tension of the crystalline cellulose. The fragmentation of the cellulose particles, with the formation of short insoluble fibers under the action of cellulases, has been observed (46-48) and is a piece of evidence in favor of this hypothesis; although the authors ascribed this phenomenon to a hypothetic C_1- or C_2-enzyme.

B. Productive and Nonproductive Adsorption of Cellulolytic Enzymes on Cellulose

In studying how the concentration of adsorbed enzymes influenced the rate of splitting of the insoluble cellulose, we found that the rate of splitting was proportional to the amount of endoglucanase fixed on the substrate. This proportionality also was observed on a nonlinear part of the adsorption isotherm, as shown in Fig. 18. However, as was discussed in the previous section, the irreversible adsorption of enzymes on insoluble substrates is very characteristic of nonspecific interaction between a protein molecule and a solid interface. Therefore, it is logical to conclude that many of the enzyme molecules are not connected with the insoluble substrate through an active center. Only some of the enzyme molecules, which at a given moment form a productive complex, do have this mode of binding. The existence of these different modes of binding was demonstrated by passing a solution of CM cellulose through a column charged with cellulolytic enzymes adsorbed on insoluble microcrystalline cellulose. We observed an effective splitting of the polymeric substrate without desorption of the enzymes (Fig. 20). Since hydrolysis of CM cellulose is controlled mainly by endoglucanase and exoglucosidase, our results showed that although these enzymes were adsorbed on the insoluble substrate, they were degrading it and thus were capable of hydrolyzing a soluble substrate at the same time. The passage of soluble CM cellulose at concentrations above those of the Michaelis constant through the column did not lead to desorption of the enzymes from the insoluble cellulose. So, interaction between the enzymes and the insoluble substrate must have been stronger than the interaction of the same enzymes with the soluble substrate.

The fact that an adsorbed enzyme appears to be capable of cleaving a soluble substrate is evidence that the active center of many of the enzyme molecules does not interact with the insoluble substrate and is accessible to soluble CM cellulose.

Thus, not all of the molecules of the adsorbed enzymes are bound on cellulose in a productive manner. Possibly, nonproductive insoluble enzyme substrate complexes are formed as a result of the chemical transformation of the productive complexes; and later on they may become productive again. In other words, such complexes may be nonproductive only temporarily.

We discuss now the adsorption of β-glucosidases of the cellulase complex, i.e. cellobiase and aryl-β-glucosidase, on cellulose. As seen in Fig. 20, these enzymes can be adsorbed almost completely by cellulose under certain conditions. However, contrary to endoglucanase, they do not directly effect the insoluble substrate because there is no correlation between the amount of adsorbed enzyme and the yield of the end product. Thus, the adsorption of β-glucosidase is nonproductive. At the same time, their binding is very similar to that of endoglucanase because they are not removed from the surface of cellulose by a soluble substrate, cellobiose, or by an inhibitor, glucose. The results in Fig. 20 show that the adsorbed cellobiase can hydrolyze cellobiose to glucose in a manner similar to the hydrolysis of CM cellulose produced by nonproductively adsorbed endoglucanase.

What is the role that β-glucosidases may play in the hydrolysis of the insoluble substrate? Adsorbed cellobiase may be a factor which reduces an inhibitory effect of the intermediate product, i.e. cellobiose, on the endoglucanase or on the cellobiohydrolase. It must be noted, however, that the catalytic activity of adsorbed cellobiase decreases considerably. As follows from Fig. 22, the rate of glucose formation in the enzymatic hydrolysis of cellulose decreases 3 fold after the removal of the soluble cellobiase from the reaction mixture, despite the presence of adsorbed cellobiase of up to 85% of the total soluble plus adsorbed cellobiase. Therefore, we can conclude that a strong and selective adsorption of cellobiase and aryl-β-glucosidase, i.e. components of cellulase complex which do not directly effect native cellulose, can

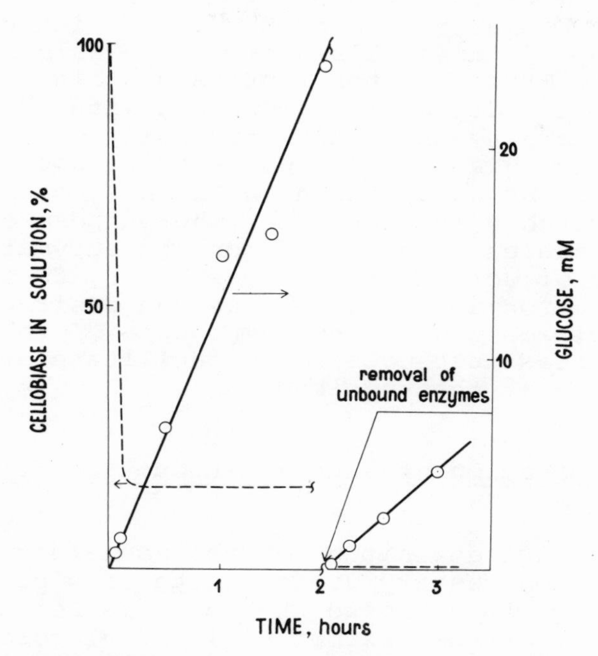

Fig. 22. Kinetics of interaction of a crude cellulase complex from *G. candidum* at 6 g/l with milled cotton linters in a batch reactor. Conditions: cotton linters 15%, pH 4.5, 40°C.

considerably alleviate the process of degradation of the insoluble substrate via the mechanism of adsorption induced strength lowering, i.e. Rebinder effect.

VI. ENZYMATIC PRODUCTION OF GLUCOSE FROM NATIVE CELLULOSE

The purpose of the part of our program which carries a high practical importance is the development of a method to produce a commercial glucose from cellulosics, particularly for the food industry. A low grade of cotton linters was selected as the raw cellulosic material; this is available normally as waste from cotton processing plants. Selection of this cellulose source was based on two factors. First, cotton linters contain cellulose to a greater degree in comparison with other cellulose containing materials, typically over 90%. Second, low grade linter waste still does not have a significant application in the national economy and is notably inexpensive. The purpose of the initial stages of our study on the enzymatic hydrolysis of cotton linters was to try to optimize this reaction and to determine the desirable levels of activities, which the components of the cellulase complex must possess to facilitate substrate hydrolysis at the required rate.

A. Optimization of Glucose Production from Cotton Linters

A strict definition of the goal that is desired is a necessary prerequisite of a project of this kind. In our case the goal was defined as the attainment of the maximum yield of glucose during 8 hr of continuous enzymatic hydrolysis of milled cotton linters. At this point the minimization of the production costs involved in glucose production was not part of our goal. A series of preliminary experiments was conducted to select the main variables which controlled the rate of enzymatic glucose production. The first choice of variables included the initial concentration of enzyme, amount of primary substrate, pH, temperature, ionic strength of the reaction medium, and the duration of the vibro energy milling of the native cotton linters. If we could modify only one variable at a time, then at four runs for each test condition our six variables would require 4^6 or 4096 runs for the optimization. Therefore, a mathematical optimization of the enzymatic hydrolysis of cellulose was selected as the only practical approach.

The optimization was made using the cellocandine G3X cellulase preparation; this is a crude dry commercial material isolated from *G. candidum* (Tables 1 and 2). The choice of the preparation was somewhat arbitrary and was based mainly on the fact that this cellulase complex contained all four of the cellulase components necessary for the effective hydrolysis of cellulose. The cotton linter was ground in a high energy vibro mill, designed and built in the Department of Chemistry of Moscow State University. The vibro mill consists of four 48 mm diameter by 60 mm long barrels, each having the capacity to grind 4 g of dried linters. The concentration of D-glucose in the reaction mixture was determined by the glucose oxidase peroxidase method (21).

The experimental scheme for optimization of the enzymatic production of glucose was developed from the Box-Wilson method (50). This method is based on approximating the relationship between the dependent and independent variables. A detailed description of the experiments and the mathematical analysis of the results are given elsewhere (51).

The purpose of the first step of the optimization was to determine the regression coefficients of the linear model:

$$y = b_0+b_1x_1+b_2x_2+b_3x_3+b_4x_4+b_5x_5+b_6x_6 \quad \text{(Eq. 33)}$$

where y is the concentration of glucose after 8 hr of continuous enzymatic hydrolysis and x_1, x_2, ... X_6 are the variables. The coefficients obtained are used later for a modified model to further optimize the conditions of the reaction. Here, x_1 is the concentration of enzyme in g/l, x_2 is the concentration of substrate in wt %, x_3 is the pH, x_4 is the temperature, x_5 is the duration of milling time in min, and x_6 is the ionic strength given as the concentration of NaCl in mole/l.

The first step in the optimization was carried out with the six variables covering the following ranges: x_1 40-60 g/l enzyme, x_2 12.5 - 17.5 wt % substrate, x_3 pH 3.5 - 5.5, x_4 45-55 °C, x_5 5-15 min milling, and x_6 0.1 - 0.4 mole/l sodium chloride. The following values were obtained for the coefficients (Eq. 33): $b_o = 2.7$, $b_1 = 0.075$, $b_2 = 0.188$, $b_3 = -1.012$, $b_4 = 0$, $b_5 = 0.15$, $b_6 = 0.05$. Of these six regression coefficients for the linear model, only the ones describing the effects of substrate concentration and pH showed useful changes (51). The negative sign of the pH coefficient shows that a lower pH of 3.5 was more favorable in comparison with a higher pH of 5.5. Independent experiments, however showed that pH values of 3.5 and lower could not be optimal because they bring about rapid denaturation of the cellulase. The insignificant effects of temperature and of ionic strength on the glucose production were noted. The latter was verified by additional experiments, so that subsequent work was done at 0.1 M NaCl. In any case, the results of the first step of the optimization could not be regarded as optimal since the maximum concentration of glucose in this series was 4.0 - 4.1 g/l, well below the maximum concentrations of 4.5 - 5.0 g/l observed during preliminary experiments under one variable optimization.

The second step of the optimization was carried out in terms of a second order polynominal. The range of variables for the second as well as the third steps of the optimization are given in Table 13. The measured glucose concentration ranged from 0.3 to 5.15 g/l. After mathematical analysis (51) and exclusion of insignificant independent coefficients, the regression equations assumed the form:

$$y = 5.163 + 0.283 X_2 + 0.822 X_3 + 0.167 X_4 - 0.486 X_1^2 - 0.336 X_2^2 - 0.586 X_3^2 - 0.786 X_4^2 + 0.356 X_3X_4 \quad \text{(Eq. 34)}$$

In this case pH showed the strongest effect on the concentration of glucose. The effect of temperature was less pronounced. The remaining variables over the ranges tested did not significantly change the concentration of glucose.

TABLE 13

RANGE OF VARIABLES FOR SECOND AND THIRD OPTIMIZATION TRIALS FOR THE PRODUCTION OF GLUCOSE FROM COTTON LINTERS

	Range of Variables	
Variable	Second Trial	Third Trial
X_1 (g/l enzyme)	30-50	50-70
X_2 (wt % substrate)	15-20	10-15
X_3 (pH)	3.5-4.5	3.5-4.5
X_4 (°C)	45-65	35-55
X_5 (min milling)	3-7	10-20
X_6 (moles/l NaCl)	0.1	0.1

Using a simplex technique to find an extremum of the function (51), we determined the maximum possible concentration of glucose to be 5.7 g/l. The values of the factors to attain this concentration were $X_1 = 0.028$, $X_2 = 0.529$, $X_3 = 0.940$, $X_4 = 0.346$, $X_5 = -0.955$. From these results we concluded that the maximum concentration of glucose of 5.7 g/l that could be expected after 8 hr of continuous enzymatic hydrolysis of milled cotton linters using cellocandine G3X could be attained by operating at 60 g/l enzyme preparation, 14 wt.% substrate, pH 4.5, 48.5°C, and 10 min of vibro energy milling of the linters. These values were then verified by a series of short one variable tests of four runs at each set of conditions, with the remaining four variables fixed at the estimated optimum levels.

Recent experiments with a commercial crude cellulase preparation cellokoningin P10X (Table 1)

allowed us to obtain 10% D-glucose syrups, with practically no contamination by other reducing sugars, including cellobiose after 48 hr of continuous enzymatic hydrolysis. The conditions for this involved 10 min vibro energy milled cotton linters at 20 wt.%, pH 4.5, 50°C, and 60 g/l enzyme.

B. Quantitative Criteria of an Optimal Cellulase Complex

Before concluding we would like to introduce the concept of an optimal cellulase complex, as one with controlled properties. Trial and error has been the only available technique up to now to produce cellulase preparations under laboratory conditions. Even after measuring the activities of the preparations, by relation to soluble substrates, it has been practically impossible to predict the activity of the complexes with insoluble cellulosics. This unpredictability of enzymes behavior was actually a source of the C_1-enzyme hypothesis. Our data, however, suggest that it may be possible to develop quantitative relationships between the activities of cellulases with relation to soluble polymeric substrates and to insoluble substrates. This would provide us with a tool to predict the reactivity of the complex at large in hydrolytic reactions of native cellulosics, using the values of the specific activities of the individual components of the complex. Moreover, on the basis of such regularities we might be able to plan the rate of hydrolysis of the native cellulose by controlling the compositon of the cellulase complex both in terms of quality and quantity.

In the context of our study, an optimal cellulase complex should produce 12% glucose syrups containing less than 5% cellobiose per 48 hr of continuous hydrolysis of milled cotton linter.

A start in this direction is the linear correlation between the activities of endoglucanase, determined viscometrically in the hydrolysis of CM cellulose, and the stationary rates of glucose formation from native cellulose. This holds also for the concentrations of cellulase preparations and cotton linters which are sufficiently high to be technologically practical. An example of the correlation of this kind is given by the following formula:

$$[G] = 2.8 \times 10^{-4}(t-2) \quad A \qquad \text{(Eq. 35)}$$

where [G] is the concentration of product glucose in g/l formed for each g/l of cellulase preparation, t is the duration of the reaction in hrs, and A is the specific activity of the endoglucanase of the cellulase complex expressed in I.U. (micromoles of glucoside bonds of CM cellulose cleaved per min per g of preparation). This correlation is observed under the following conditions: milled cotton linter concentration 12.5%, pH 4.2, 45°C, 0.1 M NaCl, 20 min milling of the substrate, and 3-60 g/l cellulase preparation. The conversion of the substrate does not exceed 60%. Good agreement between the calculated (Eq. 35) and experimentally observed values is illustrated in Table 14. Use of Eq. 35 does not work for those complexes which contain only small concentrations of cellobiase. For example, the absence of cellobiase in the cellulase complex of Rapidase results in a fast accumulation of cellobiose, which may be as high as 0.1 M during the hydrolysis of milled cotton linters. This in turn brings the whole system out of steady state operation, due to very effective inhibiton by cellobiose; and the reaction practically stops.

However, the approach exemplified by Eq. 35 seems to have a reasonable potential for establishing quantitative requirement for development of an effective cellulase complex with controlled reactivities with relation to native cellulose.

TABLE 14

GLUCOSE PRODUCTION FROM COTTON LINTERS UNDER THE ACTION OF CELLULASE COMPLEXES OF DIFFERENT SOURCES

Source	Endoglucanase Activity (I.U./g)	Cellulase Concentration (g/l)	Glucose Theory [b]	Concentration [a] Experiment [c]
Asp. foetidus	44	60	4.5	4.9
T. lignorum	80	60	8.2	9.5
T. koningii	180	60	18	16
T. lignorum [d]	1100	30	56	55
G. candidum [d]	5000	3	25	25

[a] g/ℓ after 8 hr of enzymatic hydrolysis

[b] Eq. 35

[c] Conditions described in text

[d] Purified preparations

REFERENCES

1. REESE, E. T. In "Cellulose as a Chemical and Energy Resource", (C. R. Wilke, ed.) John Wiley & Sons, New York, 1975, p. 77.
2. KLYOSOV, A. A. & RABINOWITCH, M. L. In "Enzyme Engineering and Bioorganic Catalysis" (V. L. Kretovitch and I. V. Berezin, eds.) (Russ.) Viniti Press, Moscow, 1978, p. 49.
3. Enzyme Nomenclature (Recommendations of the IUPAC and the IUB), Elsevier, Amsterday, 1973.
4. Enzyme Nomenclature, Suppl. 1, *Biochim. Biophys. Acta 429*: 31, 1976.
5. SELBY, K. In "Cellulases and Their Applications" (R. F. Gould, ed.) American Chemical Society, Washington, D.C., 1969, p. 34.
6. REESE, E. T. In "Enzymatic Conversion of Cellulosic Materials: Technology and Application" (E. L. Gaden, M. H. Mandels, E. T. Reese and L. A. Spano, eds.) John Wiley & Sons, New York, 1976, p. 9.
7. ENARI, T.-M. & MARKKANEN, P. "Advances in Biochemical Engineering," vol. 5 (T. K. Ghose, A. Fiechter, and N. Blakebrough, eds) Springer-Verlag, Berlin, 1977, p. 1.
8. GHOSE, T. K. "Advances in Biochemical Engineering", vol. 6 (T. K. Ghose, A. Fiechter, and N. Blakebraugh, eds.) Springer-Verlag, Berlin, 1977, p. 39.
9. GHOSE, T. K. & GHOSH, P. *J. Appl. Chem. Biotechnol. 28*: 309, 1978.
10. REESE, E. T. *Recent Adv. Phytochem. 11*: 311, 1977.
11. RABINOWITCH, M. L., KLYOSOV, A. A. & BEREZIN, I. V. *Bioorg. Khim.* (Russ.) *3*: 405, 1977.
12. KLYOSOV, A. A. In "Proc. 2nd Joint US/USSR Enzyme Engineering Seminar of US/USSR Working Group on Production of Substances by Microbiological Means," Corning, 1976, p. 263.
13. ALMIN, K. E. & ERIKSSON, K.-E. *Biochim. Biophys. Acta 139*: 238, 1967.
14. ALMIN, K. E. & ERIKSSON, K.-E. *Biochim. Biophys. Acta 139*: 248, 1967.
15. ALMIN, K. E., ERIKSSON, K.-E. & PETTERSSON, B. *Eur. J. Biochem. 51*: 207, 1975.

16. TSCHETKAROV, M. & KOLEFF, D. *Monatsh. Chemie 100*: 986, 1969.
17. HULME, M. A. *Arch. Biochem. Biophys. 147*: 49, 1971.
18. KLYOSOV, A. A., RABINOWITCH, M. L., SINITSYN, A. P. & CHURILOVA, I. V. *Bioorg. Khim.* (Russ.) (in press).
19. RABINOWITCH, M. L., KLYOSOV, A. A. & BEREZIN, I. V. *Dokl. Akad. Nauk SSSR* (Russ.) (in press).
20. REESE, E. T. In "Cellulases and their application" (R. F. Gould, ed.) American Chemical Society, Washington, D. C., 1969, p. 26.
21. BEREZIN, I. V., RABINOWITCH, M. L. & SINITSYN, A. P. *Biokhimiya* (Russ.) *42*: 1631, 1977.
22. KLYOSOV, A. A. & CHURILOVA, I. V. *Biokhimiya* (Russ.) (in press).
23. KLIBANOV, A. M., MARTINEK, K. & BEREZIN, I. V. *Biokhimiya 39*: 878, 1974.
24. BEREZIN, I. V., KLIBANOV, A. M., KLYOSOV, A. A., MARTINEK, K. & SVEDAS, V. K. *FEBS Lett. 49*: 325, 1975.
25. KLYOSOV, A. A., RABINOWITCH, M. L. & BEREZIN, I. V. *Bioorg. Khim.* (Russ.) *2*: 795, 1976.
26. EASTERBY, J. S. *Biochim. Biophys. Acta 293*: 552, 1973.
27. ERIKSSON K.-E. & HOLLMARK, B. H. *Arch. Biochem. Biophys. 133*: 233, 1969.
28. KLOP, W. & KOOIMAN, P. *Biochim. Biophys. Acta 99*: 102, 1965.
29. KANDA, T., NAKAKUBO, S., WAKABAYASHI, K. & NISIZAWA, K. *J. Biochem. 84*: 1217, 1978.
30. NISIZWA, K., KANDA, T., SHIKATA, S. & WAKABAYASHI, K. *J. Biochem. 83*: 1625, 1978.
31. KLYOSOV, A. A. & GRIGORASH, S. Y. *Biokhimiya* (Russ.) (in press).
32. STERNBERG, D. *Appl. Environ. Microbiol. 31*: 648, 1976.
33. GONG, C. S., LADISCH, M. R. & TSAO, G. T. *Biotechnol. Bioeng. 19*: 959, 1977.
34. SINITSYN, A. P., RABINOWITCH, M. L., KALNINA, I. A. & KLYOSOV, A. A. "Abst. 2nd Nat. Symp. Microbial Enzymes", vol. 1 (Russ.) Minsk, 1978, p. 181.

35. KING, K. W. *J. Ferment. Technol.* *43*: 79, 1965.
36. CHURILOVA, I. V., MAKSIMOV, V. I. & KLYOSOV, A. A. *Biokhimiya* (Russ.) (in press).
37. WOOD, T. M. & MCCRAE, S. I. *Biochem. J.* *171*: 61, 1978.
38. WOOD, T. M. & MCCRAE, S. I. *Biochem. J.* *128*: 1183, 1972.
39. RODIONOVA, N. A., RUMYANTSEVA, T. N., TIUNOVA, N. A., MARTINOWITCH, L. I. & BAKHTADZE, L. N. *Biokhimiya* (Russ.) *42*: 43, 1977.
40. BUCHT, B. & ERIKSSON, K.-E. *Arch. Biochem. Biophys.* *129*: 416, 1969.
41. REBINDER, P. A. & SHCHUKIN, E. D. *Uspekhi Fisicheskich Nauk (Adv. Physical Sciences)* *178*: 3, 1972.
42. PERTSOV, N. V. In "Surface Effects in Crystal Plasticity" (R. M. Latanision and J. F. Fourie, eds.) NATO Advanced Study Institutes, Series E: Applied Sciences No. 17, Noordhoff-Leyden, 1977, p. 863.
43. MANDELS, M., KOSTICK, J. & PARIZEK, R. *J. Polymer Sci.* *36*: 445, 1971.
44. HUANG, A. A. *Biotechnol. Bioeng.* *17*: 1421, 1975.
45. PEITERSEN, N., MEDEIROS J. & MANDELS, M. *Biotechnol. Bioeng.* *19*: 1091, 1977.
46. HALLIWELL, G. & GRIFFIN, M. *Biochem. J.* *169*: 713, 1978.
47. HALLIWELL, G. *Biochem. J.* *95*: 270, 1965.
48. HALLIWELL, G. *Biochem. J.* *100*: 315, 1966.
49. RODIONOVA, N. A., TIUNOVA, N. A., FENIKSOVA, R. V., KUDRYASHOVA, T. I. & MARTINOWITCH, L. I. *Dokl. Akad. Nauk SSSR* *214*: 1206, 1974.
50. BOX, G. E. P. & WILSON, K. B. *J. Roy. Statist. Soc. Ser. B* *13*: 1, 1951.
51. KLYOSOV, A. A., AGAFONOV, M. N. & GRANOVSKY, YU.V. *Biokhimiya* (Russ.) (in press).

IMMOBILIZED AMYLOGLUCOSIDASE: PREPARATION, PROPERTIES, AND APPLICATION FOR STARCH HYDROLYSIS

L. A. Nakhapetyan and I. I. Menyailova

All-Union Scientific Research Institute
of Biotechnology
Kropotkinskaya Street, Moscow, USSR

In recent years, there has been increasing interest in glucose syrups due to the development of large scale processes for the enzymatic isomerization of glucose. Starch is now the cheapest and the most available material for the production of glucose; and this process, using soluble amyloglucosidase, has been studied well and has been used in industry for a long time in many countries.

Recently, a significant number of papers dealing with the production, investigation of properties, and use of immobilized amyloglucosidase for starch hydrolysis have been published. The problems associated with the immobilized enzyme process have not been completely solved; and the process for starch hydrolysis by immobilized amyloglucosidase has not been utilized by industry.

Although there are many publications concerning the process of amyloglucosidase immobilization, it is usually difficult to compare the data obtained because of the diversity of the enzyme preparations and enzyme sources, the variety of conditions for immobilization, and the different methods of determining the activity and stability of the soluble and immobilized enzymes. In this connection we decided to carry out a systematic study of amyloglucosidase immobilization on different organic and inorganic carriers, with most of the testing being done with amyloglucosidase from *Endomycopsis bispora*. In our work we have paid particular attention to the influence of the procedures and conditions for the immobilization on the activity and stability of the resulting enzyme preparations. The purpose of our investigation was to develop a method of immobilized enzymatic hydrolysis of liquefied starch, including the preparation of the immobilized amyloglucosidase, the design of the enzymatic reactor, and the selection of optimal reaction conditions. In this paper the results of the work of amyloglucosidase immobilization on Soviet-made inorganic carriers are presented. In addition, the main results of our experimental runs in a 50 liter reactor with immobilized amyloglucosidase are described.

I. IMMOBILIZATION OF AMYLOGLUCOSIDASE ON INORGANIC CARRIERS

A. Silanization Conditions

As possible carriers for immobilized amyloglucosidase, various porous inorganic carriers manufactured by Soviet industry as well as by local laboratory techniques have been studied. The modification of the support surface was carried out by heating the carrier in aqueous or acetone solutions of γ-aminopropyltriethoxysilane. The efficiency of the silanization was estimated by determining the number of residual amino groups on the carrier by titration. We have shown that while modifying the carrier in aqueous silane solution, the latter should not be acidified, as recommended elsewhere (1), since the number of amino groups on the carrier would be reduced. This perhaps could result from hydrolysis of the ethoxy groups of the silane in the acidic medium. Silanization from aqueous solution produced 1.2 to 1.6 times more amino groups than silanization from an acetone solution (Table 1). The temperature at which the silanization was conducted had practically no influence on the number of amino groups (Table 1).

TABLE 1

NUMBER OF AMINO GROUPS ON THE CARRIER SURFACE FOR DIFFERENT SILANIZATION TEMPERATURES AND SOLVENTS

5% Silane in Acetone		5% Silane in Water	
Temperature (°C)	Amino Groups, (meq/g)	Temperature (°C)	Amino Groups, (meq/g)
50	0.31	50[a]	0.48
100	0.36	100	0.44
140	0.38	140	0.46

(a) silanization carried out at 100 mm Hg

B. Influence of Carriers

To select the optimal characteristics of the carrier structure, the number of amino groups and the activity of amyloglucosidase immobilized on various inorganic carriers were determined. In this work a unit of amyloglucosidase activity is defined as micromoles of glucose obtained per min per g of enzyme support from a 1% solution of soluble starch at 30°C and pH 4.7. From the results listed in Table 2 the number of amino groups and the activity for one m^2 of carrier surface area were largest for porous glass. This material is known to have the narrowest distribution of pore diameters and has been better studied as compared to silochrome[20] or to silica gel. There were approximately the same number of amino groups on the surface of silochrome and silica gel; but

TABLE 2

AMYLOGLUCOSIDASE IMMOBILIZATION VIA GLUTARALDEHYDE ON VARIOUS INORGANIC CARRIERS(a)

Type of Carrier	Mean Diameter of Pores (Å)	Specific Surface Area (m^2/g)	Amino Groups		Activity of Immobilized Amylo glucosidase	
			(meq/g)	(meq/m^2)	(units/g)	(units/m^2)
Silochrome(b)						
396	2200	30	0.22	0.0073	127.0	4.24
SX-402	900	70	0.42	0.0060	143.0	2.04
394	700	100	0.60	0.0060	130.0	1.30
C-50	632	89	0.55	0.0062	134.0	1.51
C-80	500	80	0.40	0.0050	123.0	1.54
Silica Gel						
MSA-247-1	2760	12	0.17	0.0142	37.6	3.14
MSA-1	2000	13	0.18	0.0138	36.0	2.82
MSA-247-2	1600	23	0.29	0.0126	38.4	1.67
MSA-245-2	1000	29	0.20	0.0068	53.0	1.83
MSA-247-a-3	800	57	0.38	0.0067	70.0	1.23
109-11-3	700	63	0.36	0.0057	48.7	0.77
MSO-50	550	50	0.36	0.0072	36.0	0.72
KSK-1-X	480	90	0.36	0.0040	27.8	0.31
MSA-2-1Y	450	60	0.36	0.0060	25.4	0.42

TABLE 2
(cont'd)

Type of Carrier	Mean Diameter of Pores (Å)	Specific Surface Area (m^2/g)	Amino Groups (meq/g)	Amino Groups (meq/m^2)	Activity of Immobilized Amylo glucosidase (units/g)	Activity of Immobilized Amylo glucosidase ($units/m^2$)
Porous Glass						
44-v	2800	10	0.20	0.020	96.0	9.60
70-b	1200	43	0.56	0.013	207.0	4.82
90-3-g	1070	59	0.71	0.012	240.0	4.07
92-2-v	500	140	0.92	0.007	240.0	1.71
58-8	250	96	0.95	0.010	102.0	1.06

(a) Enzyme from *Endomyeopsis bispora*

(b) Reference (20) for description

the activity of amyloglucosidase immobilized on silochrome was slightly higher than it was on silica gel. However the mechanical strength and chemical stability of porous glass is much lower than for silochrome or silica gel.

It is interesting to study the influence of the structural parameters of the carrier on the activity of immobilized amyloglucosidase. Data from the literature refer to porous glasses as varying both in specific surface area and in pore diameter (1,3). Carriers varying in specific surface areas and having the same mean pore diameters were obtained. In this case the activity of immobilized amyloglucosidase was shown to be proportional to the specific surface area (2). Since the specific surface area increased with smaller pore size, we found the limiting value of the pore diameter at which the diffusional limitation of starch became noticeable on the overall rate of hydrolysis. The amyloglucosidase activity per g of carrier and per m^2 of surface area is presented in Fig. 1 for different pore diameters. The decrease of the specific activity of amyloglucosidase immobilized on carriers with a mean pore diameter of less than about 550 Å suggests a significant diffusional limitation below this range of pore diameter (see Curve 2). The sharp drop in the activity of amyloglucosidase immobilized on carriers with pore diameters of more than 800 Å, is due to the corresponding decrease in the specific surface area (see Curve 1). On the basis of these results, silica gel with a pore diameter of 600 to 700 Å and a specific surface area of no lower than 60 m^2/g has been recommended for use as a carrier in commercial production and application of amyloglucosidase.

C. Influence of Chemical Methods and Conditions of Immobilization

The immobilization of amyloglucosidase on amino carriers has been carried out using three different methods: treatment with glutaraldehyde,

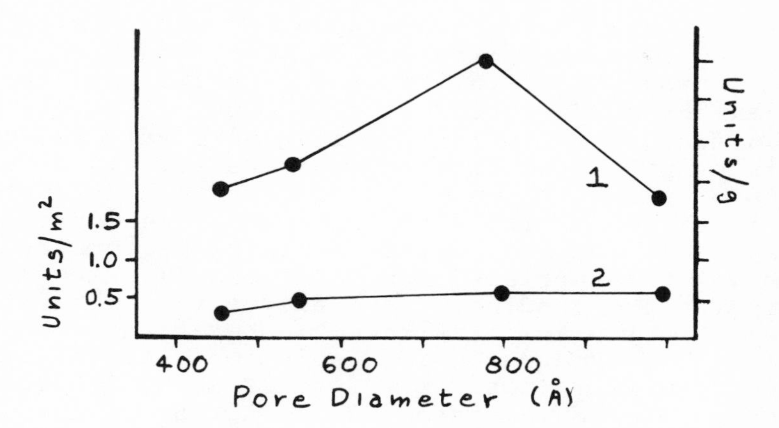

Fig. 1. Effect of pore diameter on the activity of amyloglucosidase from *End. bispora* immobilized on silica gel via glutaraldehyde. Curve 1: activity of the preparation in units/g of carrier; Curve 2: activity of the preparation in units/m^2 of carrier surface area. Enzyme activity was determined with 1% soluble starch at 30°C and pH 4.7.

reaction with azo coupling, and binding of pre-oxidized amyloglucosidase. These methods differ in principle. In the first case the binding is effected via the amino groups of the enzyme; in the second case the binding is via the benzene ring of the tyrosine moiety; and in the third case the linkage is through the aldehyde groups of the carbohydrate part of the enzyme. In all cases the binding does not involve the amyloglucosidase carboxyl groups, which are parts of the active center of the enzyme (4). It should be pointed out, that in spite of the different ways of chemical binding and the participation of various functional groups of the enzyme in the process of immobilization, both the activity and the thermostability of all the preparations were quite close; this is described later in more detail.

While preparing the inorganic carriers containing aromatic amino groups we found, that when the nitroaryl derivative of the carrier was reduced by an aqueous solution of sodium thiosulphate, a significant precipitation of sulphur was observed. This sulphur precipitation was due to acidification of the reaction medium by sodium bisulphite formed as a result of the reaction. We could not completely get rid of the liberated sulphur by washing the carriers, and the residual sulphur hampered the binding of the enzyme to the carrier. The sulphur precipitation could be avoided by conducting the reduction in a solution buffered at pH 7.0 and with an ionic strength of about 0.5 N. By such reduction we succeeded in increasing the activity of immobilized amyloglucosidase by 25%.

The conditions for treating the amino derivative of the carrier with glutaraldehyde also were specified (1,7). We treated the carriers with glutaraldehyde in phosphate buffer, pH 7.0, to prevent polymerization of the aldehyde near the positively charged carrier surface. We also studied the influence of the ionic strength of the buffer solution on the activity of the immobilized preparations. It was found that by increasing the ionic strength of the buffer from 0.1 to 0.5 N, the activity of the immobilized amyloglucosidase was increased by 20%.

While looking for the optimal conditions for amyloglucosidase immobilization on modified inorganic carriers, we studied the effect of pH, the duration of the binding reaction, and the initial amount of the native enzyme in relation to the amount and activity of the bound protein. The amount of protein bound to the carrier was determined directly by a modification of the method of Hartree (18).

In the case of azo coupling the relationship between the immobilized enzyme activity and the pH

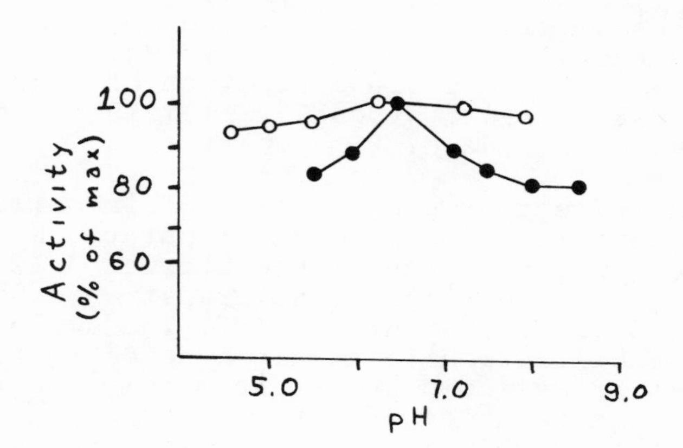

Fig. 2. Relationship between the activity of amyloglucosidase from *End. bispora* immobilized on silochrome C-80 and the pH of the medium in which the binding was carried out. Data represent binding by glutaraldehyde (O) and by azo coupling (●); lines were drawn merely by connecting the points. Enzyme assays carried out same as for Fig. 1.

of the medium in which the binding was carried out exhibited a maximum at pH 6.5 (Figure 2), as compared to published recommendations to carry out azo coupling at pH 8.0 to 8.5 (8) in the range favorable for diazo groups.

The pH of the medium in which the binding was carried out was not so important when the amyloglucosidase immobilization was carried out using glutaraldehyde (Figure 2); and a pH of 4.0 to 7.0 could be used.

While optimizing the conditions for binding preoxidized amyloglucosidase, we found that it was preferable to oxidize amyloglucosidase at pH 3.0

to 4.0, since in this case the greatest part of the initial activity of amyloglucosidase was retained (Table 3). It should be noted that instead of glucose, which was recommended by Christinson (9) for the elimination of excess sodium periodate, it was advisable to use a solution of liquefied starch, as it had a strong stabilizing effect on the amyloglucosidase.

The effect of the duration of immobilization on the activity of the bound preparations also was studied. The reaction of the carrier diazo groups with the benzene ring of the tryosine moiety was completed within two hr at 5°C. The amyloglucosidase immobilization to an aldehyde derivative of the carrier was complete at room temperature in two hr; and the reaction of oxyamyloglucosidase aldehyde groups with the amino groups of the carrier ended in one hr.

The influence of the initial amount of the enzyme on the activity of the immobilized preparations is shown in Fig. 3,4, and 5 for azo, glutaraldehyde, and preoxidation coupling, respectively. In the case of azo coupling and glutaraldehyde coupling, the units of activity of immobilized enzyme per mg of immobilized protein decreased as the initial amount of the native enzyme was increased at the same time the activity per g of enzyme-carrier increased with the added amount of native enzyme. The decrease in specific activity of the bound protein with increase in the initial amount of the enzyme was most likely due to interaction between protein molecules, as suggested by Poltorak (10). The amount of protein immobilized was determined as the difference between the starting protein and uncoupled protein in the solution after the immobilization was completed. From these data the practical ratio between the initial amount of native enzyme and the amount of carrier could be selected based on cost data.

With the preoxidized enzyme, the activity of the immobilized enzyme per g of enzyme-support and the specific activity of the bound protein both

TABLE 3

DETERMINATION OF pH OPTIMUM FOR OXIDATION OF AMYLOGLUCOSIDASE

pH	Initial Activity of Native Amyloglucosidase (units)	Activity of Preoxidized Amyloglucosidase (units)	Activity Retained after Oxidation (%)	Activity of Immobilized Amyloglucosidase (units)
7.0	560	375	66.0	81.0
6.0	560	342	61.0	79.5
5.0	560	358	64.0	32.5
4.0	560	450	80.5	85.5
3.2	560	450	80.5	88.2

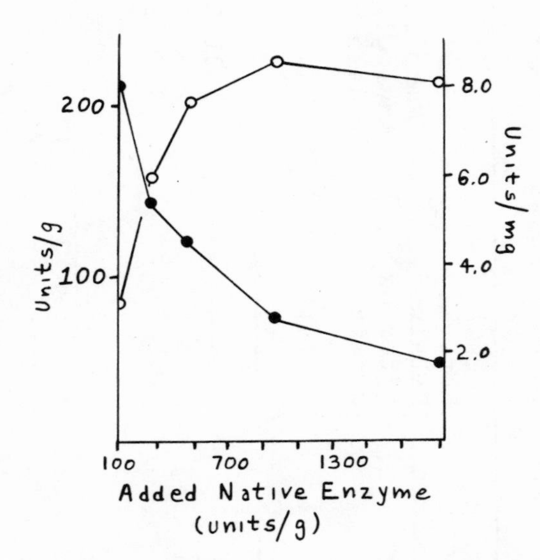

Fig. 3. Relationship between the activity of immobilized protein and the total units of native amyloglucosidase from *End. bispora* added per g of support for immobilization via azo coupling. Results are expressed as activity in units/mg of immobilized protein (●) and activity in units/g of enzyme-carrier preparation (O) (13).

were greatest with the same initial amount of native enzyme (Fig. 5). In spite of the fact that up to 80% of the initial activity of the enzyme was retained after oxidation, the retention of activity after immobilization did not exceed 45 to 47%, probably, because the carbohydrate part of amyloglucosidase consists of three saccharides and up to 80 carbohydrate residues (11). With such a large number of binding sites the amyloglucosidase molecule could become rigid enough to affect the activity of the immobilized form. However, in

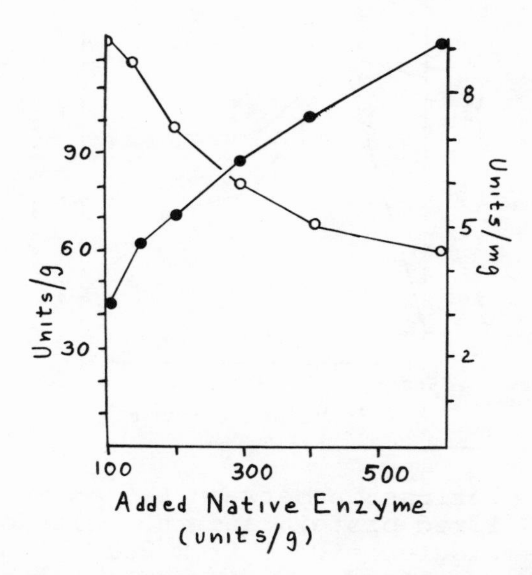

Fig. 4. Relationship between the activity of immobilized protein and the units of native amyloglucosidase from *End. bispora* added per g of support for immobilization via glutaraldehyde coupling. Results are expressed as activity in units/mg of immobilized protein (O) and activity in units/g of enzyme-carrier preparation (●).

spite of some differences in the degree of activity after immobilizing amyloglucosidase by means of a variety of methods, the activities of the immobilized preparations were rather close.

II. PROPERTIES OF NATIVE AND IMMOBILIZED AMYLOGLUCOSIDASE

It is necessary to study the properties of immobilized amyloglucosidase in order to select a preparation for commercial application and to estab-

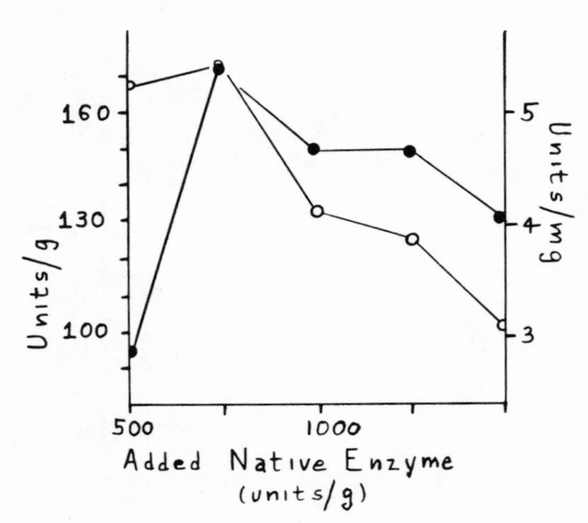

Fig. 5. Relationship between the activity of immobilized protein and the units of native amyloglucosidase from *End. bispora* added per g of support for preoxidation and then immobilization on aminosilochrome C-80. Results are expressed as activity in units/mg of immobilized protein (O) and activity in units/g of enzyme-carrier preparation (●).

lish the scientific foundations for the technology of starch hydrolysis to glucose using the immobilized enzyme.

The pH and temperature optima for the enzyme action were studied. The activities of native and immobilized preparations as a function of pH are shown in Fig. 6. With azo coupling, the pH optimum was shifted in the alkaline direction to pH 6.0. With glutaraldehyde coupling the pH optimum was at 4.7; and for immobilized preoxidized amyloglucosidase the pH optimum was at 4.0. The temperature optimum for the native enzyme and for azo or glutaraldehyde coupled enzyme on silichrome C-80

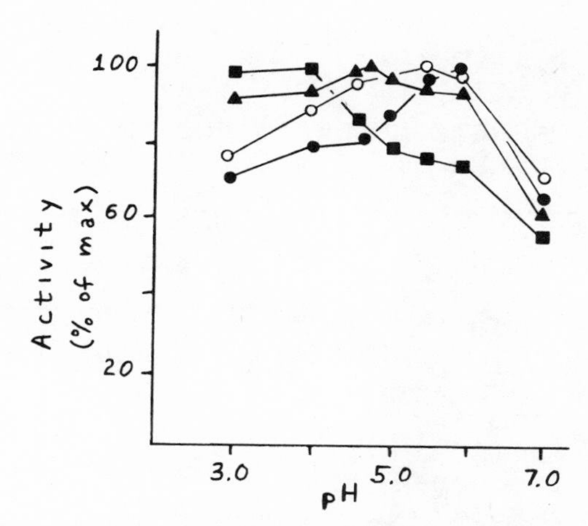

Fig. 6. pH optima of amyloglucosidase preparations from *End. bispora* immobilized on silochrome C-80 by azo coupling (●), glutaraldehyde coupling (▲), preoxidized enzyme (■), and native enzyme (O) (13).

was at 50°C for all three preparations, with rapid fall off to 10% of maximum activity at about 30° and 65°C.

The stability of the immobilized amyloglucosidase was studied at 40° to 70°C in acetate buffer at pH 4.7. The preparations were held at 40° to 60°C for four hr and at 70°C for one hr. The rate constant for thermal inactivation k_{in}, was calculated from Eq. 1:

$$k_{in} = \frac{2.303}{t} \log \left[\frac{E_o}{E_t}\right] \qquad \text{(Eq. 1)}$$

TABLE 4

INACTIVATION RATE CONSTANT FOR AMYLOGLUCOSIDASE IN ACETATE BUFFER AT pH 4.7 WITHOUT SUBSTRATE[(a)]

Amyloglucosidase Preparation	Rate Constants[(a)]			
	40°	50°	60°	70°
A. Enzyme from *End. bispora* immobilized on silochrome C-80 via:				
Azo Coupling	0.028	0.10	0.33	2.9
Glutaraldehyde	0.028	0.08	0.31	2.3
Preoxidized enzyme	0.027	0.10	0.32	2.8
B. Amyloglucosidase from *Asp. niger* immobilized on silochrome C-80 via glutaraldehyde	0.022	0.076	0.27	2.3
C. Native amyloglucosidase from:				
Asp. niger	-	0.082	-	-
End. bispora	-	0.12	-	-

(a) Rate constant in hr^{-1} after 4 hr at 40° to 60° or 1 hr at 70°C.

where E_o represents the initial activity of the native or immobilized enzyme in units/g, E_t represents the activity after the elevated temperature exposure for t hours in units/g, and t represents the duration of the elevated temperature exposure in hr. The results, presented in Table 4, show a slight increase in stability for the immobilized preparations as compared with the native enzyme. Also, the preparations had low stability at 60°C and particularly at 70°C, with inactivation half times of approximately two hr and 15 min, respectively. It should be noted that the thermostability of amyloglucosidasee from *Aspergillus niger* and from *Endomycopsis bispora* were rather close for both the soluble and immobilized preparations.

Stabilization of enzymes by the substrate has been known since 1890, when O'Sullivan and Thompson discovered a strong stabilizing effect of sucrose on invertase (12). A comparison of the thermal stability of immobilized amyloglucosidase in the presence of substrate at 50°C showed an increase in stability by a factor of 3 to 4 times (13,14). At higher temperatures the stabilizing effect of the starch was even more marked (19).

The kinetic constants for different amyloglucosidase preparations were determined for starch and maltose hydrolysis at 30°C and low substrate concentrations. The values of the Michaelis constant K_m and the apparent maximum velocity V_{max} were obtained from Lineweaver-Burk double reciprocal plots of the reaction velocity and the substrate concentration. The results, along with the overall k_{cat}, are shown in Table 5. K_m and V_{max} are only apparent values for the immobilized enzyme due to the resistance to diffusion of the substrate and products through the carrier pores. It should be noted that the k_{cat} values were similar for the different preparations, while the K_m values with azo coupling may have been somewhat lower than those with the other coupling methods. This may be due to binding via the benzene ring of the tyrosine moiety of the enzyme being more

TABLE 5

KINETIC CONSTANTS FOR DIFFERENT AMYLOGLUCOSIDASE PREPARATIONS

Preparations	Maltose as Substrate			Starch as Substrate		
	K_m (mmol)	V_{max} $\left(\frac{mmol}{l,\ sec}\right)$	k_{cat} (a)	k_m (% x 10^3)	V_{max} $\left(\frac{mmol}{l,\ sec}\right)$	k_{cat} (a)
A. Amyloglucosidase from *End. bispora* immobilized on silochrome C-80 by:						
Azo coupling	0.96	23	3.7	0.11	0.15	0.26
Glutaraldehyde	1.6	22	3.3	0.17	0.12	0.21
Preoxidized enzyme	--	--	--	0.24	0.10	0.17
B. Amyloglucosidase from *End. bispora* immobilized by glutaraldehyde on:						
Porous glass	--	--	--	0.14	0.11	0.19
Silochrome C-80	--	--	--	0.17	0.12	0.21

TABLE 5
(cont'd)

Preparations	Maltose as Substrate			Starch as Substrate		
	K_m (mmol)	V_{max} $\left(\frac{mmol}{l,\ sec}\right)$	k_{cat} (a)	k_m ($\% \times 10^3$)	V_{max} $\left(\frac{mmol}{l,\ sec}\right)$	k_{cat} (a)
C. Amyloglucosidase from *Asp. niger* immobilized on silochrome C-80 by:						
Glutaraldehyde	0.82	22	3.3	--	--	--
Azo coupling	0.71	24	3.3	--	--	--
D. Native amyloglucosidase from:						
Asp. niger	1.8	8.4	4.6	--	--	--
End. bispora	2.2	7.7	3.2	0.40	0.078	0.065

(a) Dimensions of k_{cat} are mmol/(l, sec, U); where U is unit of enzyme activity.

favorable for a higher activity molecular conformation of the active site. Supporting evidence is found in 85% retention of enzyme activity bound in this way against 44 to 47% retention for coupling via glutaraldehyde or the preoxidized enzyme.

III. HYDROLYSIS OF CONCENTRATED LIQUEFIED STARCH BY IMMOBILIZED AMYLOGLUCOSIDASE

The hydrolysis of concentrated solutions of liquefied starch by different immobilized amyloglucosidase preparations was studied a) in a batch reactor mounted on a thermostated shaker and b) under continuous processing in a column type reactor. For starch hydrolysis by native and immobilized amyloglucosidase, during the first two hr the degree of hydrolysis reached 80 to 85% after which the rate of reaction declined sharply (Fig. 7).

One of the important advantages of the enzymatic hydrolysis of starch is the possibility of carrying out a high degree of hydrolysis with attainment of dextrose equivalents (DE) of 95 to 97% (15). As our experiments have shown (16), the immobilized amyloglucosidase in contrast to the native enzyme makes it possible to obtain glucose syrups with a DE content of 92 to 93%. The DE content of glucose in the pores probably is higher due to diffusional limitations.

The influence of the liquefied starch concentration and the temperature of hydrolysis on the degree of substrate conversion was studied in shake flasks under batch conditions. The higher the starch concentration the lower was the degree of hydrolysis at all temperatures tested. This may be explained by the fact that as the starch concentration increased the viscosity of the liquified starch solution also increased, causing a greater diffusional limitation. In addition the proportionality between conversion and temperature for the three concentrations of substrate indicated that inactivation due to substrate probably was

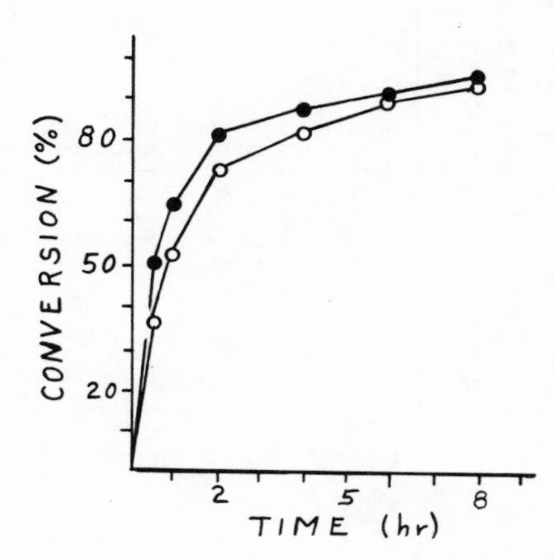

Fig. 7. Cumulative hydrolysis of 30% starch solution by amyloglucosidase from *End bispora* immobilized on silochrome C-80 via glutaraldehyde. Hydrolysis carried out at 50°C and pH 4.7, using 50 units of immobilized enzymes (●) or of soluble native enzyme (O) for each g of dry starch.

absent during the first hour of hydrolysis, even at 60°C.

The influence of flow rate on starch conversion was studied in a thermostated glass column 50 cm long and 2 cm inner diameter. From the data presented in Table 6 one can see that when the flow rate of the substrate was diminished, the amount of hydrolyzed starch per unit of enzyme activity also diminished. This can be explained by postulating significant product inhibition of the reaction at high degrees of starch conversion. In addition, the data show that with an increase in the starch concentration, the degree of substrate conversion was reduced, probably due to the greater viscosity of the higher concentrations.

TABLE 6

RELATIONSHIP BETWEEN THE DEGREE OF STARCH CONVERSION AND THE RATE OF SOLUTION FLOW THROUGH THE COLUMN

Concentration of starch (%)	Flow rate, (ml/min)	Starch Conversion (%)	Specific Productivity of Immobilized Enzyme (g of glucose/unit, hr)
20	3.1	97.0	0.0167
20	4.0	93.5	0.0208
20	5.0	92.0	0.0264
35	1.6	97.2	0.0153
35	2.3	61.0	0.0222
35	3.8	55.0	0.0348
50	0.9	73.5	0.0125
50	2.2	54.8	0.0292

The retention of immobilized amyloglucosidase activity during continuous operation of the flow reactor also was studied. A 35% solution of liquified starch was passed through the column of immobilized enzyme at 50°C for a month or longer. The time over which the immobilized preparation lost half of its initial activity, i.e. the half life, was calculated from the inactivation rate constant k_{in}. The data presented in Table 7 show that amyloglucosidase immobilized on silochrome C-80 via glutaraldehyde coupling retained the greater activity during long term operation. Amyloglucosidase bound via azo coupling to porous glass had a low half life of 15.6 day, which was rather close to the half life of 22.6 day for the analogues preparation obtained by Weetall (1). However, the same preparation immobilized on silochrome C-80 was more stable, with a half life of 124 day. Unfortunately, the half life of amyloglucosidase immobilized on silochrome via glutaraldehyde declined sharply as the temperature of hydrolysis was raised above 50°C (Fig. 8).

Since the activity of the immobilized enzyme decreased with time, we attempted to obtain a constant yield of glucose by use of three columns connected in series. Amyloglucosidase immobilized on silochrome via glutaraldehyde and having a total activity of 517 units was packed into the first column. A 35% solution of starch was passed through the column for seven days at 50°C. After that, the second column containing a total activity of 517 units was connected to the first; and after seven more days of continuous operation the third column containing 517 units was connected to the outflow of the second column. The rate of substrate flow through the columns was maintained constant during the whole experiment. On the 21st day of the experiment the activity of the immobilized preparations in the first, second, and third columns was 89.0, 95.5, and 110.0 units/g, respectively. During the experiment the degree of starch hydrolysis was 95.5 to 94.4%. The decrease in activity of the immobilized enzymes in the columns is shown in Fig. 9. These data show that

TABLE 7

STABILITY OF IMMOBILIZED AMYLOGLUCOSIDASE AT 50°C DURING HYDROLYSIS OF 35% STARCH SOLUTION IN A COLUMN REACTOR

Preparations	E_o (units/g	E_t (units/g)	Run Time (day)	k_{in} (day^{-1})	Half Life (day)
A. Amyloglucosidase from *Asp. niger* immobilized by glutaraldehyde on:					
Silochrome C-80	77	62	44	0.0051	135
Silica Gel MSA	59	47	44	0.0062	112
B. Amyloglucosidase from *End. bispora* immobilized by glutaraldehyde on:					
Silochrome C-80	70	43	92	0.0054	128
Porous Glass 90-3-g	136	84	37	0.13	54

TABLE 7
(cont'd)

Preparations	E_o (units/g)	E_t (units/g)	Run Time (day)	k_{in} (day^{-1}	Half Life (day)
C. Amyloglucosidase from *End. bispora* immobilized on silochrome C-80 by:					
Azo Coupling	89	72	37	0.0056	124
Preoxidized Enzyme	84	34	37	0.025	28
D. Amyloglucosidase from *End. bispora* immobilized on porous glass 90-3-g by azo coupling	191	134	8	0.0445	15.6

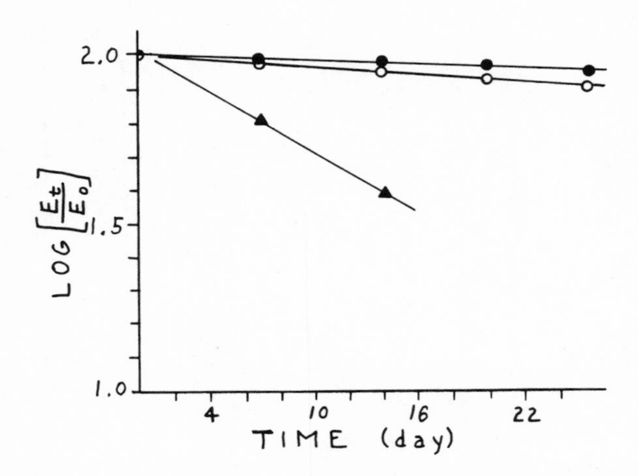

Fig. 8. The decrease in activity of amyloglucosidase from *End. bispora* immobilized on silochrome C-80 via glutaraldehyde, during hydrolysis of 35% starch solution at 45° (●), 50° (O), and 55°C (▲) (21).

the decrease in activity proceeds similarly and independently of the position of the column in the series.

On the basis of these laboratory tests conditions for producing various glucose syrups by starch hydrolysis with the aid of amyloglucosidase from *Aspergillus niger* immobilized via glutaraldehyde have been developed. The experimental testing of the processes was carried out in a pilot plant designed for this purpose and included a stirred tank reactor (17) and a column reactor. Hydrolysis of 25 to 30% maize starch solution preliquefied by acid up to a DE of 35 to 40% was studied in a 50 liter reactor. Over long time continuous operation, glucose syrups containing 88 to 93% glucose were obtained.

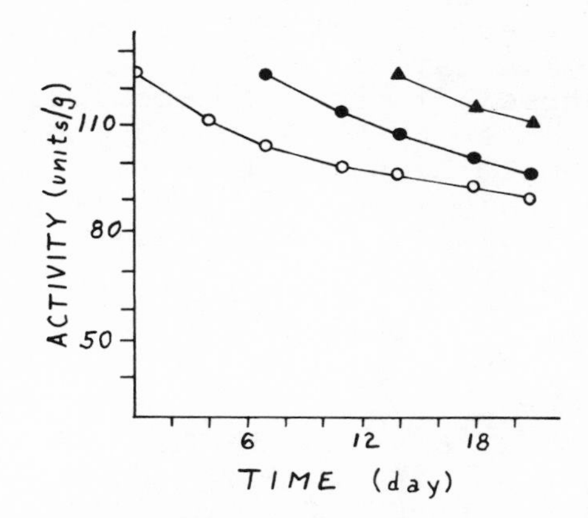

Fig. 9. The decrease in activity of immobilized amyloglucosidase from *Asp. niger*, using three flow through columns connected in series and operating at 50°C with 35% starch solution as feed to the first column. The data represent the activity in columns 1 (O), 2 (●), and 3 (▲).

The hydrolysis of 35 to 40% potato starch preliquefied by acid to 40% DE also was carried out in a 50 liter reactor but under batch conditions.

Samples of glucose pellets containing 78% dry matter and 60% reducing compounds were obtained by vacuum concentrating of glucose syrups prepared with immobilized amyloglucosidase. These pellets were used in confectionary applications.

ACKNOWLEDGMENTS

The authors are thankful to S. P. Zhdanov and Ye. V. Karomaldy of the Institute of Silicate Chemistry of the USSR Academy of Sciences and to A. V. Kiselyov and Yu. S. Nikitin of the Department of Chemistry of Moscow State University for preparation of numerous samples of porous glass, silochrome, and silica gel and for helpful discussions on the interpretation of the experimental results. The authors also are grateful to A. I. Kostner and K. E. Pappel of the Polytechnical Institute of Tallinn and to Z. M. Borodina and T. A. Shockol of the All-Union Research Institute of Starch Products for help in conducting the pilot runs of starch hydrolysis by immobilized amyloglucosidase in 50 liter reactors.

REFERENCES

1. WEETALL, H. H. & HAVEWALA, N. B. In "Enzyme Engineering" (L. B. Wingard, Jr., ed.) John Wiley, New York, 1972, p. 241.
2. NAKHAPETYAN, L. A., MENYAYLOVA, I. I., ANTONOV, V. C., ZHDANOV, S. P. & KAROMALDY, E. V. *Fermet Spirt. Prom 1*: 37, 1975.
3. MESSING, R. A. *Enzymologia 39*: 12, 1970.
4. GRAY, C. J. & JOLLEY, M. E. *FEBS Lett. 29*: 197, 1973.
5. WEIBEL, M. K., WEETALL, H. H. & BRIGHT, H. J. *Biochem. Biophys. Res. Comm. 44*: 347, 1971.
6. CAMPBELL, D. H., LUESCHER, E. & LERMAN, L. S. *Proc. Nat. Acad. Sci. USA 37*: 575, 1951.
7. ROBINSON, R. J., DUNNIL, P. & LILLY, M. D. *Biochim. Biophys. Acta. 242*: 659, 1971.
8. CAMPBELL, D. H. & WELIRY, N. "Methods in Immunology and Immunochemistry" vol. 1 (C. A. Williams & H. W. Chase, eds.) Academic Press, New York, 1967, p. 374.
9. CHRISTINSON, J. *Chem. Ind. 4*: 215, 1972.
10. POLTORAK, O. M. & VOROBJEVA, E. S. *Vestn Mosc. Univ. Chem. 6*: 17, 1966.
11. LINEBACK, D. R. & BAUMANN, W. E. *Carbohyd. Res. 14*: 341, 1970.
12. O'SULLIVON, C. & THOMPSON, F. W. *J. Chem. Soc. 57*: 834, 1890.
13. NAKHAPETYAN, L. A., MENYAYLOVA, I. I., ZHDANOV, S. P. & KAROMALDY, E. V. *Ferment. Spirt. Prom. 5*: 35, 1976.

14. MENYAYLOVA, I. I., NAKHAPETYAN, L. A., KOZLOV, L. V., ANTONOV, B. K., ZHDANOV, C. P. & KAROMALDY, E. V. *Tez. Dokl. Vses. Symp. on Production and Application of Immobilized Enzymes*, Tallinn, 1974, p. 37.
15. LADUR, T. A. & BORODINA, Z. M. & ZNIITEIPISH-CEPROM, E. I. *Krakhm. Patoch. Prom. 1*: 11, 1973.
16. MENYAYLOVA, I. I., KLEYMENOVA, G. B., NAKHAPETYAN, L. A., LADUR, T. A., BORODINA, Z. M. & SHOCKOL, T. A. *Tez. Dokl. Vses. Symp. on Production and Application of Immobilized Enzymes*, Abovian, 1977, p. 145.
17. PAPPEL, C. E., SJIMER, E. CH., KOSTNER, A. I., KULIKOVA, A. K. & TICHOMIROVA, A. S. *Trudy Tall. Polytech. Inst. 424*: 19, 1977.
18. HARTREE, G. F. *Anal. Biochem. 48*: 422, 1972.
19. SINITSYN, A. P., KLIBANOV, A. M., KLYOSOV, A. A. & MARTINEK, K. *Prikl. Biokhim. Mikrobiol. 14*: 236, 1978.
20. BEBRIS, N. K., KISELEV, A. V., MOKEEV, V. Y., NIKITIN, Y. S., YASHIN, Y. I. & ZAIZEVA, G. E. *Chromatographia 4*: 93, 1971.
21. NAKHAPETYAN, L. A. & MENYAYLOVA, I.I. *Ferment. Spirit. Prom. 2*: 41, 1976.
0.2 min^{-1} (5).

SUBSTRATE STABILIZATION OF SOLUBLE AND IMMOBILIZED GLUCOAMYLASE AGAINST HEATING

A. A. Klyosov, V. B. Gerasimas and
A. P. Sinitsyn

Department of Chemistry
Moscow State University
Moscow, USSR

Glucoamylase (1,4-α-D-glucan glucohydrolase, E.C. 3.2.1.3) catalyzes the hydrolytic cleavage of glucose units from the non-reducing end of starch and the products of partial starch hydrolysis. In the last decade this enzyme has received considerable attention owing to its great industrial usefulness in the soluble form and to its potential use in the immobilized form for production of glucose from starch maltodextrines. We have been especially interested in the production of immobilized glucoamylase that possesses high thermostability at 60 to 65°C. We believe that immobilized glucoamylase that has a half-life of about 3 to 4 weeks at 65°C could be considered a technologically feasible preparation with respect to thermostability. However, in spite of a great number of papers published on the immobilization of this enzyme, the half-life of most stable insoluble preparations at 65°C does not exceed 5 to 7 days (1,2). The question arises as to whether or not this value is characteristic of a certain limit of thermostability of glucoamylase. The answer to this question is important; because from both the theoretical and practical viewpoints it is very important to study the principles of thermostability of glucoamylase and to reveal possibilities for increasing the resistance of this enzyme to thermal action.

In the course of our studies we concentrated on the stabilization of glucoamylase against heating, using as substrate, maltodextrines. Moreover, we found that the stabilization of the enzyme by the substrate considerably exceeded the stabilization achieved by immobilization of glucoamylase on solid inorganic supports (2,3). Therefore, the first part of this paper deals with 1) a detailed kinetic analysis of the substrate stabilization of the enzyme as measured by the thermoinactivation of glucoamylase, 2) a comparison of the relative roles of immobilization and substrate stabilization on the thermoresistance of glucoamylase, and 3) the effect of substrate structure on the thermostabilization. The second part of the paper describes the effect of the immobilization technique on the stability of the glucoamylase and the degree of attainable substrate stabilization of the enzyme.

I. THEORETICAL ASPECTS

A simple scheme for thermoinactivation of an enzyme in the presence of a substrate, which produces a stabilizing or destabilizing effect on the enzyme substrate complex, can be presented in the following way:

$$\begin{array}{ccccc} E + S & \overset{K_S}{\rightleftharpoons} & ES & \xrightarrow{k_{cat}} & E + P \\ \downarrow k_{in} & & \downarrow \alpha k_{in} & & \\ D & & D & & \end{array} \qquad \text{(Eq. 1)}$$

where D is thermoinactivated enzyme, $\alpha < 1$ for the stabilization of enzyme by substrate, and $\alpha > 1$ for destabilization of the enzyme by the substrate. In this case the total rate of inactivation of the enzyme is described by the following expression:

$$v_{in} = k_{in}\,[E] + \alpha k_{in}\,[ES] \qquad \text{(Eq. 2)}$$

The material balance equation for the total concentration of enzyme, $[E]_o$, in the system is shown in Eq. 3.

$$[E]_o = [E] + [ES] \qquad \text{(Eq. 3)}$$

$$K_s = \frac{[E]\,[S]}{[ES]} \qquad \text{(Eq. 4)}$$

With the dissociation constant for the enzyme substrate complex defined as in Eq. 4 and assuming that the value of the dissociation rate constant for the enzyme substrate complex by far exceeds the value of k_{cat} and that [S] is the same as the initial substrate concentration, $[S]_o$, it follows that:

$$v_{in} = k_{in(app)} \; [E]_o \qquad \text{(Eq. 5)}$$

$$\text{where:} \quad k_{in(app)} = k_{in} \frac{\left[1 + \left(\dfrac{\alpha [S]_o}{K_s}\right)\right]}{\left[1 + \left(\dfrac{[S]_o}{K_s}\right)\right]} \qquad \text{(Eq. 6)}$$

As can be seen from Eqs. 5 and 6, in the absence of a substrate the value of the apparent inactivation rate constant for the free enzyme, $k_{in(app)}$, is equal to the inactivation rate constant for the free enzyme, k_{in}. On the other hand, at a concentration of substrate exceeding by far the value of K_s, the value of $k_{in(app)}$ equals αk_{in}, as would be inferred from Eq. 1.

To analyze Eq. 6, we may transform it in the following way:

$$\frac{1}{1-\dfrac{k_{in(app)}}{k_{in}}} = \frac{1}{1-\alpha} + \frac{K_s}{(1-\alpha)\ [S]_o} \qquad \text{(Eq. 7)}$$

If the inactivation rate constant of the free enzyme (k_{in}) can be determined from an independent experiment, then a plot of

$$\left(1 - \frac{k_{in(app)}}{k_{in}}\right)^{-1} \text{ vs. } [S]_o^{-1}$$

should allow us to obtain the values of α and K_s. However, the value of k_{in} may be too high to be determined by direct experiment. Thus, k_{in}, α, and K_s all are unknown so that the above plot cannot be made. Instead, an iterative linearization can be used, where the numerical value of k_{in} is varied until the best linear fit to the experimental results is obtained. This best fit gives an approximate value for k_{in}; and from the intercept and slope the values of α and K_s can be obtained.

II. SUBSTRATE STABILIZATION OF GLUCOAMYLASE

Glucoamylase, assumed to be derived from *Aspergillus niger*, was obtained from NOVO Industries A/S, Copenhagen, Denmark. The specific activity of the enzyme with respect to soluble starch was 9,000 $\pm$ 500 micromoles/min/g (150 $\pm$ 10 microkat/g) at 25°C. Starch hydrolysate of 24 dextrose equivalent (DE) called STAR-DRI 24-E, a partially acid hydrolyzed cornstarch from A. E. Staley Company, Decatur, Illinois, was used as the maltodextrines. Controlled pore silica having an average pore diameter of 1,150 Å, porous glass with a mean pore diameter of 750 Å, and the enzyme preparation and corn syrup solids were kindly given to us by H. Weetall of Corning Glass Works.

The immobilization of glucoamylase on silica using the silane glutaraldehyde method was carried out as described elsewhere (1), unless indicated otherwise. The half-life of such preparations in the presence of 30% dextrines at 65°C and pH 4.5 was 5 to 6 days. This agreed with published data (1). The inactivation kinetics of immobilized glucoamylase were studied mostly at 75°C in a thermostatted column, 15 cm long by 0.5 cm in diameter packed with 60 to 100 mg of the immobilized enzyme. The experiment was designed in such a way that the degree of conversion of the substrate at the outlet of the column never exceeded 30%. Glucose in the reaction products was determined by the glucose oxidase peroxidase method (4). With the amount of immobilized glucoamylase ranging from 30 to 100 mg, a linear dependence of the substrate conversion on the weight of the enzyme preparation was observed. The experimental details were described previously (5).

Our studies showed in all cases, regardless of the concentration of the added substrate or effector (maltodextrines, maltose, or glucose) or the temperature, that the kinetics of thermoinactivation of soluble and immobilized glucoamylase was first order. Fig. 1A shows the dependence of the apparent rate constants of thermoinactivation for soluble and immobilized glucoamylase on the concentration of the substrate in the reaction system. As can be seen, both substrates stabilized the enzyme to a considerable degree.

Since at 75°C the rate of denaturation of soluble glucoamylase in the absence of substrate had a half-life of only about 1 min, the value of k_{in} could not be measured with high accuracy. In addition, the limiting value of k_{in} for glucoamylase at concentrations of maltodextrines above 20% could not be used directly as the value of αk_{in} in Eq. 6, since the reaction system at these concentrations was sufficiently viscous that

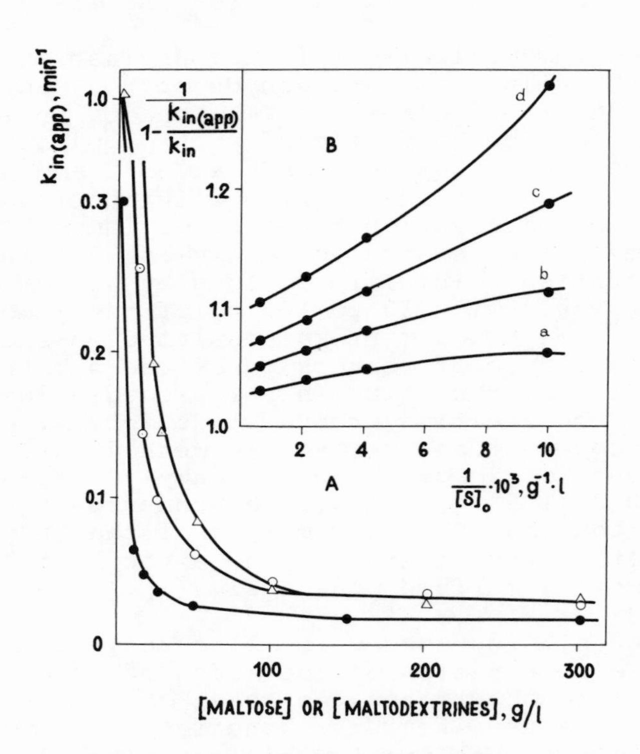

Fig. 1. Thermostability studies with glucoamylase. Part A: the apparent rate constant of thermoinactivation of soluble (Δ, o) and immobilized on silica (●) glucoamylase, with maltose (Δ) and maltodextrines (●, o) as substrates; 75°C, 0.1 M acetate buffer, pH 4.5. Part B: iterative linearization of the experimental data for thermoinactivation of immobilized glucoamylase in the presence of maltodextrines; the curves were calculated in terms of Eq. 7 for arbitrary values of k_{in} of (a) 0.7 min^{-1}, (b) 0.5 min^{-1}, (c) 0.3 min^{-1}, and (d) 0.2 min^{-1}(5).

fluid dynamic or diffusional effects became important considerations. This in turn could lead to changes in the value of the limiting inactivation rate constant. Finally according to (6), the value of the dielectric permeability of the solution could effect the rate of thermoinactivation of glucoamylase; and this variable also was difficult to take into consideration at high concentrations of the substrate. With this in mind we used the iterative linearization of our experimental data to approximate the kinetic parameters for the thermoinactivation of soluble and immobilized glucoamylase. The linearized data for the stabilization of immobilized glucoamylase by maltodextrines are shown in Fig. 1B. The best linear fit of Eq. 7 occurred at a k_{in} of 0.3 ± 0.05 min^{-1}. The values of the dissociation constant K_s for the enzyme substrate complex and of the stabilization coefficient α determined from the slope of the straight line and the intercept where 1.3 g/l and 0.06 ± 0.01, respectively. Thus, maltodextrines on being present with glucoamylase immobilized on an inorganic support give about a 15 fold increase in the thermostability of the enzyme.

The corresponding value of $k_{in(app)}$ for the immobilized glucoamylase in the presence of a saturating concentration of maltodextrines, calculated using the above constants, was 0.018 ± 0.006 min^{-1}. The experimental value of $k_{in(app)}$ in the presence of 30% cornstarch dextrines, which was more than 20 times higher than the K_s value, was 0.013 ± 0.002 min^{-1} (Fig. 1A). The good agreement of these two values suggests that the inactivation scheme of Eq. 1 may be applicable for glucoamylase and the kinetic method of analysis, as summarized in Eqs. 5 to 7, appears suitable for application to this enzyme system.

We have also studied the stabilization of soluble glucoamylase by maltodextrines and maltose. The values for the kinetic parameters are shown in Table 1. The calculated values of K_s correspond to the dissociation constants of the enzyme

TABLE 1

GLUCOAMYLASE INACTIVATION RATE CONSTANT IN THE ABSENCE OF SUBSTRATE (k_{in}), DISSOCIATION CONSTANT FOR THE ENZYME SUBSTRATE COMPLEX (K_s), AND COEFFICIENT OF STABILIZATION (α)[a][b]

	k_{in} (min^{-1})		K_s (g/l)		α	
Substrate	Sol.	Imb.	Sol.	Imb.	Sol.	Imb.
Maltodextrines[c]	1	0.30	1.6	1.3	0.025	0.06
	$\pm$ 0.1	$\pm$ 0.05	$\pm$ 0.1	$\pm$ 0.1	$\pm$ 0.002	$\pm$ 0.01
Maltose	1	-	2.6	-	0.025	-
	$\pm$ 0.1		$\pm$ 0.2		$\pm$ 0.005	

(a) 75°C., pH 4.5
(b) Sol. is soluble enzyme; Imb. is enzyme immobilized on silica.
(c) Average degree of polymerization of 7

substrate complex. This is evidenced by the fact that the value of the Michaelis constant determined in a separate experiment for hydrolysis of maltodextrines under the action of soluble glucoamylase equaled 1.3 ± 0.1 g/l at 65°C; this is the same as the K_S value of 1.6 ± 0.1 g/l and 1.3 ± 0.1 g/l for soluble and immobilized enzyme at 75°C determined by an entirely different method.

From this work we concluded first that the major role in the stabilization of soluble glucoamylase with respect to heat belongs to the substrate binding (40 fold increase in the stability on the formation of the complex both with maltodextrines and maltose) rather than to immobilization of the enzyme on the inorganic support (3 fold increase in stability). Second, the substrate stabilization is expressed to a greater degree with respect to soluble glucoamylase (40 fold) than with respect to the immobilized enzyme (15 fold). However, it should be emphasized that the limits of the thermostability increase are close to 40 to 50 times for the soluble and immobilized enzyme. Moreover, it will be shown in the following section that this limit hardly ever depends on the mode of immobilization of glucoamylase, including chemical modification of the functional groups of the enzyme, intramolecular crosslinking, and the use of various spacers between the enzyme and the support. The third conclusion is that the stabilizing effect of the substrate on glucoamylase does not depend on the length of the substrate carbohydrate chain. Maltose, that contains two glucose units, and maltodextrines, that contain on an average seven glucose units per oligosaccharide molecule, stabilized the enzyme to the same degree (Table 1). The final conclusion is that the product of the enzymatic hydrolysis, glucose, probably produces a much less pronounced stabilizing effect on glucoamylase than does the substrate. The difference in the effect of stabilization by glucose and by maltodextrines is shown in Fig. 2; although the glucose values were not measured but were calculated from data in Fig. 1 and Tables 2 and 3. The differences will be

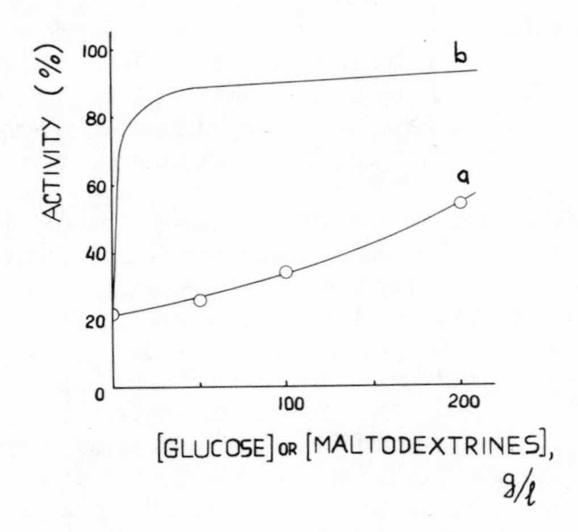

Fig. 2. Stabilizing effect of glucose and maltodextrines on the thermoinactivation of glucoamylase immobilized on silica by a silane glutaraldehyde method. Residual activity of immobilized glucoamylase was determined after incubation at 75°C for 4.5 minutes; curve a, substrate maltodextrines; curve b, theoretical results for product glucose calculated from the data of Fig. 1 and Table (5).

even greater if the concentrations on the abscissa are expressed in molar equivalents.

The data obtained in the present work have allowed us to make a number of suggestions. These pertain to the stabilizing effect of the substrate on the soluble and immobilized glucoamylase and to the character of thermoinactivation of the immobilized enzyme in reactors for production of glucose via enzymatic hydrolysis of maltodextrines. As a result of the binding of substrates maltose or maltodextrines to a certain part of the active center of glucoamylase, the latter apparently undergoes a conformational change leading to an increase in the thermostability of the enzyme.

Some ideas about the nature of the binding site can be derived from recent literature data (7-9), wherein the binding of maltose, but not glucose, induced a conformational change in glucoamylase from *Rhizopus niveus*. This conformational change in turn led to characteristic modifications in the absorption spectrum of the enzyme. The authors (7-9) suggested that the conformational change occurred with the binding of substrates or their analogs to the first subsite of the active center which contained a tryptophan residue. This led to an alteration of the microenvironment, which in turn changed the orientation of the carboxyl anion of the catalytic site of the active center. The interaction of glucose, on the contrary, occurred mostly with the second subsite of the active center, and did not induce a noticeable conformational change. The authors (7-9) did not study the thermoinactivation of glucoamylase; but their data agree basically with our evidence of the effects of maltodextrines, maltose, and glucose on the stability of the enzyme.

We hope that the elucidation of the stabilizing influence of the above effectors on soluble and immobilized glucoamylase can lead us to understand the limits in improved thermostability of the enzyme that can be achieved by this method.

The use of immobilized glucoamylase for large scale production of glucose syrups from partial hydrolyzates of cornstarch has been one of the most widely discussed possibilities of enzyme engineering during the last decade. Nevertheless, this process has not been reduced to practice so far in spite of an abundance of work on the immobilization of the enzyme and on the engineering research and development of the process. We have concluded that the moderate progress in this respect can be attributed to two factors: insufficient thermostability of the immobilized glucoamylase at 60 to 65°C and insufficient degree of enzymatic conversion of maltodextrines to glucose by the immobilized glucoamylase as compared to the soluble enzyme. For example according to literature data (1, 10-12), the degree of conversion of cornstarch dextrines into glucose at the output of a pilot plant column packed with immobilized glucoamylase did not exceed 90-92%; whereas soluble glucoamylase used for that purpose displayed a 94-95% conversion. On the other hand, discussions with corn processors indicate that the desired glucose levels should be not less than 94%, and better 95% (1).

To explain the lower degree of conversion of maltodextrines into glucose under the action of immobilized glucoamylase, as compared to the soluble enzyme, it is usually assumed (11,13,14) that glucose is present in higher concentrations in the pores of the carrier than in the bulk solution. Since glucose is the substrate for a reversion reaction that produces isomaltose and other disaccharides, the higher pore concentration of glucose would favor this reverse reaction. However, as far as we know, none of the works that have dealt with immobilized glucoamylase for production of glucose and none of the engineering studies for the reactor analysis have included the destabilization of the enzyme at relatively low substrate concentrations. As seen from Fig. 1, a marked destabilization of glucoamylase takes place at concentrations of maltodextrines of 5% and lower. Thus, in a column reactor packed with immobilized glucoamylase there is a certain distribution of the rates of inactivation of the enzyme throughout the column. The least stable enzyme should be found at the reactor outlet. After a quasi steady state distribution of the concentrations of substrate and products along the length of the column has been established, that part of the column where the degree of conversion of the substrate is 90% and higher probably contains immobilized glucoamylase that is mostly in a thermodenatured form. This effect might hinder the achievement of the high degree of conversion of the substrate in a column reactor. In the processes with the soluble enzyme, the inactivation occurs more or less uniformly throughout the volume of the reactor, which might result in a higher degree of conversion of the substrate. However, the latter case also would produce a higher total rate of inactivation of the enzyme. It seems that enzyme destabilization with lowered substrate concentrations should be taken into consideration when designing the enzyme reactors.

III. LIMIT IN THE THERMOSTABILITY OF IMMOBILIZED GLUCOAMYLASE

The data cited in the previous section and the interpretation allows one to conjecture that the thermostability of glucoamylase immobilized on inorganic supports has a certain limit which hardly ever depends on the mode of immobilization. This limit is determined mostly by the stabilization of the enzyme with cornstarch maltodextrines. The data that support this suggestion were obtained by us (2), where we studied the thermoinactivation of immobilized glucoamylase modified in 13 different ways. Apart from the enzyme immobilized on porous silica or porous glass by the silane glutaraldehyde method, we also obtained similar preparations in which the Schiff base was reduced with sodium borohydride. Preparations with glucoamylase coupled to silica or porous glass by a silane glutaraldehyde method were modified by acrolein, succinic anhydride, glutaraldehyde, dimethyladipimidate, carbodiimide/tetramethylendiamine, carbodiimide/hexamethylendiamine, carbodiimide/dodecamethylenediamine, carbodiimide/glucosamine, succinic anhydride/ glucosamine, and by maltodextrines oxidized with periodate. (Tables 2 and 3). In spite of the fact that with some of the modifications we observed a 3 to 7 fold increase in the thermostability of the immobilized preparation, some additional increase in the thermostability occurred when the substrate was added. However, in all the cases of immobilization of glucoamylase on inorganic supports that we studied, the main contribution to the increase in thermostability was as a rule made by the enzyme substrate binding. Thereby, depending on preconditioning of the enzyme

TABLE 2

THERMOSTABILITY OF IMMOBILIZED GLUCOAMYLASE FROM *Aspergillus niger* (a)

Carrier	Immobilization Technique	αk_{in} ($min^{-1} \times 10^3$)
-	Soluble glucoamylase	27 ± 3
Silica	Silane glutaraldehyde	9.0 ± 0.5
Silica	Silane glutaraldehyde + acryloylchloride	9.9 ± 0.7
Silica	Silane glutaraldehyde + 30% maltodextrines	5 ± 2
Silica	Silane gossipol (dialdehyde)	7.9 ± 0.6
Silica	Silane gossipol (in the presence of 30% maltodextrines)	7.3 ± 0.7
20% PAAG (b)	Acrolein	13 ± 1

TABLE 2
(cont'd)

Carrier	Immobilization Technique	αk_{in} ($min^{-1} \times 10^3$)
30% PAAG	Acrolein	11 ± 1
30% PAAG	Acryloylchloride	5.2 ± 0.4
30% PAAG	Acryloylchloride (in the presence of 30% maltodextrines)	5.2 ± 0.5
Silica + 20% PAAG	Silane glutaraldehyde + acryloylchloride + covalent coupling to PAAG	5.5 ± 0.5

(a) Carried out at 75°C, pH 4.5, in presence of 30% cornstarch maltodextrines
(b) Polyacrylamide gel

TABLE 3

THERMOSTABILITY OF GLUCOAMYLASE FROM *A. Niger* IMMOBILIZED ON POROUS GLASS[a]

Immobilization technique	αk_{in} ($min^{-1} x\ 10^3$)
Silane glutaraldehyde	7.2 ± 0.5
Silane glutaraldehyde + reduction with $NaBH_4$	6.4 ± 0.4
Silane glutaraldehyde + succinic anhydride	21 ± 2
Silane glutaraldehyde + acrolein	40 ± 5
Silane glutaraldehyde + glutaraldehyde	7.9 ± 0.4
Silane glutaraldehyde + dimethyladipimidate	12 ± 2
Silane glutaraldehyde + carbodiimide/tetramethylendiamine + 30% maltose	16 ± 6
Silane glutaraldehyde + carbodiimide/hexamethylendiamine	16 ± 6
Silane glutaraldehyde + carbodiimide/dodecamethylenediamine	8 ± 2

Cont'd.

TABLE 3
(cont'd)

Immobilization technique	αk_{in} ($min^{-1} \times 10^3$)
Silane glutaraldehyde + carbodiimide/glucosamine	7 ± 1
Silane glutaraldehyde + succinic anhydride + carbodiimide/glucosamine	21 ± 2
Silane glutaraldehyde + maltodextrines oxidized with HIO_4	8.7 ± 0.5

(a) Carried out at 75°C, pH 4.5, in presence of 30% cornstarch maltodextrines induced by immobilization, the substrate stabilization was higher or lower in its absolute value; but in almost all the cases a certain limit in the thermostability of glucoamylase was observed.

This conclusion seems to be very important from both the theoretical and technological points of view; for it enables one to judge the realistic possibilities for increasing the resistance of glucoamylase to thermal action. In order to obtain more evidence, we also studied the thermostability of this enzyme immobilized through various linkages on different supports. The data are summarized in Tables 2 and 3. One can see that the inactivation rate constants for the preparations are rather close; and in any case basically no different effects in the stabilization of immobilized enzyme are revealed. The additional stabilization of immobilized glucoamylase by 30% maltodextrines (2) varies for various batches of the immobilized enzyme and is not sufficiently reproducible.

In conclusion, the thermostability of the known preparations of immobilized glucoamylase appears to be restricted to a certain limit, which is determined first and foremost by the substrate binding rather than by the mode of immobilization of the enzyme. As far as we know, this is the only example in the practice of enzyme immobilization where a certain limit of thermostability has been suggested which does not depend on the mode of immobilization. The authors will be very glad if this suggested limitation in thermostability could be rejected in future studies of glucoamylase immobilization; and the data given an alternative interpretation, since the conclusions influence the practical application of fixed glucoamylase for the high temperature conversion of starch maltodextrines into glucose on an industrial scale.

REFERENCES

1. WEETALL, H. H., VANN, W. P., PITCHER, W. N., LEE, D. D., LEE, Y. Y. & TSAO, G. T. In "Immobilized Enzyme Technology, Research and Application" (H. H. Weetall and S. Suzuki, eds.) Plenum Press, New York, 1975, p. 269.
2. SINITSYN, A. P., KLIBANOV, A. M., KLYOSOV, A. A. & MARTINEK, K. *Prikl. Biokhim. Mikrobiol. (Appl. Biochem. Microbiol.) 14*: 236, 1978.
3. SINITSYN, A. P., GERASIMAS, V. B. & KLYOSOV, A. A. "Abst. 2nd Natl. Symp. on Immobilized Enzymes" (Russ.), Erevan, 1977, p. 24.
4. BEREZIN, I. V., RABINOWITCH, M. L. & SINITSYN, A. P. *Biokhimiya 42*: 1631, 1977.
5. KLYOSOV, A. A. & GERASIMAS, V. B. *Biokhimiya 44*: 984, 1979.
6. MORIYAMA, S., MATSUNO, R. & KAMIKUBO, T. *Agric. Biol. Chem. 41*: 1985, 1977.
7. OHNISHI, M., KEGAI, H. & HIROMI, K. *J. Biochem. 77*: 695, 1975.
8. OHNISHI, M. & HIROMI, K. *J. Biochem. 79*: 11, 1976.
9. OHNISHI, M., YAMASHITA, T. & HIROMI, K. *J. Biochem. 81*: 99, 1977.
10. LEE, Y. Y., LEE, D. D. & TSAO, G. T. In "Proc. 3rd Int. Biodegrad. Symp.", Kingston, R.I., 1975, p. 1021.
11. REILLY, P. J. In "Enzyme Technology Grantees-Users Conference" (E. K. Pye, ed.) University of Pennsylvania, 1976, p. 80.
12. WEETALL, H. H. In "Immobilized Enzyme for Industrial Reactors" (R. A. Messing, ed.) Academic Press, New York, 1975, p. 201.
13. LEE, D. D., LEE, Y. Y., REILLY, P. J., COLLINS, E. V. & TSAO, G. T. *Biotechnol. Bioeng. 18*: 253, 1976.
14. REILLY, P. J. In "Enzyme Technology and Renewable Resources" (J. L. Gainer, ed.) University of Virginia, 1977, p. 123.

SECTION III
BIOMEDICAL POSSIBILITIES OF ENZYME ENGINEERING

CHEMICAL ASPECTS OF ENZYME STABILIZATION AND MODIFICATION FOR USE IN THERAPY

V. P. Torchilin, A. V. Mazaev, E. V. Il'ina, V. S. Goldmacher, V. N. Smirnov, and E. I. Chazov

National Cardiology Research Center
Petroverigskiy Lane
Moscow, USSR

In recent years it has been suggested that enzyme preparations could play a greater role in practical medicine (1). Unfortunately, the everyday clinical use of enzymatic preparations has been hindered by a number of factors. One such limitation has been the short duration of enzyme activity under physiological conditions; this has increased greatly the quantity of preparation to be used in the course of treatment. Another limitation has been the antigenic properties of enzymes, which appear as foreign proteins to the organism and thus may set the maximum preparation concentration limit in the organism. Other limitations have included possible nonspecific toxicity, enzyme accessibility to the action of endogenous proteases, and the high price and shortage of particular enzymes suitable for practical use.

The problems to be solved in paving the way for wider use of enzymes in therapy are basically to decrease the total dose of the enzyme preparation and to increase the duration of its activity in the organism. In solving these problems, it may be helpful to make use of the methods worked out in the study of immobilized enzymes (2-4), particularly in regard to stabilization by different methods (5). By use of such methods it may be possible to make enzyme preparations that show prolonged activity in the organism in small doses and that produce only mild toxicity or antigenic reactions.

I. APPROACHES TO ENZYME IMMOBILIZATION FOR MEDICAL APPLICATIONS

There are two general ways of producing immobilized or modified medical enzymes: 1) water soluble enzyme derivatives or 2) water insoluble biodissolving particles. If the medical enzyme is intended for long term circulation in the organism to affect macromolecular substrates having no strict localization, it is reasonable to utilize water soluble enzyme derivatives. These derivatives need to differ from the native enzymes by having higher stability against denaturation and lower clearance rates. In those cases where the enzyme is necessary only in certain locations, for example, in local thrombosis like lesions, use should be made of immobilized enzyme biocompatible and biodissolving particles that can be easily localized in the lesion zone and made to discharge the active material at a given rate. Although the local concentration of the preparation would be high, the total dose would remain minor, thus decreasing enzyme waste as well as undesirable side reactions.

At the present time a number of attempts already have been made to develop immobilized or modified enzymes intended specially for therapeutic use. The microencapsulation of enzymes, suggested by Chang (6-8), appears to be the best explored so far. His experiments with immobilized catalase, urease, and asparaginase in animals suggest that this might be an important approach to the use of enzyme preparations for the correction of metabolic mistakes caused by low molecular weight substrates. However, the microencapsulated method is of little service in the immobilization, modification, and stabilization of enzymes intended for converting macromolecular substrates (Fig. 1.). Obviously,

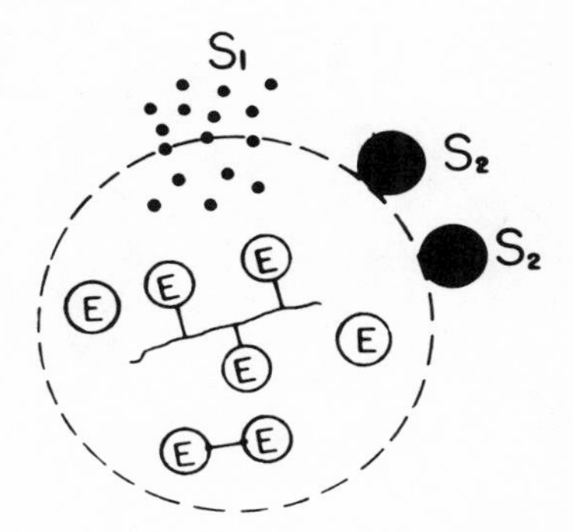

Fig. 1. Enzyme(E), native or immobilized by different methods, inside the microcapsule. Low molecular weight substrate (S_1) penetrates the membrane; high molecular weight substrate (S_2) does not penetrate the membrane (6).

other ways must be developed for large molecular weight substrates.

Work also has been done in preparing biodissolving enzyme particles for localized applications. An example is the use of thrombolytic enzymes deposited in the thrombosis zone, according to Fig. 2. The enzymes are bound covalently to a gradually destroying carrier. As a result, the enzyme is transferred into solution still covalently bound to the carrier fragments and thus possibly still stabilized by the carrier. The use of similar preparations with different carriers has been described for immunological purposes (9,10). An alternative approach is to include the enzyme in the structure of a biocompatible polymer, so that the enzyme gradually leaches out of the matrix into the surrounding solution. Such preparations, using polymer matrices of different nature and biocompatibility, were investigated by Langer and Folkman

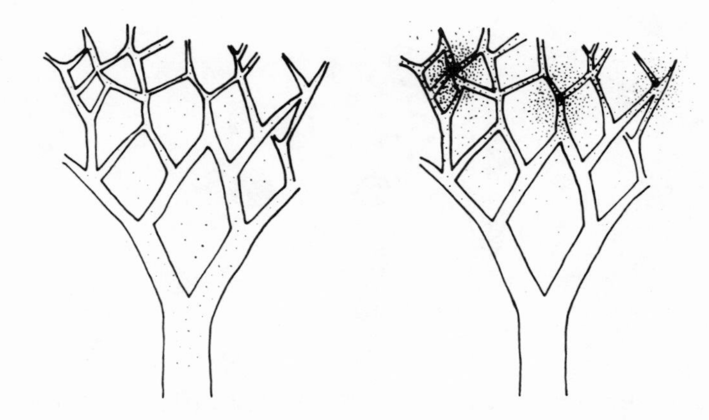

Fig. 2. Distribution of biodissolving enzyme microspheres (left) or of native enzyme (right) in a capillary vascular bed.

with soybean trypsin inhibitor and other protein inclusions (11). Various systems of polyvinylpyrrolidone gel containing chymotrypsin, which leave the polymer particle at a given rate, also have been described (12,45).

We have created a number of biodissolving carriers of different dissolution rates on the basis of partial oxidation modified Sephadex, the crosslinked derivative of the biocompatible polysaccharide dextran. Oxidized Sephadex contains reactive carbonyl groups. This material can dissolve slowly in aqueous solutions at a rate dependent on the degree of modification (13,14). The carriers were used for the immobilization of a model enzyme, α-chymotrypsin. As shown in Table I, the immobilization has little effect on the enzyme biochemical properties, but provides it with a higher thermostability both on particulate carriers and in solution. On the particulate carriers the bound enzyme is protected from inhibition under the action of macromolecular inhibitors.

TABLE 1

PROPERTIES OF BIOSOLUBLE PREPARATIONS OF IMMOBILIZED α - CHYMOTRYPSIN[a]

Preparation	Parameters of N-Acetyl-L-Tyrosine Ethyl Ester Hydrolysis		Enzyme Activity with Protein Trypsin Soy Bean Inhibition	Enzyme Activity after 6 hr at 37°C
	K_m ($M \times 10^3$)	k_{cat} (sec^{-1})	(% of original)	(% of original)
Enzyme on Aldehyde Sephadex Particles	2.5	90	95	90
Enzyme on Aldehyde Sephadex with Complete Stabilization	1.8	120	50[b]	85
Native Enzyme	1.0	160	0	0

(a) For experimental conditions see (14,22,24)
(b) Only 50% of solid matter solubilized

This is important for thrombolytic enzymes since blood contains powerful systems of protease inhibitors.

Using immobilized albumin labelled with I^{131}, we have shown the possibility of directed deposition in a coronary vascular bed. The Fig. 3 data show that while intracoronary administered protein, injected through a catheter, is distributed in the organism during an hour, the protein immobilized on carriers with different rates of solubilization stays in the myocardium for 24 hr or longer. This direction is of paramount importance for medical applications. The protein distribution time was established by the radioactivity index, defined as the ratio of the radioactivity level above the abdominal cavity to that above the heart area. The ratio became constant (relatively 1.0), when the injected protein was evenly distributed in the organism.

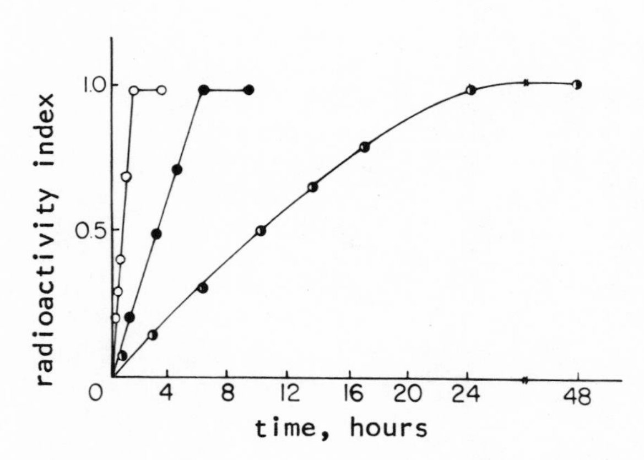

Fig. 3. Distribution of I^{131}-labelled albumin after intracoronary administration in the dog. Curves represent: (O) soluble I^{131}-albumin; (●) and (◑) I^{131}- albumin immobilized on aldehyde Sephadex; (●) carrier took 6 hr for solubilization; (◑) carrier took 24 hr for solubilization.

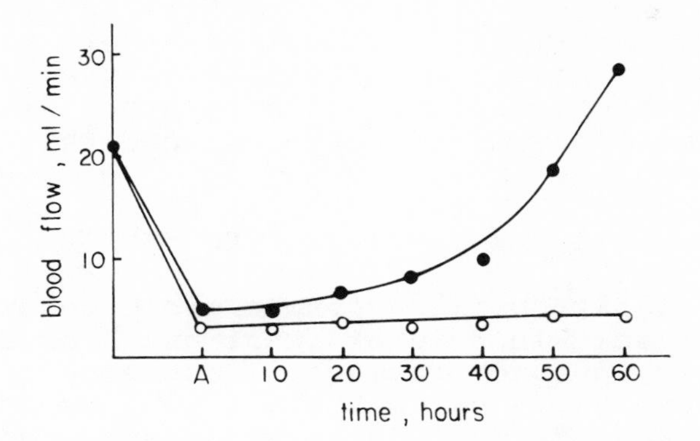

Fig. 4. Blood flow in the femoral artery of the dog after creation of an artificial thromboembolism (O), followed by treatment with a biodissolving fibrinolysin preparation (●) (16).

We also have used the enzyme fibrinolysin, as the model protein and have demonstrated for the first time its pronounced effect on a peripheral thrombus in the canine femoral artery (16). Fig. 4 shows the change in canine femoral artery blood flow after an artificial thromboembolism and subsequent administration of a suspension of microparticles of fibrinolysin immobilized on modified Sephadex. With this preparation the solubilization time was about 1 hr. It is clear that one hour after the immobilized enzyme injection, the blood flow was completely restored in the artery, where, close to the thrombus surface the preparation was injected. In the other femoral artery of the same animal the thromboembolism was induced, which served as a control; and no blood flow changes were registered. It must be pointed out that in the experiment the dose of enzyme was 100 times smaller than is necessary for lysing the same thrombus by intravenous administration of native fibrinolysin.

Thus, the very first results of studying microparticles of biodissolving immobilized enzymes demonstrate their suitability for meeting the needs described earlier. Moreover, similar preparations have the potential for use in treating not only cardiovascular but also other diseases, such as oncology or endocrine disorders.

II. WATER SOLUBLE STABILIZED ENZYMES

Generally speaking, there are a considerable number of soluble polymers that can bind compounds and thereby provide biological activity to polymers (17). Dextrans, due to their high biocompatibility and easy utilization by organisms, seem to be the most preferable carriers at present. For example, α-amylase or catalase, bound to dextrans activated by the traditional cyanogen bromide method, have significantly increased circulation times in experimental animals as compared to the unbound enzymes (18). Poly-N-vinylpyrrolidone (19) and polyethyleneglycol (20) enzyme adducts can be used to enhance the circulation time of the enzymes. Besides, in some cases a decrease in the antigenic properties of the bound albumin or enzyme has been demonstrated (21). In developing water soluble carriers special importance must be given to obtaining the maximal biocompatibility of the carriers, the binding of maximal quantities of protein, and to maintaining the specificity of the protein.

Special consideration was given to these problems in our experiments. Thus, we did not employ the cyanogen bromide method of dextran activation because of high toxicity intermidiate products but instead used oxidative activation of the carbonyl groups (22). It also turned out that the immobilized α-chymotrypsin preparations obtained according to this method showed noticeably higher thermostability, with their activity and specificity towards low or high molecular weight substrates practically unchanged (Table 2).

TABLE 2

PROPERTIES OF WATERSOLUBLE IMMOBILIZED α-CHYMOTRYPSIN(a)

Preparation	Parameters of N-Acetyl-L-Tyrosine Ethyl Ester Hydrolysis		Enzyme Activity with Protein Trypsin Soy Bean Inhibition	K_i	Enzyme Activity after 6 hr at 37°C
	K_m ($M \times 10^3$)	K_{cat} (sec^{-1})	(% of original)	($M \times 10^9$)	(% of original)
Enzyme on Soluble Aldehyde Dextran	1.3	100	25	7.8	70
Enzyme on Soluble Acrylic Acid Copolymer Without Complexing	1.2	60	0	1.5	65
Enzymes on Soluble Acrylic Acid Copolymer With Complexing	1.2	50	0	1.5	85
Native Enzyme	1.0	160	0	1.2	0

(a) Experimental conditions see (22-24).

In the course of immobilizing enzymes on synthetic polymers it was discovered that the introduction of reactive ionogenic groups into the inert polymer matrix under certain conditions gave preparations capable of electrostatic complexation with enzymes. Some of these preparations were able to bind greater quantities of enzyme than usual and to produce enzyme preparations of higher thermostability; at the same time the activity and specificity towards low or high molecular weight substrates was preserved (23,24). This is shown in Table 2.

The preparation of enzymes modified by polymer carriers, both of polyelectrolytic and non-polyelectrolytic origin, may be useful in the creation of new types of medical agents or for studying the mechanisms of certain biochemical processes taking place in the human organism. For example, we studied the effect of insulin, modified by soluble polymers of different origin and different molecular weight, on the blood glucose level in animals in comparison with the effect of unmodified insulin (25). The results showed that the insulin effect was not dependent on the polymer molecular weight; this confirmed earlier suggestions that the insulin receptors are localized on the outer surface of the cell membrane. If insulin needed to get inside the cell to modify the glucose utilization mechanism, then the molecular weight of the insulin carrier would have affected the membrane penetration rate and most likely shown up as a difference in the level of blood glucose.

III. BIOLOGICALLY ACTIVE CARRIERS

In contrast to the use of biologically inert carriers, it may be feasible to use carriers that by themselves show biological activity or are able to accentuate the action of the bound enzyme. Examples of such carriers include fibrin (26) and collagen (27), which are essentially void of antigenic properties, and the anticoagulant heparin.

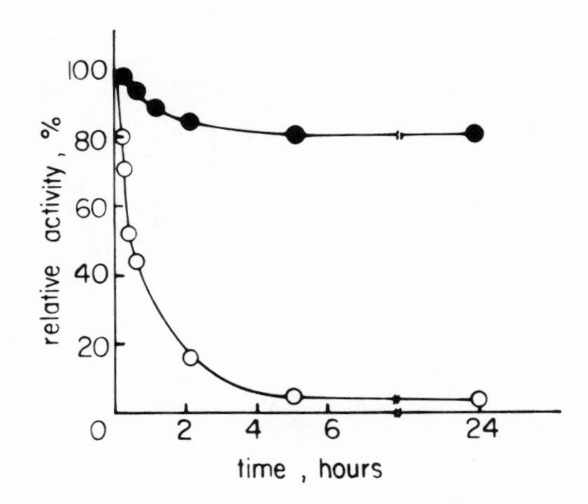

Fig. 5. Activity of native (O) and heparin-bound (●) α-chymotrypsin at pH 7.5 and 37°C (28).

Then natural anticoagulant heparin is a rigid macromolecule containing reactive carboxy groups that can be used for enzyme binding via activation with water soluble carbodiimide (28). Heparin may be suitable for the immobilization of enzymes intended for thrombolytic therapy, since the patients undergoing such therapy often are given heparin. Table 3 shows some properties of heparin immobilized chymotrypsin; while Fig. 5 shows its thermostability qualities as compared to those of the native enzyme. For the native and immobilized enzyme, the properties change little while the thermostability increases sharply on coupling to heparin. Certainly, heparin is only one of the possible biologically active carriers, and further investigations in this field are necessary, especially with enzymes of greater therapeutic potential.

TABLE 3

PROPERTIES OF HEPARIN-BOUND α-CHYMOTRYPSIN[a]

Preparation	Parameters of N-Acetyl-L-Tyrosine Ethyl Ester Hydrolysis		K_i	Optimum
	K_m ($M \times 10^3$)	k_{cat} (sec^{-1})	($M \times 10^9$)	pH
Enzyme on Heparin	1.3	80	1.2	9.0
Native Enzyme	1.0	160	1.2	8.0

(a) Experimental conditions see (24).

IV. OTHER APPROACHES TO ENZYME STABILIZATION AND IMMOBILIZATION

In many cases the presence of an insoluble or even soluble carrier matrix may become an insuperable obstacle for the enzyme to perform its function in the organism. In such cases investigators have used bifunctional reagents to stabilize enzymes by intramolecular crosslinking (29,30). As a rule these attempts have had only limited success, based on trials with many enzymes, for example galactosidase (31). The matter is complicated by the fact that the bifunctional reagent can interact with the enzyme in three different ways: one point modification, intermolecular linking, and intramolecular crosslinking. The observed stabilization effect can be explained easily by a simple modification to the enzyme or by changes in the microenvironment as a result of the modification (32). There is no necessity to speak of the possibility of intermolecular crosslinkage.

Evidently, the success or failure of stabilization by bifunctional reagents depends at least in part on the conformity between the size of the bifunctional reagent and the distance between the possible linkage centers in different regions of the enzyme globule. In addition the length of a bifunctional reagent to be in conformity with the distance between the centers to be linked may differ quite a bit for different enzymes. We tried to find the criteria for selection of bifunctional reagents for enzyme stabilization, using as a model α-chymotrypsin stabilization by intramolecular linkages. We used aliphatic diamines with hydrocarbon chain lengths of 0 to 12 methylene groups, interacting with enzyme carboxy groups preactivated by water soluble carbodiimide (33). The dependence of the crosslinked chymotrypsin inactivation rate constant at 60°C on the number of methylene groups in the diamine molecule is discussed in the paper by K. Martinek, V. V. Mozhaiv, and I. V. Berezin in this volume. The crosslinking has a sharp stabilization maximum with 1,4-tetramethylenediamine. When the enzyme was enriched with carboxy groups as a result of treating its amino groups with succinic anhydride, we observed a shift of the thermostabilization maximum to ethylenediamine (only two methylene groups) and a greater stabilization effect. Thus, we have formulated a reasonable approach to the choice of a bifunctional reagent to get stabilized enzyme preparations without using a polymer carrier, but by means of intramolecular crosslinkages.

An ideal varient would be to use a mixture of compounds having potentially reversible linkages but of different chain length for enzyme stabilization by intramolecular crosslinking. In this case, the reversibility of linking might allow the enzyme in time to select the proper length of intramolecular linkages for each region of the enzyme globule. However, some later process may be needed to fix the desired bifunctional linkages after they are formed. We have managed to realize this situation employing the thiol disulfide interchange reaction (34,35). To attain this, activated disulfide groups were introduced into the chymotrypsin molecule. After this, the enzyme was treated with dithiols of different length or with a mixture of dithiols. The modified enzyme was tested against the denaturing action of guanidine hydrochloride. The results are shown in Fig. 6, where $h_{1/2}$ is the guanidine concentration that produced a 50% decrease in enzyme activity. Thus, the enzyme preparations having larger values of $h_{1/2}$ were the more stable preparations. From Fig. 6A it is apparent that the 1,5-dithiol (1,5-pentamethylenedithiol) gave a pronounced increase in stability. We attribute this increase to the formation of intramolecular crosslinks.

The modification of therapeutic enzymes by low molecular weight intramolecular crosslinking agents may be applicable to a wide range of enzymes; however at present it is not possible to predict the stabilization effect resulting from the modification. Our results do suggest that only a small number of enzyme reactive groups are

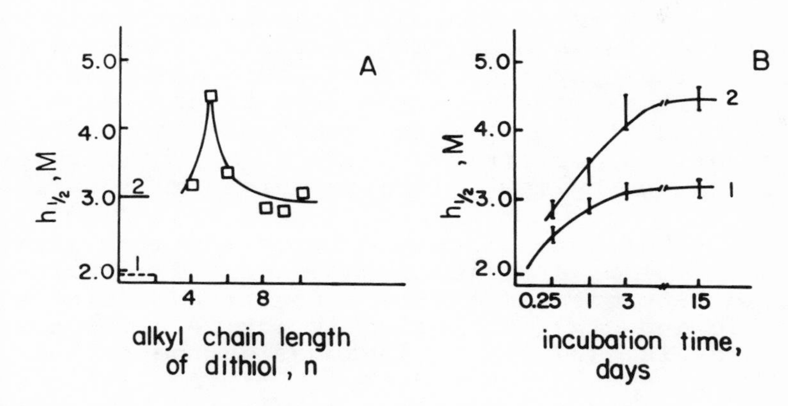

Fig. 6. Stabilization of α-chymotrypsin by dithiols. Part A: the dependence of the maximal thermostabilization effect of α-chymotrypsin, modified with different dithiols against the denaturing action of guanidine hydrochloride, on the alkyl chain length of the dithiol. No. 1 represents the stability of native enzyme and No. 2 the stability of the enzyme modified with monofunctional mercaptoethanol; the data points represent the stability of the dithiol modified enzyme. Part B: the increase in the stabilization effect as a function of time after enzyme treatment with an equimolecular mixture of different dithiols; curve 1 data were obtained in the absence of 1,5-pentamethylenedithiol; curve 2 data were obtained with a mixture containing 1,5-pentamethylenedithiol. In both parts the vertical axis represents the concentration of guanidine denaturant needed to produce a 50% loss in enzyme activity. (From V. P. Torchilin and K. Martinek, Enzyme and Microbial Technology 1: 74, 1979, by permission of the publishers, IPC Business Press Ltd. © .)

responsible for the enhanced stabilization of the active conformation of the enzyme (36). Our studies also have shown, that the modification of α-chymotrypsin amino groups by alkylation or succinylation takes place at a certain degree of conversion (see paper by Martinek, Mozhaev, and Berezin, this volume). The results have given us an opportunity to explain numerous experimental observations pertaining to the one point enzymes modification by low molecular weight modifiers (36).

Another form of enzyme immobilization consists of using artificial phospholipid vesicles called liposomes as carriers for transporting enzymatic and other biologically active compounds in the organism (37,38). Liposomes can be filled with enzymes and other water or lipid soluble compounds. These carriers are completely biocompatible and cause no antigenic or pyrogenic reactions, while the inserted compounds do not come in contact with blood components or undergo biodegradation. Liposomes also may be prepared with an appropriate antibody attached to the outher membrane and used for targeting of enzyme preparations in the organism (39). For example, antibodies to cardiac myosin have been isolated and have been shown to concentrate in the necrotic zone after myocardial infarction (40). It has been also shown, that neutral or positively charged liposomes accumulate in experimental infarction zones (41). Thus, the use of liposomes having proteolytic enzymes inside and an anti-cardiac myosin antibody outside might be useful for transport of specific enzymes to the necrosis zone in myocardial infarction as aids to promote healing.

When working out liposome enzyme transport systems, one has to take into consideration at least two difficulties. First, liposomes in a very short time after administration are assimilated by the reticuloendothelial system, and have no time to interact with the target organ cells. Second, the affinity proteins or antibody molecules adsorbed on preformed liposomes (42) or incorporated into the liposome membrane (43) may be insecurely bound to the liposome or undergo critical changes in binding ability. At present, there is no practical solution to the first problem. One can only suggest that the liposomes must be made of substances that are not substrates for liver phospholipases or the preformed liposomes must be covered by a blood protein to deceive the reticuloendothelial system cells. To answer the second question we have carried out a comparative study of different methods of binding chymotrypsin with liposomes. Besides adsorption or incorporation, we used covalent binding of the protein with the surface of a liposome via a spacer group (44). It follows from Fig. 7 that covalent binding via a spacer can preserve the binding ability of the immobilized protein with the antigen or substrate; besides, it may be possible in this way to increase the protein quantity on the surface. As a model of antibody-antigen interaction we studied the inhibition of liposome bound chymotrypsin by high molecular weight trypsin pancreatic inhibitor. It follows from Table 4 that covalent immobilization gives an opportunity to bind more enzyme than does adsorption or incorporation. Meanwhile, the liposome integrity, as measured by the leakage of preinserted radioactive H^3-deoxyglucose, remains almost constant. Of great importance seems to be the 100% preservation of the binding ability of the covalently coupled enzyme, via glutaraldehyde, towards the high molecular weight inhibitor; whereas the incorporated enzyme binding ability is considerably decreased.

V. CONCLUSIONS

In summary, the main problems in producing immobilized, stabilized, or modified enzymes for medical uses appear to be a) the need to find simple and safe methods for binding the enzyme to the carrier with complete biocompatibility of the preparation and no toxic intermediate products; b) the need to develop methods for enzyme immobilization on different biologically active carriers; c) the need to develop safe and theoretically sound methods of enzyme

TABLE 4

PROPERTIES OF CHYMOTRYPSIN BOUND WITH LIPOSOMES BY DIFFERENT METHODS

Method of Immobilization	Quantity of Immobilized Enzyme for Constant Quantity of Lipid (relative units)	Quantity of ^3H-Deoxyglucose in Liposomes after Immobilization (% of initial)	K_1/K_2 (a)
Adsorption	<30	70	(b)
Incorporation	59	80	5
Covalent Binding	100	80	1

(a) K_1 and K_2 are constants of inhibition of immobilized and native chymotrypsin, respectively with macromolecular pancreatic trypsin inhibitor at 37°C, pH 8.5, 0.145M NaCl.

(b) Could not measure due to low enzyme concentration

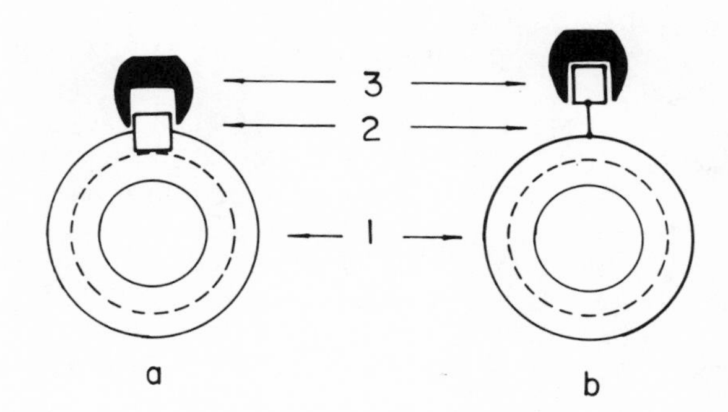

Fig. 7. Schematic picture of liposomes with immobilized enzyme or antibody on the liposome surface. Part 1 represents the liposome, showing both sides of the membrane. Part 2 represents an antibody or enzyme attached by adsorption or incorporated into the membrane in (a) and by covalent coupling through a spacer in (b). Part 3 (dark color) represents an antigen or substrate (44).

stabilization without the use of carriers; and d) the need to solve the early removal problem in enzyme targeting with liposomes.

REFERENCES

1. WOLF, M. & RANSBERGER, K. "Enzyme Therapy", Vantage Press, New York, 1972.
2. ZABORSKY, O. R. "Immobilised Enzymes" CRC Press, Cleveland, Ohio, 1973.
3. MOSBACH, K. *Meth. Enzymol. 44*: 1976.
4. BEREZIN, I. V., ANTONOV, V. K. & MARTINEK, K., eds, "Immobilizovannye Fermenty" ("Immobilized Enzymes," Russ. edition) Moscow University Press, Moscow, 1976.

5. CHANG, T. M. S., ed. "Biomedical Applications of Immobilized Enzymes and Proteins" vol. 1-2, Plenum Press, New York, 1977.
6. CHANG, T. M. S. *Meth. Enzymol.* *44*: 201, & 676, 1976.
7. CHANG, T. M. S. *Nature* *229*: 117, 1977.
8. CAMPBEL, J. & CHANG, T. M. S. In "Biomedical Applications of Immobilized Enzymes and Proteins" vol. 2 (T. M. S. Chang, ed.) Plenum Press, New York, 1977, p. 281.
9. WILSON, M. B. & NAKANE, P. K. *J. Immunol. Meth.* *12*: 171, 1976.
10. JUNOWICZ, E. & CHARM, S. E. *Biochim. Biophys. Acta* *428*: 157, 1976.
11. LANGER, R. & FOLKMAN, J. *Nature* *263*: 797, 1976.
12. TORCHILIN, V. P., TISCHENKO, E. G., SMIRNOV, V. N. & CHAZOV, E. I. *Bioorg. Khim.* (Russ. edition) *2*: 399, 1976.
13. TORCHILIN, V. P., BOBKOVA, A. S., SMIRNOV, V. N. & CHAZOV, E. I. *Bioorg. Khim.* (Russ. edition) *2*: 116, 1976.
14. TORCHILIN, V. P., TISCHENKO, E. G., SMIRNOV, V. N. & CHAZOV, E. I. *J. Biomed. Mater. Res.* *11*: 223, 1977.
15. TORCHILIN, V. P., LEBEDEV, B. S., MAZAEV, A. V., KUKHARCHUK, V. V., EVENTOV, A. Z., KRAMER, A. A., SMIRNOV, V. N. & CHAZOV, E. I. *Kardiologia* (Russ. edition) *16*(N9): 102, 1976.
16. CHAZOV, E. I., MAZAEV, A. V., TORCHILIN, V. P., LEBEDEV, B. S., IL'INA, E. V. & SMIRNOV, V. N. *Thromb. Res.* *12*: 809, 1978.
17. RINGSDORF, H. J. *Polymer Sci. Symp. No.* *51*: 135, 1975.
18. MARSHALL, J. J., HUMPHREYS, J. D. & ABRAMSON, S. L. *FEBS Lett.* *83*: 249, 1977.
19. VON SPECHT, B. U. & BRENDEL, W. *Biochim. Biophys. Acta* *484*: 109, 1977.
20. ABUCHOWSKI, A., MCCOY, J. R., PALCZUK, N. C., VAN ES, T. & DAVIS, F. F. *J. Biol. Chem.* *252*: 3582, 1977.
21. ABUCHOWSKI, A., VAN ES, T., PALCZUK, N. C. & DAVIS, F. F. *J. Biol. Chem.* *252*: 3578, 1977.

22. TORCHILIN, V. P., REIZER, I. L., SMIRNOV, V. N. & CHAZOV, E. I. *Bioorg. Khim.* (Russ. edition) *2*: 1252, 1976.
23. TORCHILIN, V. P., TISCHENKO, E. G. & SMIRNOV, V. N., *J. Solid-Phase Biochem. 2*: 19, 1977.
24. TORCHILIN, V. P., REIZER, I. L., TISCHENKO, E. G., IL'INA, E. V., SMIRNOV, V. N., & CHAZOV, E. I. *Bioorg. Khim.* (Russ. edition) *2*: 1687, 1976.
25. TORCHILIN, V. P., IL'INA, E. V., MAZAEV, A. V., LEBEDEV, B. S., SMIRNOV, V. N. & CHAZOV, E. I. *J. Solid-Phase Biochem. 2*: 187, 1978.
26. DILLON, J. G., WADE, C. W. R. & DALY, M. H. *Biotechnol. Bioeng. 18*: 133, 1976.
27. VENKATASUBRAMANIAN, K., VIETH, W. R. & BERNATH, F. R. in "Enzyme Engineering", v. 2 (E. K. Pye, and L. B. Wingard Jr., eds.) Plenum Press, New York, 1974, 439-445.
28. TORCHILIN, V. P., IL'INA, E. V., STRELTSOVA, Z. A., SMIRNOV, V. N. & CHAZOV, E. I. *J. Biomed. Mater. Res., 12*: 585, 1978.
29. SAIDEL, L. J., LEITZES, S. & ELFRING, W. H. *Biochem. Biophys. Res. Communs. 15*: 409, 1964.
30. WANG, J. H. & TU, J. *Biochemistry 8*: 4403, 1969.
31. SNYDER, P. D., JR., WOLD, F., BERNLOHR, R. W., DULLUM, C., DESNICK, R. J., KRIVIT, W. & CONDIE, R. M. *Biochim. Biophys. Acta 350*: 432, 1974.
32. ZABORSKY, O. R. In "Enzyme Engineering", vol. 1 (L. B. Wingard, Jr., ed.) John Wiley, New York, 1972, p. 211.
33. TORCHILIN, V. P., MAKSIMENKO, A. V., SMIRNOV. V. N., MARTINEK, K., KLIBANOV, A. M. & BEREZIN, I. V. *Biochim. Biophys. Acta 522*: 277, 1978.
34. TORCHILIN, V. P., MAKSIMENKO, A. V. & MARTINEK, K. *Bioorg. Khim.* (Russ. edition), *5*: 295, 1979.
35. TORCHILIN, V. P., MAKSIMENKO, A. V., SMIRNOV, V. N., BEREZIN, I. V. & MARTINEK, K. *Biochim. Biophys. Acta, 568*: 1, 1979.
36. TORCHILIN, V. P., MAKSIMENKO, A. V., SMIRNOV, V. N., BEREZIN, I. V., KLIBANOV, A. M. & MARTINEK, K. *Biochim. Biophys. Acta, 567*: 1, 1979.

37. WEISSMANN, G., COHEN, C. & HOFFSTEIN, S. *Biochim. Biophys. Acta 498*: 375, 1977.
38. GREGORIADIS, G. *New Engl. J. Med. 295*: 704 & 765, 1976.
39. GREGORIADIS, G. *Nature 265*: 407, 1977.
40. KHAW, B. A., BELLER, G. A., HABER, E. & SMITH, T. W. *J. Clin. Invest. 58*: 439, 1976.
41. CARIDE, V. J. & ZARET, B. L. *Science 198*: 735, 1977.
42. WEISMANN, G., BLOOMGARDEN, D., KAPLAN, R., COHEN, C., HOFFSTEIN, S., COLLINS, T., GOTLIEB, A., & NAGLE, D. *Proc. Nat. Acad. Sci. U.S.A. 72*: 88, 1975.
43. GREGORIADIS, G. & NEERUNJUN, E. D. *Biochem. Biophys. Res. Communs. 65*: 537, 1975.
44. TORCHILIN, V. P., GOLDMACHER, V. S. & SMIRNOV, V. N. *Biochem. Biophys. Res. Communs., 85*: 893, 1979.
45. GOLDBERG, E. In "Polymeric Drugs," Academic Press, New York, 1978, p. 239.

MODIFICATION OF TRYPSIN PANCREATIC INHIBITOR BY POLYSACCHARIDES FOR PROLONGATION OF THERAPEUTIC EFFECT

N. I. Larionova, I. Y. Sakharov, N. F. Kazanskaya, A. G. Zhuravlyov, V. G. Vladimirov, and P. I. Tolstich

Moscow State University and
N. I. Pirogov Second State Medical Institute
Moscow, USSR

A polyvalent inhibitor of proteinases from the bovine pancreas effectively blocks the action of kallikrein, chymotrypsin, plasmin, plasmin activator, blood coagulation factors, and tissue and leukocytic proteinases (1). Owing to its wide specificity, this pancreatic inhibitor has long been used in therapeutic practice (2, 3). The great variety of the modes of action of the pancreatic inhibitor on the organism is illustrated by the list of possible targets for its curative effect (Fig. 1). This inhibitor is used in treating diseases associated with activation of the kinin systems in the organism (5). Such diseases include acute pancreatitis, primary hyperfibrinolytic hemorrhage and coagulopathy, burns, nephrotic syndrome, and shock of different etiology (4,6). Pancreatic inhibitor is very effective in normalizing the indices of the kinin system when large doses of this expensive preparation are used with these diseases. Large doses are needed because the inhibitor does not persist in the blood flow. The half-life of pancreatic inhibitor is 7 to 10 min depending on the species of the animal and on the dose (4,7). Also, it is desirable to maintain

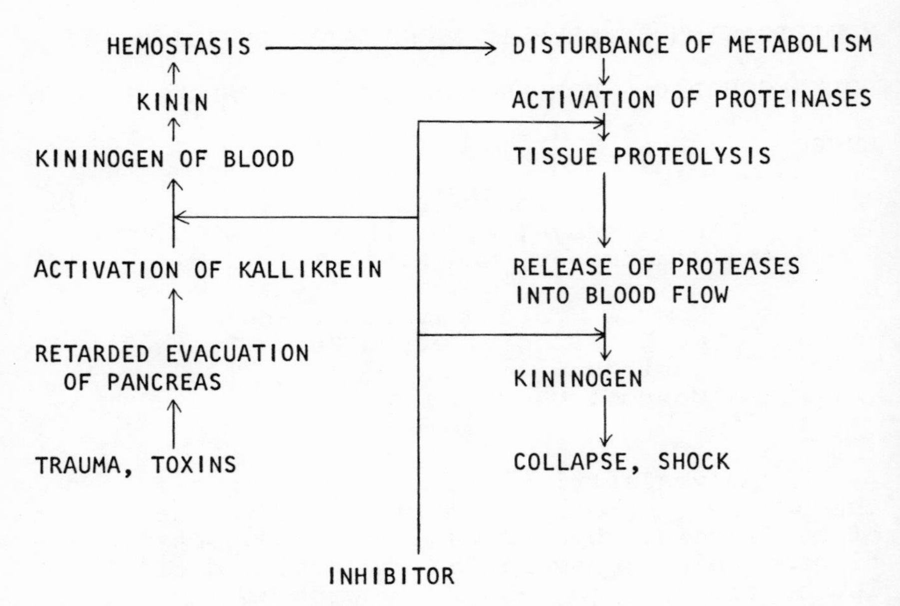

Fig. 1. Targets for trypsin pancreatic inhibitor (4).

a high concentration of pancreatic inhibitor in the blood to prolong the inhibition.

Pancreatic inhibitor usually is administered by intravenous injection. On the basis of literature data (8,9), it seems likely that the problem of repeated dosing could be solved with the help of water soluble high molecular weight derivatives, that have greater stability and longer duration in the circulation. The problem of modification of biologically important proteins by soluble high molecular weight compounds is now growing into a specific branch of enzyme engineering of importance to medicine. This paper describes the binding of pancreatic inhibitor to soluble polysaccharide carriers and discusses the resulting behavior when used *in vivo* (9,10).

I. BINDING OF TRYPSIN PANCREATIC INHIBITOR TO SOLUBLE POLYSACCHARIDE CARRIERS AND PROPERTIES OF THE CONJUGATES

Several preparations of pancreatic inhibitor bound to soluble polysaccharide carriers having different charges were synthesized (11,12). Modified dextrans and carboxymethylcellulose were used as the carriers. The molecular weight of the derivatives, the participation of protein amino acid residues in the binding to the carriers, the pH optimum, the kinetic properties, and the thermostability of the conjugates were estimated.

A special study was made of the noncovalent interaction of pancreatic inhibitor with the soluble supports. It turned out that the binding was rather strong. The complexes of pancreatic inhibitor, a basic protein with a pI of 10.5, and negatively charged polysaccharide carriers (CM cellulose and CM dextran) dissociated only in sodium chloride solutions greater than 0.3 M.

The use of binding agents having various functional groups allowed us to obtain preparations of different molecular weight. For example, activation of a carrier by the chlorine derivatives of s-triazine increased the molecular weight of the carrier, as noted by the different elution time with gel filtration (12) (Fig. 2). To interrupt the polymerization of carriers containing carboxy groups (CM dextran), soluble carbodiimide was used (Fig. 2).

Pancreatic inhibitor belongs to the group of proteins that have an amino group in the active center (13). This limits the possibilities for immobilization of pancreatic inhibitor, since many of the conventional modes for covalent binding to carriers involve the interaction with amino groups of the protein. We have succeeded in obtaining com-

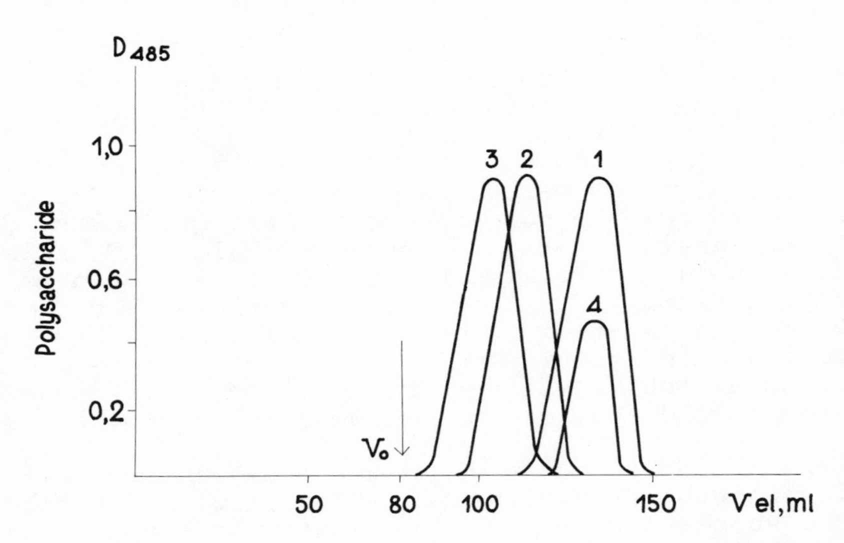

Fig. 2. Estimation of the molecular modification of polysaccharides by gel filtration; optical density of column eluant versus elution volume; V_o represents column void volume; Sepharose 6B packing in 260 ml total column volume. Curves represent: 1, polysaccharide; 2, polysaccharide activated by cyanuric chloride in a nonaqueous medium; 3, polysaccharide activated by cyanuric chloride and incubated at pH 8.3 for 18 hr; 4, preparation of pancreatic inhibitor bound to polysaccharide by carbodiimide.

pletely active preparations of pancreatic inhibitor on carriers activated by chlorine derivatives of s-triazine (Table 1). The ε-amino groups of lysine residues of the inhibitor were treated previously with citraconic anhydride. Then the modified inhibitor was reacted with the triazine activated support to give coupling via the tyrosine residues. The lysine residues then were reactivated by incubation at pH 3.0 (11). The use of a carboxyl

TABLE 1

COMPARATIVE CHARACTERISTICS OF PANCREATIC INHIBITOR BOUND TO POLYSACCHARIDE SUPPORTS BY CHLORIDE DERIVATIVES OF s-TRIAZINE

Polysaccharide	Amount of Bound Protein (mg/g preparation)	Inhibitor Activity of Bound Protein (%)
Dextran	2.6	9
Dextran	1.2 [a]	100 [a]
CM Cellulose	75	100
DEAE Dextran	0.3	-

(a) Pancreatic inhibitor added after reaction with citraconic anhydride containing carrier does not require modification of the protein amino groups, as shown by the large amount of bound inhibitor with retention of full activity (Table 1).

Another advantage of the caroboxyl containing polysaccharide became obvious when the pH dependency of the association of trypsin with preparations of pancreatic inhibitor bound to carriers was measured (Fig. 3) (14). The trypsin pH profile with pancreatic inhibitor bound to the carboxyl containing polysaccharide was much closer to that for the native inhibitor than with the other carriers. This probably was due to a buffering action of the CM cellulose around the optimum pH for the inhibitor action. As was reported elsewhere (14), the pK_{app} of the carboxy group of the CM cellulose used by us was 7.5; therefore, the buffering capacity of the carrier probably masked the sharp decrease in the degree of association for other supports with the pH different from the optimum.

For the kinetic study of the interaction of trypsin with pancreatic inhibitor bound to soluble carriers, the association and dissociation rate constants were defined as

$$E + I \underset{k_d}{\overset{k_a}{\rightleftharpoons}} EI \qquad \text{(Eq. 1)}$$

with the equilibrium constant K_i defined as k_d/k_a. The results (Table 2) showed that even a sharp increase in the molecular weight of the inhibitor did not affect the kinetic or equilibrium parameters of the inhibitor association (14). This may be explained by the suggestion that the limiting stage of inhibition of proteases by protein inhibitors is not controlled by diffusion (15); this reasoning is supported by other data not shown on the kinetics of association with increasing viscosity of the medium (in 20% sucrose solution).

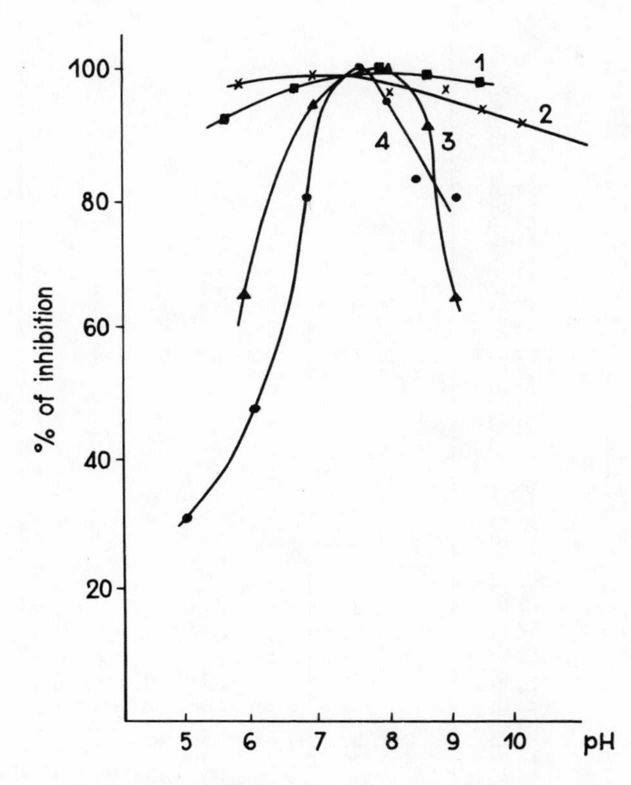

Fig. 3. pH dependence of the activity of trypsin in the presence of pancreatic inhibitor bound to carriers. Data represent: 1, native inhibitor; 2, inhibitor bound to CM cellulose; 3, inhibitor bound to dextran; 4, inhibitor bound to DEAE dextran (14).

TABLE 2

KINETIC AND EQUILIBRIUM CONSTANTS FOR THE INTERACTION OF TRYPSIN WITH TRYPSIN PANCREATIC INHIBITOR BOUND TO SOLUBLE POLYSACCHARIDE CARRIERS

Carrier	k_a (M^{-1} sec^{-1})	k_d (sec^{-1}	K_i (M)
-	1×10^6	9×10^{-5}	0.9×10^{-10}
CM Cellulose	0.6×10^6	6×10^{-5}	1.0×10^{-10}
Dextran	0.5×10^6	3.5×10^{-5}	0.7×10^{-10}
DEAE Dextran	-	-	2.0×10^{-10}

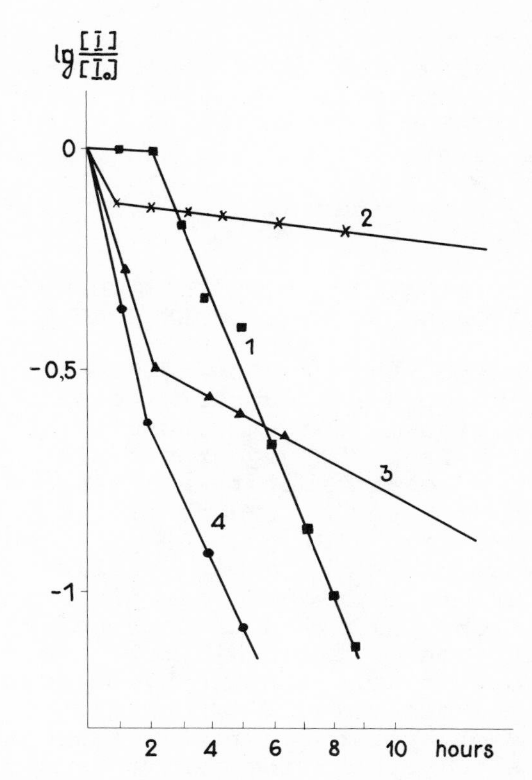

Fig. 4. Thermal stability of pancreatic inhibitor bound to carriers, expressed as the log of the ratio of actual to initial inhibitor activity, as tested at pH 4.7, 97°C, with 0.1 M NaCl. Data represent: 1, native inhibitor; 2, inhibitor bound to CM cellulose; 3, inhibitor bound to dextran; 4, inhibitor bound to DEAE dextran.

Owing to its structure, native pancreatic inhibitor has a very stable conformation and undergoes inactivation rather slowly. However, our covalent binding of pancreatic inhibitor to soluble polysaccharides gave several fractions, including some that were more stable than the native protein with respect to thermal inactivation (Fig. 4). The negative charge of the carrier apparently contributes greatly to the stabilization of the globule of the protein inhibitor. From the data of Fig. 3, it may be possible to carry out the sterilization of pancreatic inhibitor preparations at elevated temperatures.

II. *IN VIVO* BEHAVIOR OF PROTEINASE PANCREATIC INHIBITOR BOUND TO SOLUBLE CARRIERS

Pancreatic inhibitor bound to a negatively charged polysaccharide support was used for studying the distribution of the inhibitor in organs and excretion from the organism. A solution of the bound or free inhibitor was injected into the caudal vein of a rat at a dose of 1,000 Kallikrein Inactivator Units (KIU) per 100 g of body weight. The rats were killed by administration of hexinal, a barbiturate that produced a rapid stoppage of respiration and which was assumed not to influence the distribution of the inhibitor; and the level of inhibitory activity in the blood serum, kidney, and urine was determined. Comparison of the residence times of native pancreatic inhibitor with that of its polysaccharide derivatives shows that the rate of removal of the modified inhibitor from the blood was almost 10 times lower than that for the native material (Fig. 5). Similar results have been reported for other enzymes bound to soluble carriers (16,17).

The data in Fig. 6 demonstrate that the binding of pancreatic inhibitor with a negatively charged carrier greatly decreases the accumulation of the inhibitor in the kidney, but does not markedly affect its excretion into the urine, for the rat. The relatively fast decrease in the serum level of native inhibitor may have been due to rapid diffusion into the extracellular tissue fluid (2,7), with later diffusion back into the

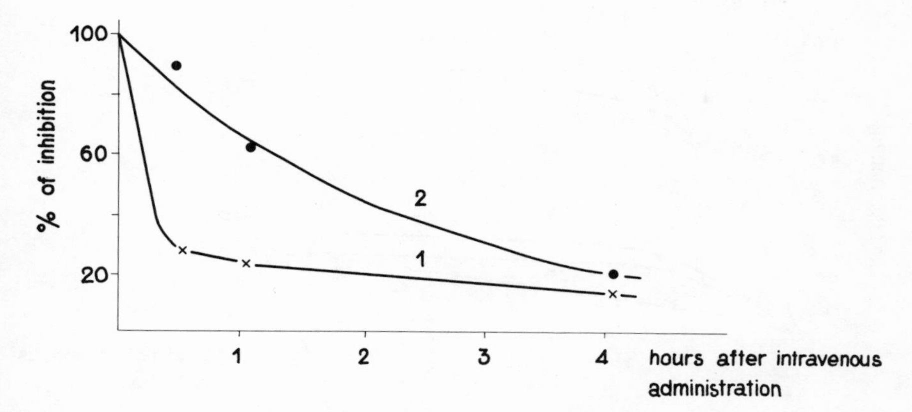

Fig. 5. Disappearance of pancreatic inhibitor from the circulation of rats after intravenous administration: Data represent: 1, native inhibitor; 2, inhibitor bound to a negatively charged polysaccharide support.

blood and discharge through the kidney. Interaction of pancreatic inhibitor with a polysaccharide support carrying a negative charge causes a sharp decrease in the isoionic point of the inhibitor (pI 4.6); this in turn might explain the observed decrease in the level of inhibitor in the kidney.

Changes in the properties of pancreatic inhibitor with binding to a carrier may also change the rate of its removal by endocytosis (18). Thus, all of the factors suggested above in theory could lead to an increase in the time of circulation of pancreatic inhibitor, bound to a soluble carrier, in the organism.

The inhibitor bound to CM dextran was tested for its effect on acute pancreatites in dogs.

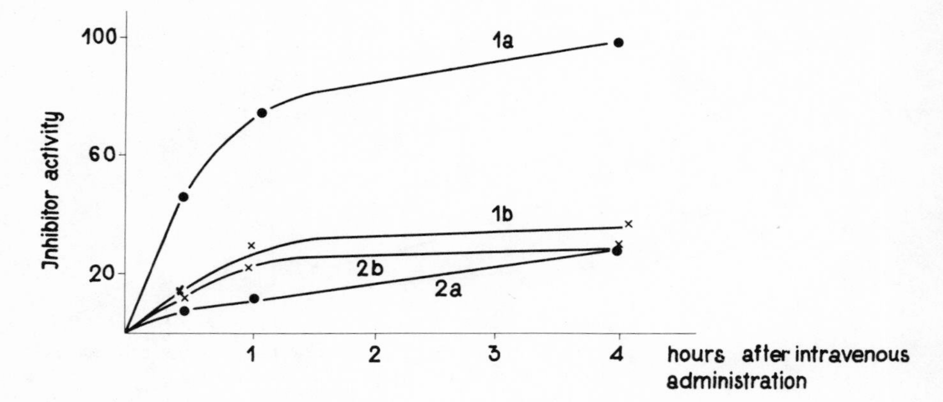

Fig. 6. Accumulation of pancreatic inhibitor in kidney, 1, and urine, 2. Data represent: a, native inhibitor; b, inhibitor bound to a negatively charged polysaccharide support. Vertical axis is expressed as KIU.

Acute pancreatitis was stimulated by introducing a preincubated mixture of gall with trypsin into the main pancreatic duct, at a dose of 0.25 ml per kg of animal weight. Infusion of the inhibitor was carried out 3 hr after the introduction of the mixture, at a dose of 5000 KIU per kg of body weight. Blood samples were taken from the fermoral vein until pancreatitis was manifested and at 3, 24 and 72 hr thereafter. The results of the experiments are given in Table 3. Three hours after administration of gall and trypsin and pancreatitis developed; and all the untreated animals perished. Administration of the inhibitor produced a positive effect. The use of the high molecular weight derivative of the inhibitor increased the number of surviving animals by 1.3 fold and the time following induction of pancreatitis for the animals that died by 2 fold. The above data suggest that more effective use of this pancreatic inhibitor

TABLE 3

PANCREATIC INHIBITOR IN TREATMENT OF ACUTE PANCREATITIS IN DOGS

Preparation of Inhibitor	Number of Animals	Dose (KIU/kg)	Survival (%)	Time[a] (days)
-	11	5000	0	0.5
Native	10	5000	63	7
CM Dextran	10	5000	80	14

(a) Average time dogs lived following induction of pancreatitis and administration of pancreatic inhibitor for therapy might be obtained by using preparations of the inhibitor bound to soluble carriers. However, much testing remains to be carried out before the practical use of this method can be ascertained.

REFERENCES

1. VOGEL, R., TRAUTSCHOLD, I. & WERLE, E. "Natural Proteinase Inhibitors", Academic Press, New York, 1968.
2. MARX, R., IMDAHL, H. & HABERLAND, G. L. "Neue Aspekte der Trasylol Therapie", Schattauer-Verlag, Stuttgart, 1968.
3. WERLE, E., In "Proceedings of the International Research Conference on Proteinase Inhibitors" (Ed. H. Fritz and H. Tschesche, eds.) Walter de Gruyter, Berlin, 1971, p. 23.
4. VEREMEENKO, K. N. "Enzymes of Proteolysis and Their Inhibitors in Medical Practice" (Russ.), Kiev, 1971.
5. PASKHINA, T. S. & MENSHIKOV, V. V. "Kinins and Kinin Blood System: Sbornik Nauchnik Trudov" (Russ.), First I. M. Sechenov Med. Inst., Moscow, 1976.
6. PASKHINA, T. S., DOLGINA, N. I., NARTIKOVA, V. F., KRINSKAJA, A. B., MOROSOVA, N. A. & ROSSINSKAYA, E. V. *Med. Chimii* (Russ.) *23*: 689, 1977.
7. FRITZ, H., OPPITZ, K.-H., MECKL, D., KEMKES, B., HAENDLE, H., SCHULT, H. & WERLE, E. *Hoppe-Seyler Z. Physiol. Chem.* *350*: 1541, 1969.
8. BROUN, G., In "Enzyme Engineering" vol 2 (E. K. Pye and L. B. Wingard, Jr., eds) Plenum Press, New York, 1974, p. 433.
9. MARSHALL, J. J. *Trends Biochem. Sci.* *3*: 79, 1978.
10. BEREZIN, I. V., ANTONOV, V. K. & MARTINEK, K. "Immobilizovannye Fermenty" ("Immobilized Enzymes"), Moscow University Press, Moscow, 1975.

11. LARIONOVA, N. I., KAZANSKAYA, N. F. & SAKHAROV, I. Y. *Biochemistry* (Russ.) *42*: 1237, 1977.
12. LARIONOVA, N. I., KAZANSKAYA, N. F. & SAKHAROV, I. Y. *Biochemistry* (Russ.) *43*: 880, 1978.
13. CHAUVET, J. & ACHER, R. *J. Biol. Chem. 242*: 4274, 1976.
14. LARIONOVA, N. I., KAZANSKAYA, N. F. & SAKHAROV, I. Y. *Biochemistry* (Russ.) *44*: 350, 1979.
15. LUTHY, J. A., PRAISSMAN, M., FINKENSTADT, W. R. & LASKOWSKI, M. JR. *J. Biol. Chem. 248*: 1760, 1973.
16. ABUKHOVSKY, A., VAN ES, T., PALCZUK, N. C. & DAVIS, F. F. *J. Biol. Chem. 252*: 3578, 1977.
17. SHERWOOD, R. F., BAIRD, J. K., ATKINSON, T., WIBLIN, C. N., RUTTER, D. A. & ELLWOOD, D. C. *Biochem. J. 164*: 461, 1977.
18. PRATTEN, M. K., DUNCAN, R. & LLOYD, J. B. *Biochim. Biophys. Acta 540*: 455, 1978.

ENZYMATIC MODIFICATION OF β-LACTAM ANTIBIOTICS: PROBLEMS AND PERSPECTIVES

V. K. Svedas, A. L. Margolin and
I. V. Berezin

Moscow State University,
Moscow, USSR

β-Lactam antibiotics make up a large group of physiologically active compounds including natural and semisynthetic penicillins, cephalosporins, and cephamycins. Many of them are characterized by unique antimicrobial properties, bacteriocidal action, and low toxicity and are widely used for practical applications in medicine. Because of their practical effectiveness, the synthesis of β-lactam antibiotics is a major problem of present interest. Many fine mechanisms of the molecular action of penicillins and cephalosporins already have been investigated (1,2). But despite great success in this field, it is difficult to select the structure and predict the properties of new β-lactam antibiotics. In connection with this one cannot but mention the great variety of these compounds, several tens of thousands at present.

Synthesis of new antibiotics proceeds, as a rule, according to the following scheme:

natural or biosynthetic
β-lactam compound

↓

semiproduct
(6-aminopenicillanic, 7-aminocephalosporanic
or 7-aminodeacetoxycephalosporanic acid)

↓

semisynthetic antibiotic

For the first step certain natural or modified natural β-lactam compounds when hydrolysed by penicillin amidase (EC 3.5.1.11) form a semiproduct such as 6-aminopenicillanic acid (6-APA) or 7-aminodeacetoxycephalosporanic acid (7-ADCA). Hydrolysis of this kind is well known now and is used practically every time such semiproducts are needed (3-8). 7-Aminocephalosporanic acid (7-ACA) is obtained without use of enzymes (9-10). For the second step the semiproducts form new "semisynthetic" antibiotics. As a rule, modification of semiproducts is carried out by chemical means in a multistep process under special conditions of low temperatures, organic solvents, etc. All this makes the chemical steps difficult and labor-consuming (11).

It is possible to obtain semisynthetic antibiotics from semiproducts by enzymatic means. However, the potential for enzymatic synthesis of penicillins and cephalosporins is obscure because no adequate quantitative description of the process is as yet available. Due to the large variety of these compounds, the whole problem will not be solved even if the synthesis of penicillin or cephalosporin becomes well understood. This paper discusses both the general thermodynamic and kinetic regularities of the enzymatic synthesis of β-lactam antibiotics. The creation of a scientific basis for this process will allow the prediction of the possibility and optimal conditions for synthesis of a given antibiotic.

I. THERMODYNAMIC ASPECTS OF ENZYMATIC MODIFICATION OF β-LACTAM ANTIBIOTICS

The thermodynamic study of the reversible synthesis of β-lactam antibiotics is of great value. Here the term reversibility refers only to the enzymatic hydrolysis step due to the presence of labile β-lactam rings. For example, when hydrolyzing penicillin non-enzymatically, only a small part is converted to 6-APA and respective carbonic acids (12) due to the labile nature of the rings.

For this study of the thermodynamic parameters of the hydrolysis and synthesis of penicillins and cephalosporins, both natural (benzylpenicillin) and semisynthetic (ampicillin, 7-phenylacetamidodeacetoxycephalosporanic acid (7-PADCA), cephalexin, cephalothin, cephaloridine) β-lactam compounds as well as benzylpenicilloic acid and phenylacetylglycine have been chosen. Enzymatic hydrolysis of all of the compounds mentioned above proceeds according to the following scheme:

$$\text{Antibiotic} + H_2O \rightleftharpoons \text{Acid} + \text{Semiproduct}$$

This reaction is catalyzed by the enzyme penicillin amidase and with an equilibrium constant K'_c. The semiproduct refers to 6-APA, 7-ACA, 7-ADCA or their derivatives.

$$K'_c = \frac{[\text{Acid}]\ [\text{Semiproduct}]}{[\text{Antibiotic}]\ [H_2O]} \qquad \text{(Eq. 1)}$$

Fig. 1 shows the general scheme of the reversible hydroylsis of these compounds exemplified with cephalothin. The structural formulas of some β-lactam antibiotics are given in Table 1, with experimental values of the equilibrium constants K'_c in Table 2. Using these data the change in Gibbs free energy $\Delta G^{o\prime}_c$ was calculated using the expression

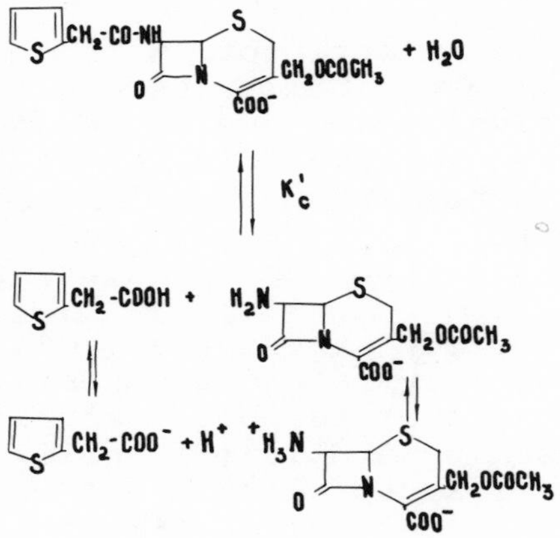

Fig. 1. A scheme of hydrolysis-synthesis of cephalothin catalysed by penicillin amidase.

$$\Delta G_c^{o\prime} = -RT \ln K_c^{\prime} \qquad \text{(Eq. 2)}$$

Also calculated was the Gibbs energy change due to hydrolysis, with respect to the non-ionized forms of the reagents (ΔG_c^o), and the pH-dependence of $\Delta G_c^{o\prime}$ (Table 2 and Fig. 2), using our own experimental data and literature data for the ionization constants of the products of the reaction. The symbols, units and terms are in accord with standard practice (13). It is of interest to analyze the hydrolysis step, i.e. the course of the reaction from left to right, to estimate what concentration of 6-APA it is possible to obtain in practice.

It is known that when the concentration of the initial substrate benzylpenicillin is increased, the yield of 6-APA begins to decrease (14). However, the mechanism of this decrease in yield is

TABLE 1

STRUCTURAL FORMULAS OF SOME β-LACTAM ANTIBIOTICS

Compound	Formula[(a)]
benzylpenicillin	C_6H_5-CH_2-CONH–[penam ring: S, $C(CH_3)_2$, N, C=O, COOH]
7-PADCA	C_6H_5-CH_2-CONH–[cephem ring: S, N, C=O, CH_3, COOH]
cephalothin	[2-thienyl]-CH_2-CONH–[cephem ring: S, N, C=O, CH_2-$OCOCH_3$, COOH]
cephaloridine	[2-thienyl]-CH_2-CONH–[cephem ring: S, N, C=O, CH_2-$\overset{+}{N}$(pyridinium), COO-]

cont'd

TABLE 1 (cont'd)

Compound	Formula[a]
ampicillin	C_6H_5-CH(NH$_2$)-CONH- [β-lactam/thiazolidine ring: S, C(CH$_3$)$_2$, CH-COOH, N, C=O]
cephalexin	C_6H_5-CH(NH$_2$)-CONH- [β-lactam/dihydrothiazine ring: S, CH$_2$, C-CH$_3$, C-COOH, N, C=O]
benzylpenicilloic acid	$C_6H_5-CH_2$-CONH- [HOOC-CH; thiazolidine ring: S, C(CH$_3$)$_2$, CH-COOH, NH]
N-phenyl-acetylglycine	$C_6H_5-CH_2-CONH-CH_2COOH$

(a) Not all hydrogens are shown

TABLE 2

THERMODYNAMIC PARAMETERS FOR HYDROLYSIS OF β-LACTAM ANTIBIOTICS (a)

Compound	K_c' (b) (kJ/mol)	$\Delta G_c^{o\prime}$ (b) (kJ/mol)	$\Delta G_c^{o\prime}$ (c) (kJ/mol)	ΔG_c^{o} (kJ/mol)
benzylpenicillin	0.00290	14.38	4.14	20.13
7-PADCA	0.000680	17.97	8.13	24.10
cephalotin	0.00685	12.29	2.38	18.60
cephaloridine	0.0144	10.45	0.96	17.18
ampicillin	0.0106	11.20	2.05	28.01
cephalexin	0.0108	11.16	2.38	28.30
benzylpenicilloic acid	0.55	1.45	1.54	28.20
N-phenylacetylglycine	1.2	1.71	2.01	26.00

(a) 25°C, in 0.1M, KCl; (b) pH 5.0; (c) pH 7.0

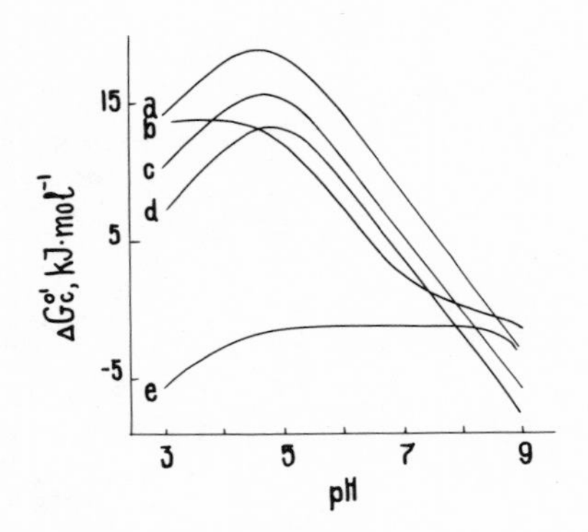

Fig. 2. pH-Dependences of standard Gibbs energy changes in enzymatic hydrolysis of β-lactam antibiotics: a, 7-PADCA; b, ampicillin and cephalexin; c, benzylpenicillin; d, cephalothin and cephaloridine; e, benzylpenicilloic acid.

not understood; and hence it is not known how to approach the control of this phenomenon. From a detailed kinetic study, it can be shown that the decrease in benzylpenicillin conversion is independent of any effects of phenylacetic acid, 6-APA or benzylpenicillin on the catalytic properties of penicillin amidase (15). Destruction of benzylpenicillin or 6-APA under the reaction conditions also cannot account for the decrease in yield of 6-APA. However, the equilibrium position may be analyzed if the pH-dependence of the equilibrium constant of the reaction is known (16).

The degree of benzylpenicillin conversion into 6-APA depends on the concentration of the substrate and on the pH of the solution (Fig. 3). It is obvious from this figure that the degree of conversion of benzylpenicillin to 6-APA begins to decrease even at fairly low substrate concentrations in acid medium. An increase in pH leads to

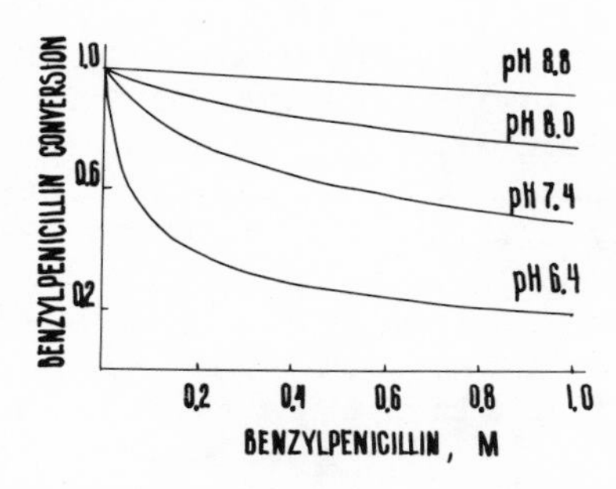

Fig. 3. Equilibrium yield of 6-APA vs initial concentration of benzylpenicillin in enzymatic hydrolysis. Conditions: 25°C, penicillin amidase 5×10^{-6} M.

a higher degree of conversion. However, under a pH of approximately 7.5 commonly employed in practice, benzylpenicillin cannot be completely hydrolyzed if its concentration exceeds 0.05 M. It thus appears that the reversibility of the enzymatic hydrolysis of penicillins may have an appreciable effect even when the conversion has gone only a small fraction of the equilibrium value.

Of greater interest and importance today, however, is the study of the enzymatic synthesis of β-lactam antibiotics (17-22). Table 1 and Fig. 2 show that differences in the structure of the nucleus of an antibiotic have a slight effect on the value of the observed Gibbs energy change ($\Delta G_c^{o\prime}$) and on the value of the pH-independent component (ΔG_c^o) of the standard Gibbs free energy change. This conclusion proceeds from comparison of the data for pairs of antibiotics that have the identical acyl moiety and that differ in the structure only of the nucleus. Thus, benzylpenicillin and 7-PADCA form a pair as do cephalothin and cephaloridine or ampicillin and cephalexin. For all these pairs the difference in

$\Delta G_c^{o\prime}$ does not exceed 4.0 kj/mole (1.0 kcal/mole), the difference in ΔG_c^o being still lower. It should be noted that the value of the pH-independent component (ΔG_c^o) of the standard Gibbs free energy change is more sensitive to the structure of the acyl moiety than to the structure of the nucleus. Benzylpenicillin and ampicillin or 7-PADCA and cephalexin differ in the structure of the acyl moiety only; and an increase of approximately 6.2 kj/mole (1.5 kcal/mole) is observed. It should be underlined that the pH-dependences of $\Delta G_c^{o\prime}$ are very similar for all β-lactam antibiotics; regardless of the structure of the acyl part and the modification in the nucleus. β-Lactam compounds have a well defined thermodynamic pH-optimum for synthesis of <5.0. At the same time the splitting of the β-lactam ring, as for benzylpenicilloic acid, provides a considerable change in the pH-dependence. Over the pH range 4.0 - 8.0 the standard Gibbs free energy change for hydrolysis of benzylpenicilloic acid does not vary and is close to zero. At pH 7.0 $\Delta G_c^{o\prime}$ for benzylpenicillin hydrolysis exceeds the respective value for benzylpenicilloic acid hydrolysis by 2.6 kj/mole (0.6 kcal/mole); at pH 4.4 the difference in the corresponding Gibbs free energy changes of hydrolysis is 15.6 kj/mole (3.7 kcal/mole). Moreover, the value of the pH-independent component of the Gibbs free energy change (ΔG_c^o) of hydrolysis of benzylpenicilloic acid exceeds the ΔG_c^o of hydrolysis of benzylpenicillin by 8.1 kj/mole (1.9 kcal/mole).

These data unambiguously testify to the significant contribution of ΔG_{ion} in the process of hydrolysis of benzylpenicillin. The high value of ΔG_{ion} and a sharp pH-optimum in the $\Delta G_c^{o\prime}$ should be ascribed to the closeness of the values of the ionization constants of carboxy and amino groups in the products of hydrolysis. The pK values of the carboxy group of D-(-)-α-aminophenylacetic, 2-thienylacetic, and phenylacetic acids are equal to 1.96, 4.15 and 4.20, respectively; and the pK values of the amino group in the nuclei are in the 4.6 - 4.8 range. Splitting of the β-lactam ring in

6-APA results in a sharp increase for the pK value of the amino group from 4.60 to 8.90. This is equivalent to a 23.5 kj/mole (5.6 kcal/mole) decrease in ΔG_{ion} at pH 4.4.

Thus the β-lactam ring is not only the structural element of an antibiotic which is responsible for its antimicrobial properties, but it also provides favorable thermodynamic conditions for the synthesis of penicillins and cephalosporins in aqueous solutions.

Up till now we have discussed hydrolysis-synthesis of antibiotics in which synthesis of the end product involves the use of free acid as the acylating agent. Very often acid derivatives, such as esters, amides, and N-acylated amino acids that are richer in energy, are used. The peculiarities of enzymatic synthesis in this case are of interest.

Let us analyze, by way of example, the enzymatic syntheses of benzylpenicillin using either phenylacetylglycine (Type 1) or ethyl phenylacetate (Type 2) as acylating agents. The reactions are summarized as follows:

$$\text{Phenylacetylglycine} + \text{6-APA} \xrightleftharpoons{\Delta G_1^{o\prime}} \text{benzylpenicillin} + \text{glycine}$$

$$\text{Ethyl phenylacetate} + \text{6-APA} \xrightleftharpoons{\Delta G_2^{o\prime}} \text{benzylpenicillin} + \text{ethanol}$$

For any of these reactions the total standard Gibbs free energy change ($\Delta G_{\Sigma}^{o\prime}$) will be equal to the sum of the Gibbs free energy change of hydrolysis of the phenylacetic acid derivative ($\Delta G_1^{o\prime}$ or $\Delta G_2^{o\prime}$) and the Gibbs free energy of synthesis of benzylpenicillin ($\Delta G_3^{o\prime}$) as follows:

$$\text{Phenylacetylglycine} + H_2O \overset{\Delta G_1^{o\prime}}{\rightleftharpoons} \text{phenylacetic acid} + \text{glycine}$$

$$\text{Phenylacetic acid} + \text{6-APA} \overset{\Delta G_3^{o\prime}}{\rightleftharpoons} \text{benzylpenicillin} + H_2O$$

$$\text{Phenylacetylglycine} + \text{6-APA} \overset{\Delta G_\Sigma^{o\prime}}{\rightleftharpoons} \text{benzylpenicillin} + H_2O + \text{glycine}$$

$$\Delta G_\Sigma^{o\prime} = \Delta G_1^{o\prime} + \Delta G_3^{o\prime} \qquad \text{(Eq. 3)}$$

In the case of N-acylated amino acids the value of $\Delta G_1^{o\prime}$ in the studied pH range changes little and is close to zero or somewhat lower (see Table 2). Hence, the pH dependence of the total Gibbs free energy change for a Type 1 reaction ($\Delta G_\Sigma^{o\prime}$) will be identical to the pH-dependence of the standard Gibbs free energy change of benzylpenicillin synthesis ($\Delta G_3^{o\prime}$). From the thermodynamic point of view this method of acylation has no advantages over acylation by free acids (Fig. 4).

Much more advantageous thermodynamic conditions for the synthesis of β-lactam compounds can be achieved when esters of acids are used as the acylating agents as per a Type 2 reaction. The pH-dependence of the Gibbs free energy change of hydrolysis of the ester ($\Delta G_2^{o\prime}$) is determined only by the ionization constant of phenylacetic acid, (PAA) which has a pK of 4.2. At pH >pK a sharp decrease in $\Delta G_2^{o\prime}$ is observed; and at pH 7.0 the value of $\Delta G_2^{o\prime}$ becomes -21 to -29 kj/mole or -5 to -7 kcal/mole. It shold be noted that if esters of acids are used, the thermodynamic pH-optimum of synthesis is observed at pH> 5.0 (Fig. 4) and coincides with the pH-optimum of the enzymatic

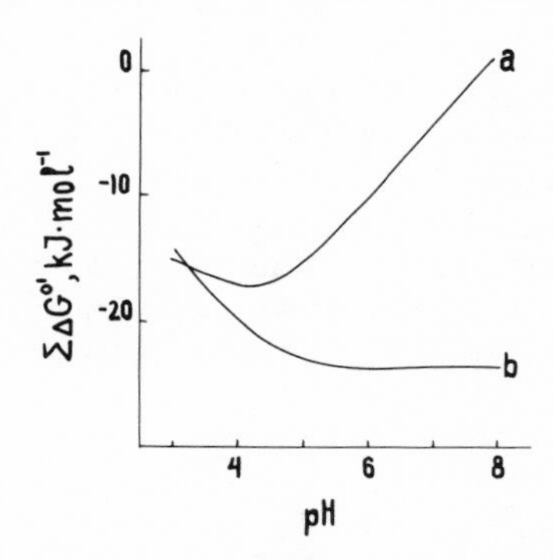

Fig. 4. pH-Dependences of standard Gibbs free energy changes ($\Delta G^{o\prime}$) calculated with Eq. 3 for synthesis of benzylpenicillin from phenylacetylglycine (a) and from ethyl phenylacetate (b). $\Delta G_1^{o\prime}$ at pH 7.0 of 0.46 kJ/mole (-0.11 kcal/mole) for (a) and $\Delta G_2^{o\prime}$ at pH 7.0 of -19.7 kJ/mole (-4.7 kcal/mole) for (b) were used in the calculations.

activity of penicillin amidase. In practice it should be taken into consideration, however, that the esters of phenylacetic and 2-thienylacetic acids have poor solubility in water; and it is therefore difficult to achieve a high concentration of acylating agent. It is much more advantageous to use esters of acids in the synthesis of ampicillin, cephalexin, and similar compounds. We have already mentioned above the essential effect of the β-lactam ring on the values of the standard Gibbs free energy change in synthesis of penicillins and cephalosporins. Analysis of the data shows that the syntheses of all β-lactam compounds have similar thermodynamic parameters. Neither modification of the acyl moiety nor its nucleus greatly effects the value of the standard Gibbs free energy change in hydrolysis of penicillins and cephalosporins. This analysis allows the conclusion to be made that hydrolysis-synthesis of the other antibiotics of this class will have a similar pH-dependence of the standard Gibbs free energy change. Consequently, there are no thermodynamic reasons preventing enzymatic synthesis of such antibiotics as cephazolin, cephtezol, cephatrizin, cephaloglycine, and others. It should be emphasized that we mean the possibility of the synthesis, i.e. the equilibrium yield of antibiotic. The practical yield of product will depend not only on the thermodynamics but also on the kinetics of the reaction.

II. KINETIC REGULARITIES OF SYNTHESIS OF β-LACTAM ANTIBIOTICS CATALYZED BY PENICILLIN AMIDASE FROM *ESCHERICHIA COLI*

Enzymatic synthesis of β-lactam antibiotics from semiproducts can be performed with the help of acylating agents, either free acids or their activated derivatives such as esters, amides, or N-acylated amino acids. Here the analysis of the kinetics is presented, with the enzymatic synthesis of benzylpenicillin as a model. Benzylpenicillin is a natural specific substrate of penicillin amidase; the kinetics of its enzymatic hydrolysis has been studied in detail in this laboratory (3,23).

It has been shown before that the transfer of the acyl group onto 6-APA in the course of the synthesis of benzyl penicillin is described by the following kinetic scheme (17):

$$
\begin{array}{ccccccc}
E + S & \underset{}{\overset{K_s}{\rightleftharpoons}} & ES & \underset{P_1}{\overset{k_2}{\longrightarrow}} & EA & \overset{K_3}{\longrightarrow} & E + P_2 \\
K_N \Big\updownarrow N & & K_N \Big\updownarrow N & & K_N \Big\updownarrow N & & \\
EN + S & \underset{K_s}{\rightleftharpoons} & ESN & \underset{P_1}{\overset{\alpha k_2}{\rightleftharpoons}} & EAN & \begin{array}{l} \overset{\beta k_3}{\longrightarrow} EN + P_2 \\ \underset{k_4}{\searrow} E + P_3 \end{array} &
\end{array}
$$

Where S is a derivative (ester, amide, N-acylated amino acid) of some carbonic acid; P_1 is alcohol, amine, or amino acid; P_2 is corresponding acid; P_3 is antibiotic; N is nucleus; and α and β are numerical coefficients.

In Fig. 5 is reported the rate dependence of the formation of benzylpenicillin on the concentration of phenylacetic acid; these data show that the experiment was performed under the conditions $[S]_o >> K_M(app)$. The rate dependence of the enzymatic reaction on the concentration of 6-APA also is shown in Fig. 5. From these data the following effective values were obtained:

$$k_{cat(app)} = 11 \text{ sec}^{-1}$$

$$K_{M(app)} = 17 \times 10^{-3} M$$

$$(k_{cat} / K_M)_{(app)} = 6.5 \times 10^2 \ (M, \text{sec})^{-1}$$

Undoubtedly, binding of 6-APA takes place. Since the reaction for the enzymatic synthesis of benzylpenicillin is reversible, its kinetic and thermodynamic parameters are connected by the Haldane ratio (Eq. 5). In this case a scheme for the reversible synthesis and the ratio between the kinetic parameters of the foward and reverse reactions can be presented in the following way:

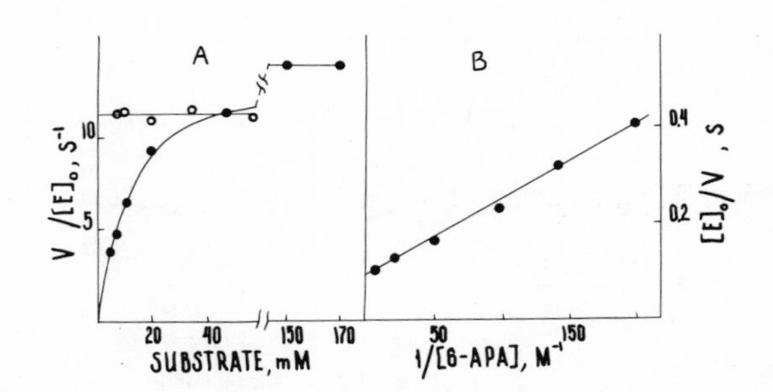

Fig. 5. Enzymatic synthesis of benzylpenicillin catalyzed by penicillin amidase. Part A: dependences of velocity of reaction on the concentration of phenylacetic acid (O) and 6-APA (●); Part B: determination of binding constant of 6-APA in this reaction. Conditions: 25°C, pH 6.0, penicillin amidase 1.95×10^{-7} M. Vertical axis S stands for sec.

$$E + I \underset{}{\overset{K_p}{\rightleftharpoons}} EI \underset{k_{-2}\ \ H_2O}{\overset{k_2}{\rightleftharpoons}} EA \underset{K_N}{\overset{+N}{\rightleftharpoons}} EAN \underset{k_{-3}}{\overset{k_3}{\rightleftharpoons}} ES \quad ES \overset{K_s}{\rightleftharpoons} E + S$$

(Eq. 4)

where I is PAA, S is benzylpenicillin, N is 6-APA, and EA and EAN are intermediate complexes. The overall equilibrium constant K_c' becomes

$$K_c' = \frac{K_p \; K_N \; k_{-2} \; k_{-3}}{k_s \; k_2 \; k_3} \qquad \text{(Eq. 5)}$$

At pH 6.0, k_c' is calculated to be 2×10^{-2}. This is based on measured values of 3.3×10^{-7} $(M, Sec)^{-1}$ for k_{-3}/K_s and 1.0×10^{-5} M for K_p. The value of $(k_3)/(k_{-2}K_N)$ for phenylacetic acid is unknown; but for estimation one can use the value of 1700 ± 150 M^{-1} (17) previously obtained for the transfer of the phenylacetic group from n-carboxy-m-nitroanilide to 6-APA. Based on these data the rate constant for acylation of the enzyme with free acid was estimated as $k_2 \sim 10$ sec^{-1}. The value of the effective catalytic constant of benzylpenicillin synthesis from phenylacetic acid is 11 sec^{-1}. The similarity of the experimental catalytic constant of synthesis and the calculated rate constant of acylation of the enzyme by free acid suggests that the acylation step evidently limits the velocity of the enzymatic synthesis. Under the same conditions of pH 6.0 and 25°, the synthesis of benzylpenicillin from phenylacetylglycine has been investigated; the value of $k_{cat(app)}$ of 25 sec^{-1} has been determined (Fig. 6).

Thus, the values of the effective catalytic constants for benzylpenicillin synthesis from free phenylacetic acid and from phenylacetylglycine are similar at the same conditions. The k_{cat} values for benzylpenicillin synthesis from phenylacetic acid and from phenylacetylglycine also are much alike and in addition are large in magnitude. Since k_{cat} for the hydrolysis of benzylpenicillin under these same conditions is 30 sec^{-1}, the synthesis of the antibiotic can proceed not only with high yield but also with a high rate when both acylating agents are involved.

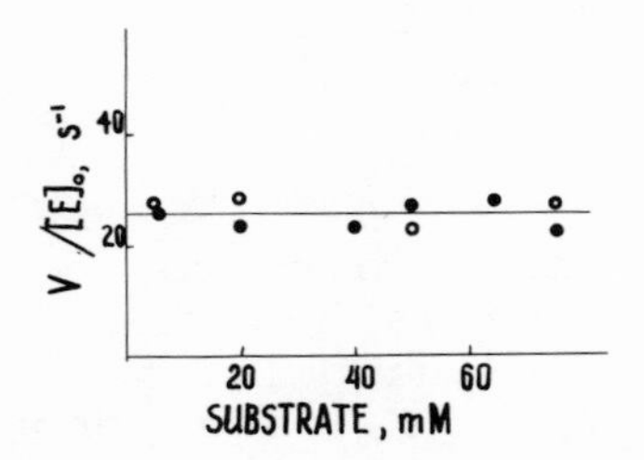

Fig. 6. Dependences of the velocity of enzymatic synthesis of benzylpenicillin on the concentration of 6-APA (●) and phenylacetylglycine (O). Conditions: 25°C, pH 6.0, penicillin amidase 1.95 x 10^{-7} M.

III. ACCUMULATION OF BENZYLPENICILLIN

The accumulation versus time curves for β-lactam antibiotic syntheses follow the characteristic features of an enzymatic catalyzed reaction. The optimal conditions for the synthesis of the desired compounds could be determined on a scientific basis if the concentration of antibiotic could be predicted at each moment of time.

Progress curves for the accumulation of the antibiotic on enzymatic synthesis of benzylpenicillin from phenylacetylglycine are compared to analogus curves for the synthesis from phenylacetic acid. In the synthesis from phenylacetic acid, benzylpenicillin is accumulated gradually until it reaches equilibrium (Fig. 7). In the case of phenylacetylglycine, more complicated kinetic curves with a definite maximum are observed (Fig. 8). The maximum value and its position on the time axis depend on the concentration of the reagents and the pH value. The appearance of the maximum is due to the formation of the intermediate product, benzylpenicillin, which is

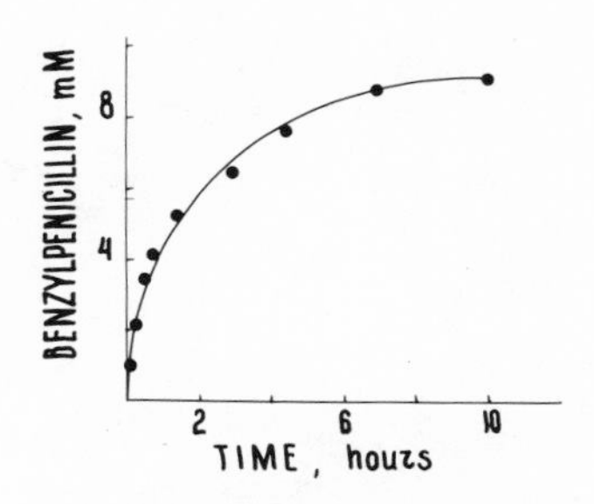

Fig. 7. The kinetics of benzylpenicillin synthesis from phenylacetic acid and 6-APA catalysed by immobilized penicillin amidase in a batch reactor. Conditions: 25°C, pH 5.0, phenylacetic acid 30 mM, 6-APA 10mM.

further hydrolyzed. This hydrolysis goes to completion at pH values greater than 7.0, while the equilibrium is observed at pH 6.0. The general kinetic scheme of the process is rather complicated. It includes hydrolysis of phenylacetylglycine, synthesis of benzylpenicillin from phenylacetylglycine and 6-APA, and the further reversible hydrolysis of benzylpenicillin to 6-APA and phenylacetic acid as follows:

$$\text{Phenylacetylglycine } (S_1) + \text{6-APA} \underset{K_n}{\overset{k_1K_1}{\rightleftharpoons}} \text{Benzylpenicillin } (S_2)$$

$$\text{Phenylacetylglycine } (S_1) \xrightarrow{k_4K_1} \text{Phenylacetic Acid + Glycine}$$

$$\text{Benzylpenicillin } (S_2) \underset{k_3K_3K_n}{\overset{k_2K_2}{\rightleftharpoons}} \text{Phenylacetic Acid + 6-APA } (N)$$

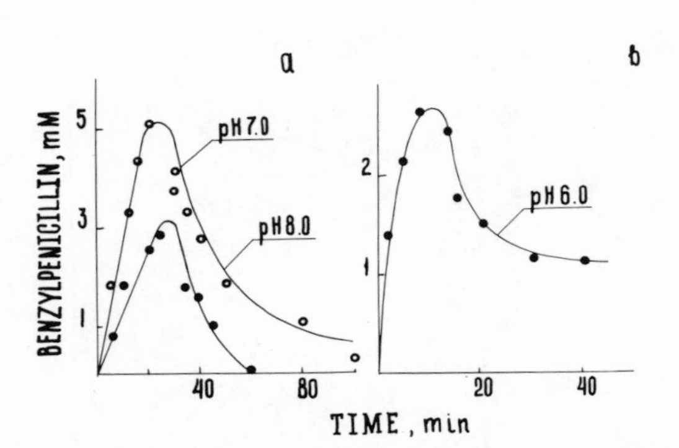

Fig. 8. The progress curve for accumulation of benzylpenicillin from phenylacetylglycine and 6-APA. Curves based on solution of Eqs. 6-9. Part a: 25°C, penicillin amidase 1.95×10^{-7} M, phenylacetylglycine and 6-APA each 20 mM; Part b: 25°C, penicillin amidase 5.85×10^{-7} M, phenylacetylglycine and 6-APA each 10.5 mM.

Each reaction of this process is shown to be catalyzed by the same enzyme, penicillin amidase, and is characterized by its Michaelis constant (K_i) and catalytic constant (k_i); where $V_m = k_i\ [E]_o$. The following two differential rate equations for substrate 1 and 2 and two material balance equations describe the whole proess:

$$\frac{d[S_1]}{dt} = \frac{V_1\ [S_1]\ /(1 + K_N/\ [N]\ )}{[S_1]\ +\ K_1K_N/(\ [N]\ +\ K_N)} - \frac{V_4\ [S_1]}{K_1\ + [S_1]} \quad \text{(Eq. 6)}$$

$$\frac{d[S_2]}{dt} = \frac{V_1\ [S_1]\ /(1 + K_N\ \ [N]\ )}{[S_1] +\ K_1K_N/(\ [N]\ +\ K_N)} +$$

$$\frac{V_3[P]/1 + K_N/[N]}{[P] + K_3K_N/([N] + K_N)} - \frac{V_2[S_2]}{[S_2] + K_2(1 + [P]/K_p)}$$

(Eq. 7)

$$[N]_o = [N] + [S_2] \qquad \text{(Eq. 8)}$$

$$[S_1]_o = [S_1] + [S_2] + [P] \qquad \text{(Eq. 9)}$$

To calculate the progress curve for the reaction according to Eqs. 6-9, numerical integration of the differential equations was used. For this calculation the pH-dependences of the kinetic parameters of hydrolysis of benzylpenicillin (k_2, K_2) as well as the binding constant of phenylacetic acid and other kinetic parameters that had been determined earlier (16,17,23) were used. At pH 6.0, 25°C, and an ionic strength of 0.1M, the values were:

$V_1 = 4.5\ 10^{-6}$ M, Sec^{-1} $\quad K_1 = 10^{-4}$ M

$V_2 = 5.8\ 10^{-6}$ M, Sec^{-1} $\quad K_2 = 2.0 \times 10^{-6}$ M

$V_3 = 2.2\ 10^{-6}$ M, Sec^{-1} $\quad K_3 = 10^{-5}$ M

$V4 = 4.8\ 10^{-6}$ M, Sec^{-1} $\quad K_N = 10^{-3}$ M

Fig. 8 shows that at these values satisfactory correlation between the experimental and calculated curves was achieved. Optimal correlation was achieved with the values of the kinetic parameters given in Table 3. It should be noted that the differences between the optimal and experimental val-

TABLE 3

KINETIC PARAMETERS OF SYNTHESIS OF BENZYLPENICILLIN (a)

pH	V_1	V_2	V_3	V_4	K_N	K_1	K_2	K_3
	(M/sec) x 10^6				(M)	(M) x 10^5	(M) x 10^6	(M) x 10^5
8.0	7.6	8.7	0	5.9	0.0039	6.0	2.0	0.9
7.0	8.6	6.9	1.2	6.9	0.0023	6.0	2.0	0.9
6.0	4.3	5.1	2.0	4.3	0.0010	3.0	2.0	0.9

(a) $[E_o] = 1.95 \times 10^{-7}$ M; 25°C

ues of the kinetic parameters was within the limits of experimental error.

A similar approach was employed to determine the unknown constants. For example, at pH 7.0 and 8.0, values of K_N and K_1 were unknown. To find the values of these constants all the other parameters in Table 3 were fixed, and the K_N and K_1 values were varied until the optimal correlation between calculated and experimental dependences was achieved. The resultant values for K_N and K_1 are shown in Table 3.

This approach allows one to find out in what way the change of the concentrations of the reagents and the catalyst will influence the yield of the product. For example, at pH 7.0 and 0.05M phenylacetylglycine, 0.02M 6-APA, and 2×10^{-7}M penicillin amidase this approach makes it possible to predict that the maximum concentration of benzylpenicillin of 4.4 mM will occur 60 min after the start of the reaction. An experiment carried out under these conditions (Fig. 9) showed that the maximum concentration of benzylpenicillin of 3.7 mM was achieved in 55 min after the process began. Such agreement seems good when it is taken into account that eight kinetic parameters, each determined with some error, are involved in the calculation. Two types of yields of the antibiotic may be obtained at pH 6.0: equilibrium and kinetic. With further decrease in pH or increase in concentrations of the reagents so as to favor the conditions for the reverse reaction, the equilibrium yield becomes more than the kinetic yield, and the specificity of the reaction for activated derivatives of acids disappears.

In summing up the results it should be said that if activated acid derivatives, specifically phenylacetylglycine, are used one can obtain the antibiotic in concentrations that essentially exceed the equilibrium ones. Besides, there is the possibility of performing the synthesis reaction in the same pH range as that of the pH-

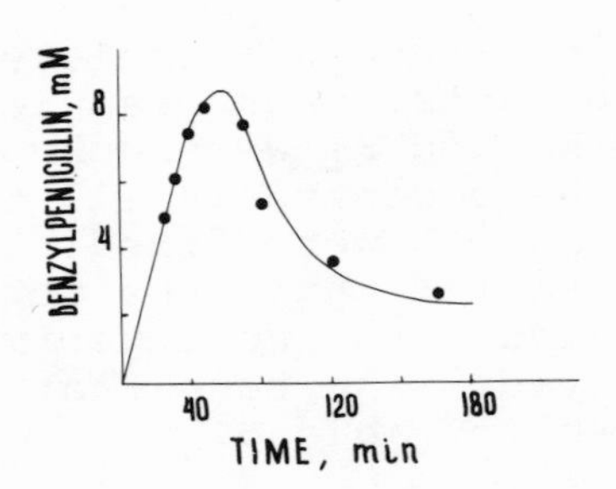

Fig. 9. The progress curve for accumulation of benzylpenicillin from phenylacetylglycine and 6-APA. Conditions: 25°C, pH 7.0, penicillin amidase 2.3×10^{-7} M, 20mM 6-APA, 50 mM phenylacetylglycine.

optimum for enzymatic activity. In order to obtain a desired compound under these conditions (for example, benzylpenicillin) the reaction should be stopped at the appropriate time.

The conclusion can be drawn (Table 3) that values of V_m influence greatly the curve for accumulation of the antibiotic much more than does the binding constant K_i. Consequently, to estimate the potential for enzymatic synthesis of new β-lactam antibiotics, one should know a small number of general kinetic parameters for the reaction.

IV. KINETIC REGULARITIES OF ENZYMATIC SYNTHESIS OF β-LACTAM ANTIBIOTICS WITH FREE AMINO GROUP

In Table 4 the values of the general kinetic parameters of enzymatic hydrolysis are presented for a number of penicillin amidase substrates. The data show that the natural substrate of penicillin amidase, benzylpenicillin, has the highest value of the bimolecular constant k_{cat}/K_m for the enzymatic reaction of all the compounds investigated. Most amide substrates of penicillin amidase, containing the phenylacetyl group in the acyl moiety, are characterized by similar values of the catalytic constants ($\sim$50 sec^{-1}) (24-27). Substitution of the 2-thienylacetyl group, as in cephalothin or cephaloridine, for the phenylacetyl group of similar structure and electron properties results in a slight decrease of k_{cat}. On the other hand, substitution of the leaving group in the benzylpenicillin structure may result in a considerable loosening of binding and a decrease in the bimolecular rate constant.

In fact, the substrate specificity of penicillin amidase in hydrolytic and synthetic reactions should be identical (28,29). This allows some assumptions concerning enzymatic synthesis to be made on the basis of the data from Table 4. According to the results obtained at low pH values, the catalytic constants of hydrolysis and synthesis of benzylpenicillin from phenylacetic acid and phenylacetylglycine are similar. Since the k_{cat} of hydrolysis of 7-PADCA, cephalothin, and cephaloridine does not differ greatly from the k_{cat} of hydrolysis of benzylpenicillin, the catalytic constants of their synthesis from acids or corresponding derivatives should be similar to each other and of high absolute values. Consequently, it should be possible to synthesize these antibiotics at high velocity.

Penicillins and cephalosporins containing a free amino group in the acyl moiety, i.e. ampicillin, cephalexin, or cephaloglycine, make up another large group of β-lactam antibiotics which are of interest. In our opinion, these compounds should be considered as nonspecific substrates of penicillin amidase. To find out the possibility for synthesis of these antibiotics, the pH dependences of the bimolecular constants of hydrolysis (k_{cat}/K_M) of two model compounds, p-nitroanilides of phenylacetic and D-(-)-α-aminophenylacetic acids have been studied. In the first case, the pH dependence is bell shaped, typical of other

TABLE 4

SUBSTRATE SPECIFICITY OF PENICILLIN AMIDASE[a]

Substrate	k_{cat} (l/sec)	K_m (M)	k_{cat}/K_m (l/(M, sec))
Benzylpenicillin	48	4.6×10^{-6}	1.0×10^{7}
7-PADCA	50	10.0×10^{-6}	5.0×10^{6}
Ethyl phenylacetate[b]	170	4.5×10^{-5}	3.8×10^{6}
p-Nitrophenyl phenylacetate[b]	170	3.1×10^{-5}	3.5×10^{6}
p-Nitroanilide of PAA	55	9.7×10^{-5}	5.7×10^{5}
p-Carboxy-m-nitroanilide of PAA[b]	112	6.0×10^{-5}	1.9×10^{6}
Phenylacetylglycine	47	8.0×10^{5}	5.9×10^{5}
Benzylpenicilloic acid	40	2.0×10^{-3}	2.0×10^{4}
Cephalothin	25	4.2×10^{-5}	0.6×10^{6}
Cephaloridine	33	1.0×10^{-4}	3.3×10^{5}

TABLE 1 (cont'd)

Substrate	k_{cat} (1/sec)	K_m (M)	k_{cat}/K_m (1/(M, sec))
Ampicillin	11	5.2×10^{-3}	2.2×10^{3}
Cephalexin	54	2.1×10^{-3}	2.6×10^{4}
p-Nitroanilide of D-(-)-α-aminophenylacetic acid	0.54	3.2×10^{-3}	157

(a) at pH 7.5 and 25°C
(b) K_m values determined at pH 6.0.

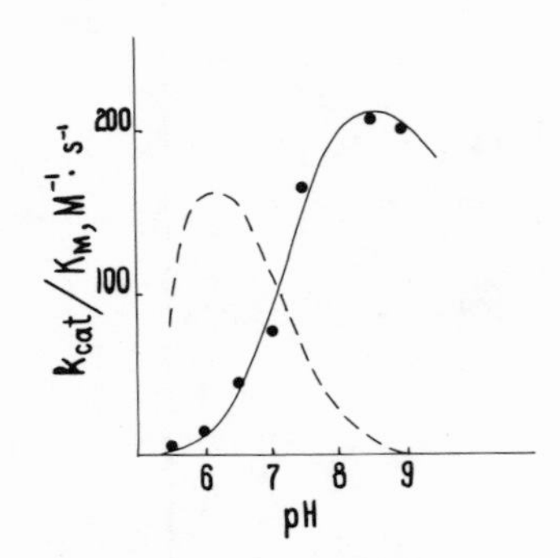

Fig. 10. pH Dependence of k_{cat}/K_M for hydrolysis of p-nitroanilide of D-(-)-α-aminophenylacetic acid. The theoretical curves are calculated with the values of pK_a 5.3, pK_b 9.6, pK_c 7.1 and k_{cat}/K_M 220 M^{-1}, sec^{-1} for binding of the deprotonated (unbroken curve) and protonated (broken curve) forms. Conditions: 25°C, penicillin amidase 5 x 10^{-8} M, substrate 0.2 - 5mM.

substrates of penicillin amidase; the dependence may be described by the pK values of the enzyme ionogenic groups equal to 5.3 and 9.6 and a k_{cat}/K_M of 57 x 10^4 M^{-1}, sec^{-1}. In the second case, the dissociation of the amino group of the substrate must be taken into account since the pK value of the amino group is close to 7, e.g. 7.02 for ampicillin and 7.1 for p-nitroanilide D-(-)-α-aminophenylacetic acid. This means that over a wide range of pH, the protonated and deprotonated forms of the substrate will be present. It has been ascertained in this laboratory that the active site of penicillin amidase is characterized by an extremely high hydrophobic nature (15,30-32). From this it could be assumed that the enzyme has the ability to bind only the deprotonated form of the substrate (Fig. 10 and Eq. 10).

$$k_{cat}/K_M = \frac{k_2/K_s}{(1 + [H^+]/K_a + K_b/[H^+])(1 + [H^+]/K_c)}$$

(Eq. 10)

On the one hand, the maximum velocity of the enzymatic reaction does not seem to depend on the form of the bound substrate. However, the Michaelis constant depends on the concentration of the deprotonated form:

$$K_{M(app)} = K_M(1 + \frac{[H^+]}{K_c})$$ (Eq. 11)

The value of $K_{M(app)}$ for ampicillin is rather high (see Table 4), even at pH 7.5; if the pH is decreased, the reaction will proceed in a bimolecular regime. This can be explained by the fact that even at a substrate concentration up to 1 M, the conditions for maximum velocity of the enzymatic process, $[S]_o >> K_M$, will not be maintained.

The example presented testifies to the fact that the efficiency of substrate binding to the enzyme is of great importance sometimes in biotechnology. Evidently, the standard catalytic parameters for enzymatic reactions cannot be employed to estimate the efficiency of penicillin amidase usage to obtain compounds containing a free amino group in the side chain. Instead, a bimolecular constant appears to be the appropriate parameter. If at pH 7.5 the ratio of bimolecular constants of enzymatic hydrolysis of both substrates is 4,000, than at pH 6.0 it may equal 34,000 and go as high as 300,000.

Now it is clear, why penicillin amidase from *E. coli* is not capable of catalyzing the synthesis of ampicillin directly from the free acid. The pH value for D-(-)-α-aminophenylacetic acid is 9.1. This means that at pH 6.0, less than 0.1% of the substrate is in the deprotonated form, which is the form involved in the reaction. Thus, the

TABLE 5

KINETIC PARAMETERS FOR HYDROLYSIS OF p-NITROANILIDES OF PHENYLACETIC AND D-(-)-α-AMINOPHENYLACETIC ACIDS

Substrate	k_{cat}/K_m (l/(M, sec))		
	pH 7.5	pH 6.0	pH 5.0
p-Nitroanilide of phenylacetic acid	57×10^4	51×10^4	45×10^4
P-Nitroanilide of D-(-)-α-amino-phenylacetic acid	150	15	1.5

effective value of the binding constant is considerably increased; and consequently, k_{cat}/K_M is decreased (Table 5). At relatively high pH (more than 8.0) and high concentration of a given acid, synthesis is impossible because of the unfavorable thermodynamic conditions. So there is only one way to synthesize ampicillin, through use of the activated derivatives of the corresponding acid. In this case synthesis may proceed at a pH close to 6.0, where the concentration of the deprotonated form of the substrate (pK_c is equal to 7.6) is rather high.

V. PERSPECTIVES ON THE USE OF ENZYMES FOR MODIFICATION OF β-LACTAM ANTIBIOTICS: DIFFICULTIES AND WAYS TO OVERCOME

As mentioned above, in aqueous solutions there exist favorable conditions for the enzymatic synthesis of β-lactam antibiotics. Thermodynamically there is a greater possibility for synthesis of amide bonds in penicillins and cephalosporins than in N-acylated amino acids, peptides, or proteins (33-36). Thus, the enzymatic synthesis of β-lactam antibiotics should be quite reasonable from a practicable standpoint. Nevertheless, only a few practical cases are known where enzymes have been used for such purposes (37-40). How can this be explained?

One of the reasons evidently is the difficulty in understanding the regularities of the enzymatic synthesis of the desired compounds, particularly for transfer of an acyl group. We have already mentioned that from a kinetic standpoint the possibilities for obtaining penicillins and cephalosporins are quite acceptable. Moreover, sometimes it is possible to achieve a concentration of the desired compound which exceeds the equilibrium value. So, the advantages in the use of enzymes as catalysts for synthetic reactions becomes more obvious in the course of a detailed investigation of the kinetic peculiarities of these reactions. However, the enzymatic synthesis of β-lactam antibiotics seems to have a bright future.

On the other hand, it should be noted that there are some difficulties and unsolved problems. First, the kinetic regularities of the enzymatic synthesis are observed only in the presence of highly specific enzymes. In the case of ampicillin or carbenicillin, penicillin amidase is not an effective catalyst; and the rate of enzymatic synthesis becomes too low to be taken into consideration. Evidently, in this case the synthesis proceeds only if the enzyme of the required specificity is involved. Similar examples are given in the literature (41,42). Our data provide evidence to say that the practical synthesis of any β-lactam antibiotic is favorable from a thermodynamic point of view. If microbiological research can provide enzymes of the required substrate specificity and rather high activity, then the only problem becomes one of quantitatively describing the kinetic peculiarities of each enzyme action. In principle, we may expect complicated kinetic regularities as well as more simple ones.

In both cases a highly specific enzyme will catalyze the conversion of the initial semiproducts into the final compounds at least to the equilibrium yield. In order to increase the degree of conversion, it is necessary to perform the enzymatic process under favorable thermodynamic conditions. It is clear now that the most sensitive parameter for this purpose is pH (14,16,19,23). As a rule the thermodynamic pH optimum for the synthesis of an antibiotic and the pH optimum for catalytic activity and stability of the enzyme used to synthesize this antibiotic differ greatly (23, 43-45).

Consequently, a search for the optimal thermodynamic as well as kinetic conditions is of great importance. In our opinion, there exist several approaches. Regulation of the pH dependent properties of enzymes due to changes in the microenvironment of their active sites is one approach (46). Shifts of the pH dependences of such properties of an enzyme as stability and activity may be in the range of several pH units. The difference between the thermodynamic pH optimum for synthesis of benzylpenicillin and the pH optimum of penicillin amidase catalytic activity is 3.6 pH units. It has been shown before, that in the case of penicillin amidase shifts of the pH dependences of activity and stability can be done using soluble complexes of enzyme and polyelectrolyte. It is possible to put such effects into practice if the shifts are observed in the case of an immobilized enzyme and in the concentrated solutions of the reagents.

It turned out that in the case of penicillin amidase, such an effect may be achieved using a complex of an enzyme and two oppositely charged polyelectrolytes. The immobilized enzyme was stable up to an ionic strength of at least 0.1 M; and an essential change in the acid branch of the pH dependency of catalytic activity was achieved (Fig. 11).

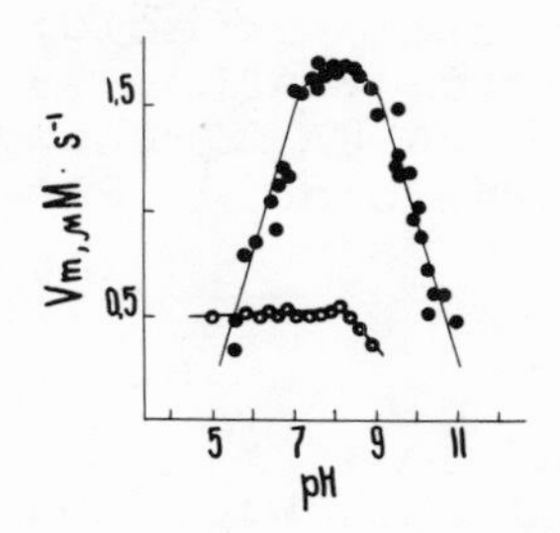

Fig. 11. pH Dependences of the catalytic activity of native penicillin amidase (●) and of penicillin amidase immobilized in polyelectrolyte gel (O). Conditions: 25°C, ionic strength 0.1 M.

Undoubtedly, one may meet other difficulties in putting into practice enzymatic methods of synthesis of new β-lactam antibiotics. The lability of the initial substances and reaction products, the difficulty in separation of the obtained reaction mixture into components, the interaction of components, and the influence of component parts of the reaction mixture on the catalyst properties are several possibilities. Nevertheless, we note that a detailed investigation of the peculiarities of the action of enzymes for modifying antibiotics makes the perspectives of their application more obvious. We believe that only the first steps have been made in this field.

SUMMARY OF ABBREVIATIONS

6-APA, 6-Aminopenicillanic acid; 7-ACA, 7-Aminocephalosporanic acid; 7-ADCA, 7-Deacetoxycephalosporanic acid; 7-PADCA, 7-Phenylacetamidodeacetoxycephalosporanic acid; PAA, Phenylacetic acid.

REFERENCES

1. STROMINGER, J. L., WILLOUGHBY, E., KAMIZYO, T., BLUMBERG, P. & YOCUM, R. *Ann. N. Y. Acad. Sci. 235*: 210, 1974.
2. DUSART, J., LEYH-BOUILLE, M., & GHUYSEN, Y-M. *Eur. J. Biochem. 81*: 33, 1977.
3. BEREZIN, I.V., KLYOSOV, A. A., NYS, P.S., SAVITSKAYA, YE. M. & SVEDAS, V. K. *Antibiotiki* (Russ.) *19*: 880, 1974.
4. SVEDAS, V. K., KLYOSOV, A. A., NYS, P. S., SAVITSKAYA, YE. M. & BEREZIN, I. V. *Antibiotiki* (Russ.) *21*: 698, 1976.
5. WARBURTON, D., BALASINGHAM, K., DUNNILL, P. & LILLY, M.D. *Biochim. Biophys. Acta 284*: 278, 1972.
6. CARRINGTON, T. R. *Proc. Roy. Soc. B 179*: 321, 1971.
7. DINELLI, D. *Process Biochem. 7*: 9, 1972.
8. VANDAMME, E. J. & VOETS, J. P. *Adv. Appl. Microbiol. 17*: 311, 1974.
9. LODER, B. & NEWTON, G. G. F. *Biochem. J. 79*: 408, 1961.
10. SVEDAS, V. K. In "Immobilizovannye Fermenty" ("Immobilized Enzymes", Russ.) I. V. Berezin, V. K. ntonov, and K. Martinek eds.) Moscow University Press, Moscow, 1975, p. 221.
11. MANHAS, M. S. & BOSE, A. K. "Synthesis of Penicillin, Cephalosporin and Analogs," Marcel Dekker, New York, 1969.
12. BATCHELOR, F. R. & CAMERON-WOOD, J. *Nature 195*: 1000, 1962.
13. Anon., Recommendation for Measurement and Presentation of Biochemical Equilibrium Data" *J. Biol. Chem. 251*: 6879, 1976.

14. SVEDAS, V. K., MARGOLIN, A. L., GALAEV, I. YU. & BEREZIN, I. V. In "Proceedings Third Joint US/USSR Enzyme Engineering Seminar US/USSR Joint Working Group on Production of Substances by Microbiological Means", 1978, p. 679.
15. SVEDAS, V. K., "Kinetics and Mechanism of the Action of Penicillin Amidase," Candidate of Chemical Sciences Dissertation, Moscow, 1976.
16. BEREZIN, I. V., KLYOSOV, A. A., MARGOLIN, A. L., NYS, P. S., SAVITSKAYA, YE. M. & SVEDAS, V. K. *Antibiotiki* (Russ.) *21*: 511, 1976.
17. KLYOSOV, A. A., MARGOLIN, A. L. & SVEDAS, V. K. *Bioorg. Khim.* (Russ.) *3*: 654, 1977.
18. BEREZIN, I. V., MARGOLIN, A. L. & SVEDAS, V. K. *Dokl. Acad. Nauk. USSR* (Russ.) *5*: 1127, 1977.
19. MARGOLIN, A. L. & SVEDAS, V. K. In "Second National Symposium on Production and Use of Immobilized Enzymes," Abstracts (Russ.), Abovyan, 1977, p. 34.
20. MARGOLIN, A. L., SVEDAS, V. K., NYS, P. S., KOLTSOVA, YE. V., SAVITSKAYA, YE. M. & BEREZIN, I. V. *Antibiotiki* (Russ.) *23*: 114, 1978.
21. MARGOLIN, A. L. & SVEDAS, V. K. In "Second National Symposium on Microbial Enzymes", vol. 1, Abstracts (Russ.), Minsk, 1978, p. 158.
22. MARGOLIN, A. L. "Kinetic and Thermodynamic Regularities of Enzymatic Synthesis of β-Lactam Antibiotic Catalyzed by Penicillin Amidase", Candidate of Chemical Sciences Disservation, Moscow, 1979.
23. BEREZIN, I. V., KLYOSOV, A. A., MARGOLIN, A. L., NYS, P. S., SAVITSKAYA, YE. M. & SVEDAS, V. K., *Antibiotiki* (Russ.) *21*: 411, 1976.
24. SVEDAS, V. K., MARGOLIN, A. L., SHERSTIUK, S. F. & BEREZIN, I. V. In "Third National Symposium on Structure and Functions of Active Sites of Enzymes", Abstracts (Russ.) Pushino, 1976, p. 24.
25. SVEDAS, V. K., MARGOLIN, A. L., SHERSTIUK, S. F. & BEREZIN, I. V. *Dokl. Acad. Nauk SSSR* (Russ.) *232*: 1127, 1977.
26. SVEDAS, V. K., MARGOLIN, A. L., SHERSTIUK, S.

F. & BEREZIN, I. V. *Bioorg. Khim.* 3: 546, 1977.
27. MARGOLIN, A. L. & SVEDAS, V. K. In "Second National Symposium on Production and Use of Immobilized Enzymes", Abstracts (Russ.) Abovyan, 1977, p. 154.
28. COLE, M. *Biochem. J.* 115: 741, 1969.
29. COLE, M. *Biochem. J.* 115: 747, 1969.
30. SVEDAS, V. K., KLYOSOV, A. A., BEREZIN, I. V., ROTANOVA, T. V., VASILYEVA, N. V., GINODMAN, L. M. & ANTONOV, V. K. In "Third National Symposium on Structure and Functions of Active Sites of Enzymes", Abstract (Russ.), Pushino, 1976, p. 25.
31. KLYOSOV, A. A., GALAEV, I. U. & SVEDAS, V. K. *Bioorg. Khim.* (Russ.) 3: 663, 1977.
32. KLYOSOV, A. A., SVEDAS, V. K. & GALAEV, I. U. *Bioorg. Khim.* (Russ.) 3: 800, 1977.
33. CARPENTER, F. H. *J. Amer. Chem. Soc.* 82: 1111, 1960.
34. FERSHT, A. R. & REQUENA, Y. *J. Amer. Chem. Soc.* 93: 3499, 1971.
35. DYACHENKO, YE. D., KOZLOV, L. V. & ANTONOV, V. K. *Bioorg. Khim.* (Russ.) 3: 99, 1977.
36. ANTONOV, V. K., GINODMAN, L. M. & GUROVA, A. G. *Molecul. Biol.* (Russ.) 11: 1160, 1977.
37. OKACHI, R., MISAWA, M., DEGUCHI, I. & NARA, T. *Agr. Biol. Chem.* 36: 1193, 1972.
38. MARCONI, W., BARTOLI, F., CECERE, F., GALLI, G. & MORISI, F. *Agr. Biol. Chem.* 39: 277, 1975.
39. SHIMIZU, M., MASUIKE, I., FIJITA, H., IIDA, I., KIMURA, K. & NARA, T. *Agr. Biol. Chem.* 39: 1745, 1975.
40. FUJII, I., MATSUMOTO, K. & WATANABE, I. *Process Biochem.* 11: 21, 1976.
41. OKACHI, R. & NARA, T. *Agr. Biol. Chem.* 37: 2797, 1973.
42. SHIMIZU, M., OCACHI, R., KIMURA, K. & NARA, T. *Agr. Biol. Chem.* 39: 1655, 1975.
43. SVEDAS, V. K., KLYOSOV, A. A., NYS, P. S., SAVITSKAYA, YE. M. & SINITSIN, A. P. In "First National Symposium on Production and Use of Immobilized Enzymes", Abstracts (Russ.), Tallin, 1974, p. 82.

44. BEREZIN, I. V., KLIBANOV, A. M., KLYOSOV, A. A., MARTINEK, K. & SVEDAS, V. K. *FEBS Lett.* *49*: 325, 1975.
45. NYS, P. S., SAVITSKAYA, YE. M., KLYOSOV, A. A., SINITSIN, A. P., SVEDAS, V. K. & BEREZIN, I. V. *Antibiotiki* (Russ.) *23*: 46, 1978.
46. GOLDSTEIN, L., LEVIN, Y. & KATCHALSKI, E. *Biochemistry* *3*: 1913, 1964.
47. STRELTSOVA, Z. A., SVEDAS, V. K., MAKSIMENKO, A. V., KLYOSOV, A. A., BRAUDO, YE. YE., TOLSTOGUZOV, V. B. & BEREZIN, I. V. *Bioorg. Khim.* (Russ.) *1*: 1464, 1975.

MODIFICATION OF ENZYMES WITH WATER SOLUBLE POLYMERS

I. M. Tereshin and B. V. Moskvichev

Laboratory for Enzyme Immobilization
Technological Institute of Antibiotics
Leningrad, USSR

Quite a number of works are known which describe the chemical modification of enzymes, using either low or high molecular weight compounds.

The use of low molecular weight compounds has been discussed in a recent comprehensive review (1); and although these smaller compounds often are of limited utility, in some situations they can serve a very useful purpose. For example, Berezin et al. (2) showed that simple acylation with low molecular weight agents can have a stabilizing effect on some enzymes. Modification with small bi or poly functional compounds also may provide an intermediate stage in the preparation of immobilized enzymes (3).

The first studies with high molecular weight enzyme modifiers was by Katchalski (4,5), who obtained high molecular weight polypeptide derivatives of trypsin and chymotrypsin. Subsequently, proteinase derivatives based on copolymers of ethylene and maleic anhydride were studied (4-10), as well as derivatives of other enzymes modified with DEAE dextran (11,12). The main result of these studies was the discovery of the interrelationship between the electrostatic potential of the polymeric chain and the displacement of the optimal pH of the enzyme when acting on low molecular weight substrates. In other work on chymotrypsin binding with copolymers of the vinyl series (13), the influence of electrostatic interaction of the polymer and enzyme on their covalent binding was found. Of great interest also was the elucidation of the effect of polymeric modification on the characteristics of catalase (14). For example, modification of catalase with methoxypolyethylene glycol was shown to prolong the effect of this enzyme in the blood stream. Other studies include the effect of polymeric modification on the stability of protein inhibitors of proteinases (15); and the retention of the catalytic activity of chymotrypsin modified with polycarboxylic acids and then adsorbed on gelatinized ionite (16).

From the above studies as well as our own work, it seemed possible that the modification of enzymes might be a suitable approach for the development of biocatalysts for technological or medical use. With this aim in sight, we have been studying the modification of proteinases with water soluble polymers. The objective of this paper is to summarize our results.

I. ENZYME MODIFICATIONS

Most of our work was done with the proteinases trypsin and terrilytin. The latter is a proteolytic enzyme from *Aspergillus terricola* and is produced on an industrial scale at the Mosmedpreparaty works. Some of the modified enzyme preparations are listed in Table 1. Modification of the proteinases with copolymers of vinyl pyrrolidone and acrolein, as well as with oxidized dextran or activated albumin, was realized by using the reaction between the carboxylic groups of the polymer and the amino groups of the enzyme to form a linkage in the Schiff bases. Subsequent treatment of such products with sodium bisulphite resulted in the formation of a polymeric matrix having a negative charge. Treatment of the Schiff bases with sodium boron hydride reduced the azomethine linkage (-CH=N-) between the polymer and protein to the linkage ($-CH_2-NH-$), whereas excessive aldehyde

TABLE 1

MODIFICATION OF PROTEINASES USING WATER SOLUBLE POLYMERS

Enzyme	Polymeric Soluble Matrix	Fixing Modifier	Symbol
I. Terrilytin	-	-	Ter
	Human Serum Albumin	$NaBH_4$	AlbTerN
	Copolymer of Vinyl Pyrrolidone and Acrolein	$NaHSO_3$	VPyrAS
	Same	$NaBH_4$	VPyrAN
	Same	NH_4Cl	VPyrAB
	Dextran (Rheopolyglukin)	$NaHSO_3$	RheoTerS
	Same	$NaBH_4$	RheoTerN
	Same	NH_2OH	RheoTerB
II. Trypsin	-	-	Tr
	Human Serum Albumin	$NaHSO_3$	AlbTrS
	Same	$NaBH_4$	AlbTrN
	Copolymer of Vinyl Pyrrolidone and Crotonic Acid	-	VPyrCr
	Copolymer of Vinyl Pyrrolidone and Cinnamic Acid	-	VPyrCin
	Dextran (Rheopolyglukin)	$NaHSO_3$	RheoTrS
	Same	$NaBH_4$	RheoTrN

groups on the polymer were reduced to alcohols. The enzyme modification with copolymers of carboxylic acids was performed using water soluble carbodiimide to form the amide linkage (-CO-NH-) between the carboxylic acid residues of the polymer and the amino groups of the protein.

II. PROPERTIES OF MODIFIED ENZYMES

The polymeric derivatives of the enzymes of course had greater molecular weights as compared to those of the native enzymes and thus had lower rates of diffusion in solution or in a porous medium. Thus, in medical applications where the modified enzyme might be injected into the blood stream, the slower diffusion of the modified product through the capillary membranes might help to ensure a lower rate of enzyme elimination from the body (17). Fig. 1 shows gel chromatograms of native and modified enzymes; the results are indicative of the 2 to 4 fold increase in molecular weight, while the diffusion coefficients showed a 2 to 3 fold decrease.

The specific activity of the enzymes modified with polymers were determined, using as a high molecular weight substrate, casein in solution or fibrin free of profibrinolysin impurity and in the form of artificial clots or fibrinous films. The caseinolytic and fibrinolytic activities of the immobilized terrilytin and trypsin are shown in Table 2. It can be seen that in some cases enzyme activation with respect to fibrin was observed. Such a phenomenon was characteristic of a number of polymeric derivatives of trypsin, where the fibrinolytic activity was higher as compared to that of the native enzyme (Table 2). This was true of trypsin covalently bound to human serum albumin or to copolymers of carboxylic acids. Such an effect might be due to a definite affinity of albumin for fibrin or with polycarboxylic acids similar to that of heparin. An appreciable effect of the polymeric matrix on the biochemical characteristics of trypsin can be seen from Fig. 2, which shows the time

TABLE 2

ENZYMATIC ACTIVITY OF POLYMERIC DERIVATIVES OF TERRILYTIN AND TRYPSIN

Enzyme	Specific Activity (PU/mg of Enzymic Protein)[a]	Relative Fibrinolytic Activity[b]	Symbol
I. Terrilytin	5.0	1.0	Ter
	4.5	0.5	AlbTerN
	4.3	0.2	VPyrAS
	5.0	0.7	VPyrAN
	4.4	0.4	RheoTerS
	5.0	1.0	RheoTerN
II. Trypsin	4.7	1.0	Tr
	3.9	16.0	AlbTrS
	0.8	15.0	AlbTrN
	1.0	2.2	VPyrCr
	1.4	4.2	VPyrCin
	0.9	2.0	RheoTrS
	1.0	2.0	RheoTrN

(a) PU units of proteolytic activity measured per (18) with 1% casein in 0.1M tris-HCl at pH 8.0 for trypsin and pH 7.6 for terrilytin.
(b) With relation to native enzyme, measured with fibrinous films

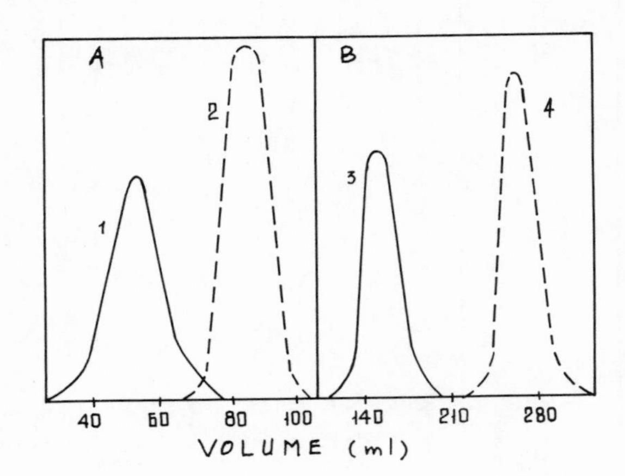

Fig. 1. Gel chromatograms of trypsin and terrilytin. Curves represent: 1, RheoTrN; 2, native trypsin; 3, VPyrAN; 4, native terrilytin. Part A: chromatography on Sephadex G-75; Part B: chromatography on Sephadex G-100. For curves 3 and 4, k_{dif} was 5.9×10^{-7} and 14.6×10^{-7} cm^2/sec, respectively.

of human plasma recalcification according to the level of modified enzyme. In case of the native trypsin, fibrin clots were formed earlier than with the control; the latter was given as 100%. As a result of the enzyme modification with albumin, the nature of the trypsin effect on the plasma recalcification time was changed greatly.

The polymeric microenvironment of the enzyme influences other characteristics of the enzyme as well as activity. For example, inactivation of the polymeric derivatives of trypsin, as with the native enzyme, followed second order kinetics. It was found that the modified trypsin had at least two fractions, one labile and one stable. The inactivation rate of the labile trypsin fraction

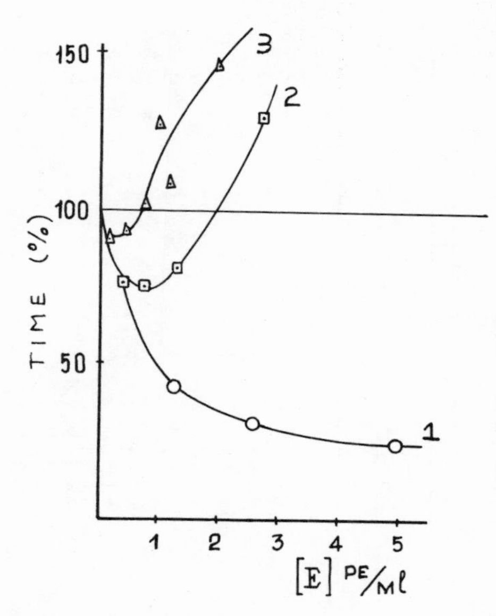

Fig. 2. Dependence of relative time of human plasma recalcification on the concentration of native or modified trypsin. Curves are for: 1, native trypsin; 2, AlbTrN; 3, AlbTrS.

(Table 3) was lower than that of the native trypsin. Under similar conditions the stable fraction retained practically its full activity on a steady level for a very long period of time. The lower rate of autolytic denaturation of the modified trypsin might be determined by 1) the formation of two or more cross linkages between the molecule of polymer and that of enzyme, which would make branching of the polypeptide chain difficult, 2) shielding by the polymer, so that the branched macromolecule of enzyme would not be as good a substrate for itself, and 3) blocking of the main amino acid residues of lysine as a result of the modification. Inactivation of terrilytin under

TABLE 3

RATIO OF RATE CONSTANTS FOR DENATURATION OF NATIVE ENZYMES TO ENZYMES MODIFIED WITH POLYMERIC MATRICES

Enzyme[a]	$k_{native}/k_{modified}$	
	Labile Fraction	Stable Fraction
Terrilytin	1.0	1.0
VPyrTerS	1.8	0
VPyrTerN	8.9	0
RheoTerN	10.2	0
RheoTerS	1.4	0
Trypsin	1.0	1.0
VPyrCr	10.0	0.5
VPyrCin	10.0	0.5
RheoTrS	17.0	0.8
RheoTrN	20.0	0.8
AlbTrS	2.5	0.5
AlbTrN	2.8	0.6

(a) Conditions for terrilytin enzymes where pH 8.0 and 25°C.; conditions for trypsin enzymes were pH 7.8 and 50°C.

similar conditions followed first order kinetics. The polymeric forms of terrilytin were rather homogeneous with respect to denaturation; and the rate constants for inactivation were lower than those for native terrilytin. Table 3 shows the relative constants for terrilytin and trypsin thermodenaturation obtained under the conditions listed. As a rule, the modified enzymes showed higher stability as compared to the native enzymes.

The dependence of modified enzyme stability on the number of cross linkages between the matrix and the enzyme has been discussed in the literature. However, we obtained new results showing that in the case of equal numbers of cross linkages, the enzymic stability would be determined by the nature of the electrochemical environment of the enzyme. Fig. 3 shows such an effect with terrilytin immobilized on the copolymer of vinyl pyrrolidone and acrolein or on dextran. In the neutral pH range an inert uncharged matrix ensured a 10 fold increase in terrilytin stability. Fig. 4 shows an increase in terrilytin stability in relation to another proteolytic enzyme, trypsin, as a result of terrilytin modification with the copolymer of vinyl pyrrolidone and acrolein. The inactivation curve for native terrilytin fell steeply, while the activity level of the modified enzyme decreased slowly.

The influence of the polymeric metrix on the affinity of protein inhibitors for the native and modified proteinases revealed that the affinity decreased as a result of enzyme modification. Fig. 5 shows the dependence of the residual activities of the native and immobilized trypsin and terrilytin on the concentration of the added human blood serum. In all cases the modified enzymes showed higher residual activities than did the native enzymes. Table 4 shows the decrease in the affinity of the serum inhibitors for the enzymes, as the latter are converted from the native to the modified form. The decrease was as great as 1,000 fold.

TABLE 4

INHIBITION OF TERRILYTIN AND TRYPSIN BY HUMAN BLOOD SERUM AND THE ACUTE TOXICITY

Preparation	Relative Equilibrium Binding Constants $K_{modified}/K_{native}$	Relative Acute Toxicity[(a)]
Terrilytin	1.0	1.0
AlbTer	9.4	0.8
VPyrTer	9.6	1.0
RheoTer	4.3	1.0
Trypsin	1.0	1.0
AlbTrS	3.0×10^3	0.008
AlbTrN	1.6×10^2	0.05
VPyrCr	2.1×10^2	0.05
RheoTrS	3.2×10^2	0.03
RheoTrN	3.0×10^2	0.01

(a) Shown as LD_{50} (native)/LD_{50} (modified); determined on 18-20g mice following intravenous injection of enzyme preparations.

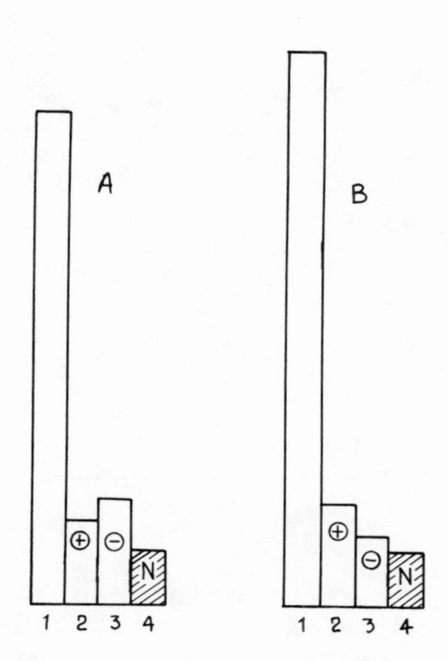

Fig. 3. Diagram of relative stability of terrilytin modified with a copolymer of vinyl pyrrolidone and acrolein (Part A) or with dextran (Part B). Bars represent: 1, VPyrAN; 2, VPyrAB; 3, VPyrAS; 4, native terrilytin. Conditions: pH 7.9, 25°C.

One might expect changes in toxicity and antigenicity associated with the conversion of the native enzymes to the modified forms because of the decrease in the affinity of the serum inhibitors and because of possible shielding of parts of the protein globule by the polymer. The data on acute toxicity of some of the enzyme preparations shown in Table 4 tended to substantiate the idea that the modification of enzymes can result in lower toxicity; the greatest reduction in toxicity was noted with the modified trypsin preparations.

Studies on the immunogenic, antigenic, and allergenic properties of the modified enzymes, pre-

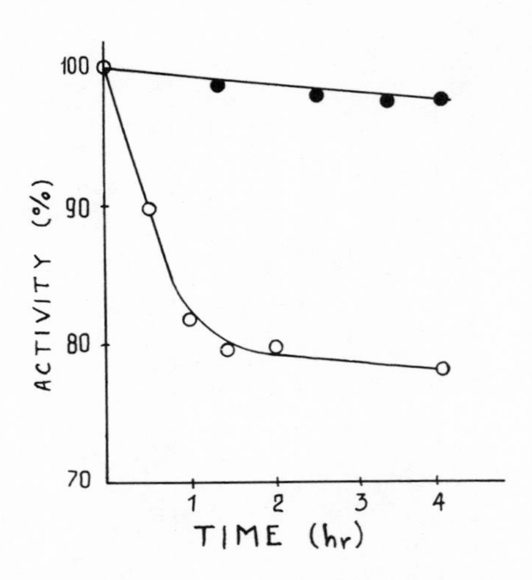

Fig. 4. Kinetics of trypsin inactivation of native (O) and modified terrilytin VPyrAN (●) at pH 8.0 and 37°C.

pared on the basis of thymus independent polymeric matrices of polyvinyl pyrrolidone and dextran, showed such modified enzymes to differ from the native ones. The results of the quantitative compliment fixation test (19,20). (Fig. 6) show that the modification to terrilytin induced a substantial decrease in its ability to react with antibodies directed against the antigenic determinants of the native enzyme. Coupling of proteolytic enzymes with polymeric matrices did not result in complete elimination of the immunogenic properties of the enzyme; however the intensity of the production of antibodies to the modified protein macromolecules was lower than to the starting enzyme. The minimum sensibilizing dose of the modified terrilytin was two orders above that of the native enzyme.

Studies on the anaphylactogenic activity of terrilytin and its modified forms in cross experi-

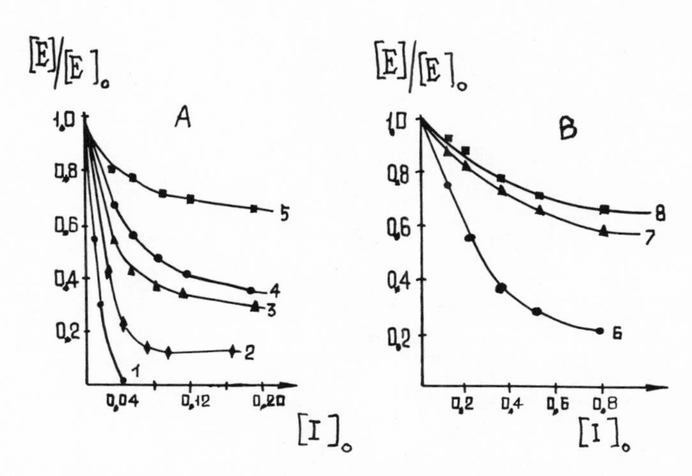

Fig. 5. Inhibition of trypsin (Part A) and terrilytin (Part B) and their modified forms by human blood serum. Curves: 1, native trypsin; 2, DexTrN; 3, AlbTrN; 4, DexTrS; 5, AlbTrS; 6, native terrilytin; 7, DexTerN; 8, DexTerS. Ordinate axis shows residual activity of enzyme $(E)/(E)_o$; abscissa axis shows inhibitor level as ratio of serum ml to 1.0 ml of incubation mixture.

ments with guinea pigs showed that the strength of the anaphylactic response in the animals sensibilized with terrilytin, and which received a challenging injection of the modified enzyme, was much less than in the animals which received the homologous native enzyme and was practically the same as in the control animals (21).

Thus, it may be concluded that coupling of enzymes with water soluble polymeric matrices makes it possible to obtain enzymic preparations having improved antigenic, immunogenic, and allergenic properties. Medical use of such preparations looks

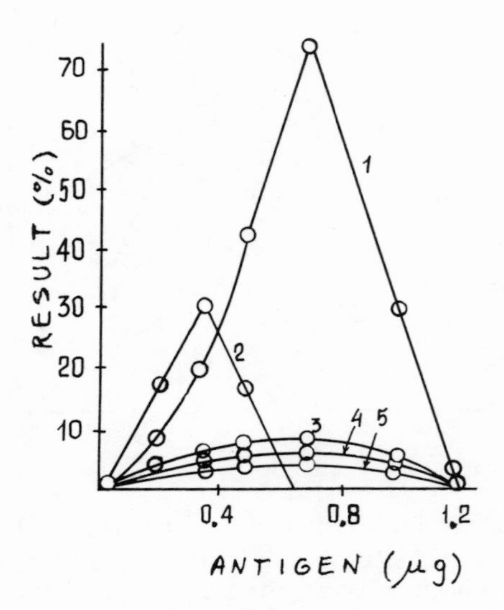

Fig. 6. Comparison of antigenic properties of native terrilytin and of its modified forms in the complement fixation test. Abscissa shows amount of antigen in μg. Curves: 1, native terrilytin; 2, RheoTerN; 3, AlbTerN; 4, $AlbTerN_2$; 5, VPyrAN.

promising for their application both as original drugs and as preparations for prolonged use in combination with the native enzyme or its various modifications. Combined use of the modified and native enzymes could be an alternative to the tendency to use combinations of enzymes having similar catalytic activities but differing in antigenic respect and prepared from different sources.

The electrochemical nature of some polymeric chains makes it possible to carry out interactions of the modified enzyme with a number of solid synthetic or mineral carriers, through structural and covalent binding or mechanical confinement in gel pores. Fig. 7 shows the wash out of native terrilytin, with a molecular weight of about 30,000, and modified terrilytin from polyacrylamide gel. It is clear that the modified enzyme was held in the gel for a longer period of time.

III. CONCLUSION

In conclusion, it should be noted that enzyme modification might be used as a complex approach for the preparation of various biocatalysts and biosorbents intended for medical or industrial use.

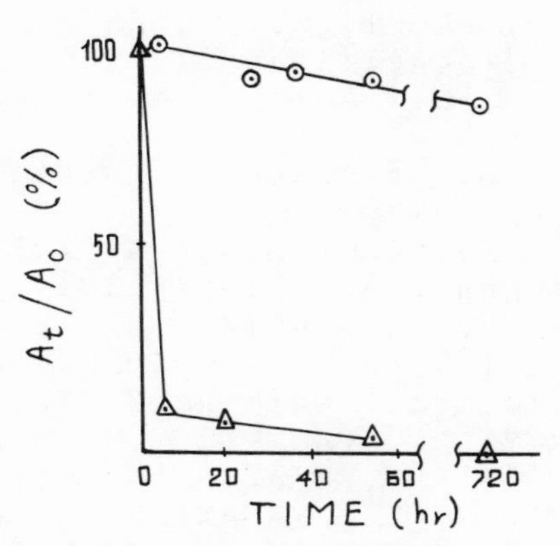

Fig. 7. Wash out of native terrilytin, triangles, and VPyrAN, circles, from polyacrylamide gel prepared from 15% acrylamide (w/v) and 2% N,N′-methylene-bis-acrylamide (w/v).

REFERENCES

1. SEVERIN, E. S., KUROCHKIN, S. N. & KOCHETKOV, S. N. *Uspekhi Biol. Khim.* *15*: 65, 1974.
2. UGAROVA, N. N., BROVKO, L. Yu., ROZHKOVA, G. D. & BEREZIN, I. V. *Biokhimiya* *42*: 1212, 1977.
3. BESSMERTNAYA, L. A., KOZLOV, L. V. & ANTONOV, V. K. *Biokhimiya* *42*: 1825, 1977.
4. KATCHALSKY, E. & SELA, M. *Adv. Protein Chem.* *13*: 243, 1958.
5. GLAZER, A. N., BAR-ELI, E. & KATCHALSKY, E. *J. Biol. Chem.* *237*: 1832, 1962.
6. EPSTEIN, C. J., ANFINSEN, C. B. & SELA, M. *J. Biol. Chem.* *237*: 3458, 1962.
7. BAR-ELI, A. & KATCHALSKY, E. *J. Biol. Chem.* *238*: 1690, 1963.
8. PECHT, M. & LEVIN, Y. *Biochem. Biophys. Res. Commun.* *46*: 2054, 1972.
9. LEVIN, Y., PECHT, M., GOLDSTEIN, L. & KATCHALSKY, E. *Biochemistry* *3*: 1905, 1964.
10. LOVENSTEIN, H. *Acta Chem. Scand.* *28*: 1098, 1974.
11. SVENSSON, B. *Biochim. Biophys. Acta* *429*: 954, 1976.
12. WYKES, G. W., DUNNIL, P. & LILLY, M.D. *Biochim. Biophys. Acta* *250*: 1971.
13. TORCHILIN, V. P., REIZER, I. L., TISHCHENKO, E. G., IL'INA, E. V., SMIRNOV, V. V. & CHAZOV, E. I. *Bioorgan. Khimiya* *2*: 1687, 1976.
14. ABUKHOVSKY, A., VAN ES, T., PALCZUK, N. C. & DAVIS, F. F. *J. Biol. Chem.* *252*: 3578, 1977.
15. LARIONOVA, N. I., KAZANSKAYA, N. F. & SAKHAROV, I. Y. *Biokhimiya* *42*: 1237, 1977.
16. KINSTLER, O. B. & KOZLOV, L. B. *Biokhimiya* *42*: 1674, 1977.
17. PEREL'MAN, A. E., VISHNEVSKY, B. I., VAVILIN, G. I., IL'INA, GL B., YAKOVLEVA, T. M., KROPACHEV, V. A. & TRUCHMANOVA, L. B. *Chim.-Pharm. Zhurnal* *11*: 63, 1977.
18. LASKOVSKY, M. *Meth. Enzymol.* *2*: 26, 1955.
19. GORYUHINA, O. A., LEONTJEVA, G. E. & KASHKIN, A. P. *Zhurn. Mikrobiol. Epidemiol. Immunobiol.* (Russ.) *11*: 91, 1976.

20. KASHKIN, A. P. & MACHOVENKO, L. V. *Vestnik Dermatol. Venerol.* (Russ.) *6*: 26, 1978.
21. KASHKIN, A. P., LEVI, E. V. & MOCHOVA, G. A. in Abstracts, "All-Union Conf. Design. Technol. Prod. Polyene. Antibiotics Microb. Enz. Med. Appl.," Leningrad, USSR, 1976.

IMMOBILIZED ENZYMES AND OTHER MATERIALS FOR THE STUDY OF MAMMALIAN CELL SURFACES

L. B. Wingard Jr.

University of Pittsburgh Medical School
Pittsburgh, Pennsylvania, USA

The branched chain sugar molecules that extend from the surface of mammalian cells appear to play key roles in the regulation of cell growth and cell differentiation, the immune response, and trans-membrane as well as intercellular communication. These sugar chains, known as glycolipids or glycoproteins, extend from protein or lipid moities that are embedded in the ordered lipid bilayer membrane that encloses the cell. In addition, these lipid or protein moieties are capable of undergoing lateral and rotational diffusion within the bilayer membrane. This relative movement of the membrane components gives rise to changing localized surface concentrations of specific cell surface sugars (1,2). The purpose of this short chapter is to point out a few of the possibilities for the use of immobilized enzymes and other compounds in the study and modification of cell surface constituents.

I. IMMOBILIZED ENZYMES

Hydrolytic enzymes such as trypsin, neuraminidase, and galactosidase have been used in soluble form for the modification of the surface constituents of mammalian cells. In a particularly instructive example, trypsin and a few other proteases were shown to initiate the proliferation of cultured quiescent chick enbryo fibroblases (3). However, there was considerable controversy as to whether or not the trypsin acted at the cell surface or inside the cells. By immobilizing the trypsin on solid supports that could not enter the cells, it was reasoned that this question of a surface versus an intracellular site of action could be answered. Commercial preparations of trypsin immobilized on Sepharose or acrylamide, using cyanogen bromide coupling, were shown to give a slow release of trypsin; thus, with these preparations it was not possible to differentiate between surface and intracellular sites of action. However, trypsin immobilized on polystyrene beads via carbodiimide coupling gave stable attachment after an initial pre-test conditioning with chick fibroblasts (3). Tests with the carbodiimide coupled trypsin led to the conclusion that the cell proliferation was initiated by trypsin action at the cell surface. This study pointed up the unique possibilities as well as one of the major problems in the use of immobilized enzymes for the study of cell surface phenomena. The need for extensive control experiments to insure that none of the immobilized enzyme becomes detached from the support is made evident in this example. Immobilized enzyme release catalyzed by cell-surface proteases or esterases may be an especially difficult problem to overcome.

Neuraminidase, an enzyme that cleaves N-acetylneuraminic acid (sialic acid) from cell surface plycoproteins, also has been immobilized to several types of supports. This enzyme from *Clostridium perfringens* has been coupled to Sepharose 4B and to porous glass beads by several techniques; however, all of the coupled preparations showed a slow release of neuraminidase (4). This same enzyme from *Vibrio cholerae* has been immobilized on nylon tubes, using dimethyl adipimidate or glutaraldehyde as the coupling agents, and shown to remove sialic acid from lymphocytes of leukemic AKR mice (5). The nylon tube preparations have not been tested for possible release of immobilized neuraminidase.

Immobilized enzymes also have been used to develop special analytical methods for the study of cell surface materials. In the technique for attaching radiolabelled iodine to cell surface proteins, immobilized glucose oxidase and immobilized lactoperoxidase have been used. Glucose and the oxidase are used to generate hydrogen peroxide; this in turn reacts with radiolabelled iodide and the peroxidase to reduce the iodide to iodine, which reacts with protein to form the radiolabelled protein. In another technique immobilized β-galactosidase has been used to develop an immunoassay for the determination of surface immunoglobulin (Ig) receptors on β lymphocytes. This type of cell contains easily detectable surface Ig receptors, for which anti-Ig can be made. By coupling β-galactosidase to the anti-Ig and incubating the lymphocytes with the enzyme-anti-Ig material, the B cells become labelled with galactosidase at the surface Ig receptors. The subsequent addition of fluorogenic substrates for β-galactosidase results in the formation of fluorescence in the vicinity of the surface Ig receptors. This technique can be used to identify β lymphocytes or to assess the surface concentration of Ig receptors (6).

II. OTHER IMMOBILIZED COMPOUNDS

These same immobilization techniques can be utilized for coupling non-enzymatic probes to enzymes, polymers, or solid supports for the study of cell surface constituents. Considerable use has been made already for attaching lectins to solid supports for studies on the binding of lectins to specific cell surface sugars (7), with newer supports and coupling techniques being reported (8). Covalently immobilized lectins also are available commercially. The problem of the release of bound lectin must be evaluated separately for each experimental situation where the release of bound lectin could influence the results of the study.

Drugs of course represent another class of compounds that in some cases can be coupled to solid supports for the study of the drug interaction with the surface of mammalian cells. As an example of the type of coupling reactions that can be utilized, we have immobilized the antibiotic tetracycline on glass beads for a colleague to use in some studies with drug resistant bacteria (unpublished). The coupling was carried out by preparation of a diazonium salt of aminoalkane glass followed by reaction with tetracycline, to form the compound shown in Fig. 1. Other work to examine the interaction of cancer chemotherapeutic agents with mammalian cell surfaces is underway.

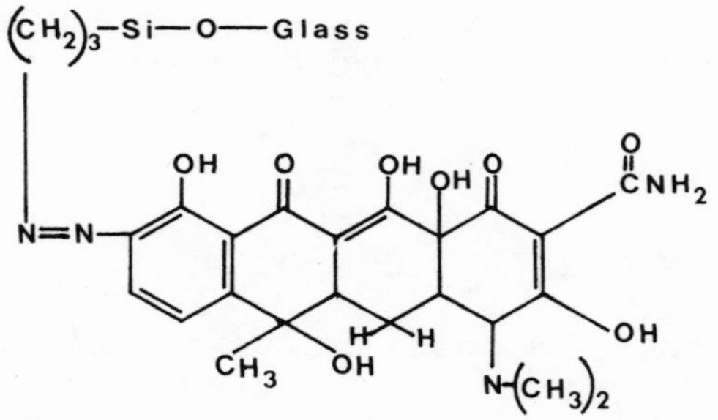

Fig. 1. Tetracycline covalently coupled to glass beads, using aminopropyltriethoxy silane and a diazonium linkage for the coupling materials.

REFERENCES

1. ROBBINS, J. C. & NICOLSON, G. L. In "Cancer," Vol. 4 (F. F. Becker, ed.) Plenum Press, New York, 1975, p. 3.
2. SCHLESSINGER, J., KOPPEL, D. E., AXELROD, D., JACOBSON, K., WEBB, W. W. & ELSON, E. L. *Proc. Nat. Acad. Sci. USA 73*: 2409, 1976.
3. CARNEY, D. H. & CUNNINGHAM, D. D. *Nature 268*: 602, 1977.
4. PARKER, T. L., CORFIELD, A. P., VEH, R. W. & SCHAUER, R. *Hoppe Seyl. Z. Physiol. Chem. 358*: 789, 1977.
5. BAZARIAN, E. R. & WINGARD JR., L. B. *J. Histochem. Cytochem. 27*: 125, 1979.
6. CAMERON, D. J. & ERLANGER, B. F. *J. Immunol. 116*: 1313, 1976.
7. KINZEL, V., KUBLER, D., RICHARDS, J. & STOHR, M. *Science 192*: 487, 1976.
8. MIRON, T., CARTER, W. G. & WILCHEK, M. *J. Solid Phase Biochem. 1*: 225, 1976.

SECTION IV
ENZYME ENGINEERING IN ENERGY TRANSFER, PHOTOGRAPHY, AND FINE CHEMICAL PROCESSING

MICROORGANISMS AS HYDROGEN AND HYDROGENASE PRODUCERS

I. N. Gogotov

Institute of Photosynthesis
USSR Academy of Sciences
Pushchino, USSR

Great attention has been paid recently to the investigation of microorganisms transforming light energy with the evolution of molecular hydrogen (1,2). However, the efficiency of hydrogen production by microorganisms *in vivo* is not usually high and depends on a number of factors (2). Therefore, for practical purposes it is probably necessary to form artificial model systems, which produce hydrogen by photodecomposition of water with much higher efficiency and which have high stability with respect to a number of factors. The functional stability of such systems will depend primarily on the stability of hydrogenase to the oxygen produced during photosynthesis and to the temperature. In this paper an attempt has been made to analyze recent literature and our own data about the efficiency and mechanism of hydrogen production and the properties of the enzymes that participate in hydrogen production.

I. HYDROGEN PRODUCTION EFFICIENCY BY MICROORGANISMS.

The organisms of different systematic groups can be characterized by their ability to evolve molecular hydrogen. There are more than twenty genera, even among bacteria, which evolve hydrogen (2). The same property is also characteristic for a number of algae (3) and some protozoa (4). There also are reports about hydrogen evolution by higher plants (5). However, the function and scale of this process is not the same in different organisms and often varies due to medium conditions.

A. Hydrogen Evolution by Chemotrophes.

Some bacteria, growing under anaerobic conditions, are able to evolve hydrogen as one of the products of normal activity (Table 1). Among them saccharolytic clostridia carrying out typical butyric acid fermentation of carbohydrates evolve hydrogen in great amounts. According to the experimental data *Clostridium butyricum* during the fermentation of one mole of glucose produces up to two or more moles of hydrogen (6). An isolated strain of *C. perfringens* growing in 10 liters of glucose medium in a glass fermentor produces hydrogen at the rate of 18.23 l/hour (7).

The second group of hydrogen producing microorganisms is represented by a number of enterobacteria, including *Escherichia coli* (Table 1) and some others (2). All these species grow under aerobic conditions, respiring with oxygen participation; but they are also able to grow under anaerobic conditions, fermenting carbohydrates and other organic compounds with hydrogen production.

The third group of hydrogen producing microorganisms is the sulfate-reducing bacteria, including two genera of *Desulfovibrio* and *Desulfotomaculum*. Some species of these bacteria are able to grow in media without sulfates and other oxidized sulfur compounds or in the presence of pyruvate, fumarate, or malate and evolve hydrogen under these conditions (8,9).

B. Hydrogen Evolution by Phototrophes.

Many representatives of purple sulfur and nonsulfur bacteria, cyanobacteria, and algae are able

TABLE 1

Hydrogen evolution and hydrogenase activity in chemotrophic bacteria

Species	Hydrogenase Activity[a]		Hydrogen Evolution Rate[a]
	MV[b]	MV + $Na_2S_2O_4$	
I. Chemoorganoheterotrophic			
A. Anaerobes			
Clostridium butyricum	2.6	5.4	2.5
C. pasteurianum 38	3.0	1.1	n.d.
B. Aerobes			
Azotobacter vinelandii	n.d.	0.2	0.3
Escherichia coli	0.5	0.8	n.d.
II. Chemolithotrophic			
A. Aerobes			
Alcaligenes eutrophus Z-1	52.3	2.4	n.d.[c]
Nocardia opaca 18	7.2	1.7	n.d.

(a) Hydrogen evolution determined in the presence of 10 mM pyruvate; units μ moles hydrogen/hr/mg protein
(b) 10 mM methyl viologen
(c) n.d. is not determined

to produce hydrogen (2). Of these, the hydrogen production by purple bacteria has been studied the most. Hydrogen production by purple bacteria takes place only under anaerobic conditions of growth in organic media or in cell suspension under anaerobic conditions by adding exogenous H-donors (Table 2). Depending on the properties of the substrate and the microbial species the evolution of hydrogen may take place in the light and dark or only during illumination of the cells. Light-dependent hydrogen evolution by purple bacteria has been shown recently to be associated mainly or completely with the action of nitrogenase (10,11); while dark hydrogen evolution has been shown to be catalyzed by hydrogenase (11,12). The amount of hydrogen evolved by purple bacteria can be varied, depending on the species and strain, the character of the substrate, and other factors of the medium (Table 2). The hydrogen evolution rate is 1-2 μmoles/hr/mg of protein in the presence of some organic substrates, e.g. pyruvate, and is 14 μmoles/hr/mg of protein in the presence of reduced methyl viologen.

In contrast to purple bacteria all other phototrophes which can evolve molecular hydrogen carry out photosynthesis with oxygen formation and are able to grow under aerobic conditions. Cyanobacteria are the most interesting among them. It is known that some species of cyanobacteria, e.g. *Anabaena cylindrica*, containing heterocysts, can simultaneously evolve hydrogen and oxygen in light (13). This is likely connected with the fact that the hydrogen photoevolution is separated in space from the oxygen production and takes place not in vegetative cells but in heterocysts. Hydrogen production by *A. cylindrica* as well as by purple bacteria is likely to occur due to nitrogenase action (14,15). The energy requirements for this reaction are supplied by the action of light or the oxyhydrogen reaction. Although hydrogenase does not take part in hydrogen photoevolution by *A. cylindrica*, this enzyme can cause hydrogen production in the presence of an exogenous H-donor and protect nitrogenase in the heterocysts from oxygen inactivation, thus supplying the nitrogenase with the

TABLE 2

HYDROGEN EVOLUTION AND HYDROGENASE ACTIVITY IN PHOTOTROPHIC BACTERIA

Species	Hydrogenase Activity[a]		Hydrogen Photoevolution with Pyruvate[a]
	MV[b]	MV+$Na_2S_2O_4$	
I. Purple bacteria			
Rhodospirillum rubrum	0.8	3.6	1.2
Rhodopseudomonas palustris	1.8	2.6	1.9
Rh. acydophila 7050	4.3	4.7	2.0
Rh. spheroides	0.5	0.3	1.4
Rh. capsulata	0.4	0.4	0.1
Rh. lovis	0.6	0.4	0.1
Rhodomicrobium vannielii	0.5	0.6	0.2
Ectothiorodospira shaposhnikovii	0.8	0.9	0.3
Thiocapsa roseopersicina BBS	3.0	13.7	1.2
II. Cyanobacteria or Blue-green Algae			
Anabaena variabilis	0.9	0.3	0.1
A. cylindrica	0.4	0.2	0.1
Spirulina platensis	1.3	0.1	0.1

(a) Units of μmoles/hr/mg protein
(b) 10 mM methyl viologen

energy due to the oxyhydrogen reaction. But even cyanobacteria have a low efficiency of hydrogen production (Table 2); so that the study of their hydrogen production mechanism will help to create more effective and more stable model systems for hydrogen production.

Green algae also are able to provide hydrogen evolution under anaerobic conditions (Table 3). The process begins only after adaptation of the cells in an atmosphere of inert gas and lasts for a few minutes to 10-15 hours and even for several days (16,17). Hydrogen production by green algae takes place in the dark and in the light; but in the last case much more intensively (18). Moreover, hydrogen photoevolution by some species of green algae increases in the presence of exogenous organic compounds, such as glucose and acetate (3). It also has been shown that in the light as well as in the dark only the carbon monoxide inhibited hydrogenase of the green algae *Chlorella vulgaris* takes part in hydrogen evolution. In spite of a rather high efficiency of hydrogenase activity of 13.4 μmoles hydrogen/hr/mg of protein, *Chlamydomonas reinhardii* cells stop producing hydrogen even with an apparently insignificant amount of oxygen in the gas phase.

II. THE PATHWAYS OF MOLECULAR HYDROGEN PRODUCTION

The reactions with direct hydrogen evolution are known to be very few. Hydrogen evolution by *Clostridium* and other strict anaerobs (Fig. 1a,b) in many cases is connected with pyruvate decomposition with ferredoxin, through acetyl phosphate into acetate, carbon dioxide, and hydrogen (19). As a result of such reactions with ferredoxin participation, hydrogen evolution is likely to take place during the fermentation of some other substrates (20). It has been shown for the *C. kluyveri* cell-free extract that NADH or NADPH can be the substrates for hydrogen evolution by this organism; and the process is carried out through the formation of the reduced ferredoxin (21).

TABLE 3

HYDROGEN EVOLUTION AND HYDROGENASE ACTIVITY IN ANAEROBICALLY ADAPTED CELLS OF VARIOUS ALGAE SPECIES

Species	Hydrogenase Activity[(a)]		Hydrogen photo-evolution[(a)]	References
	MV[(b)]	$MV+Na_2S_2O_4$		
Chlorella vulgaris	46.0	21.1	0.3	(31)
Ch. vacuolata	n.d.	n.d.	15.1	(1)
Ch. protothecoides	n.d.	n.d.	16.4	(1)
Ch. fusca	n.d.	n.d.	2.2	(1)
Ch. sorokiniana	n.d.	n.d.	14.5	(1)
Chlamydomonas reinhardi		13.4		our data
Coelastrum proboscideum	n.d.	n.d.	24.6	(1)
Scenedesmus obliquus	n.d.	n.d.	20.2	(1)
Ankistrodesmus braunii	n.d.	n.d.	11.1	(1)
Kirchneriella lunaris	n.d.	n.d.	32.1	(1)

(a) Units of μmoles/hr/mg chlorophyll; 10 mM pyruvate
(b) 10 mM methyl viologen
(c) n.d. not determined

Facultative anaerobic bacteria, for example *E. coli*, evolve hydrogen through formate breakdown (8) with formic hydrogenlyase and a low redox potential cytochrome c_3 participation (Fig. 1b,f).

The sulfate-reducing bacteria can evolve hydrogen by pyruvate as well as formate breakdown (9). However, the hydrogen production by these bacteria takes place with cytochrome c_3 participation but not with ferredoxin (22,23).

Purple bacteria also produce hydrogen in the dark either by pyruvate decomposition with ferredoxin participation or by formate breakdown (24,25) The latter hydrogen production is likely to take place with cytochrome c_3 participation (11).

Another mechanisms can act to give hydrogen production. According to one point of view (26) hydrogen photoevolution by purple bacteria takes place as the result of non-cyclic electron transport at the reaction center where bacteriochlorophyll participates. According to a second point of view (19), hydrogen evolution by purple bacteria takes place under the action of a reverse electron transfer system; this view is supported by the suppression of this process by photophosphorylation uncouplers. It has been shown also that purple bacteria preparations, containing chromatophores, are able to evolve hydrogen from NAD(P)H by adding an ATP-generating system, pyrophosphate, or during illumination (25). The process depends on the presence of ferredoxin (27).

There are few data about the partways of hydrogen production by cyanobacteria. In the dark some cyanobacteria are able to evolve hydrogen from pyruvate (28) or formate (29) decomposition. However, cyanobacteria containing heterocysts, i.e. *A. cylindrica*, are able to produce simultaneously hydrogen and oxygen in light (13). Since hydrogen photoevolution is not suppressed by diurone, but is inhibited by phosphorylation uncouplers (15), hydrogen evolution occurs not as the result of water photolysis but apparently from endogenous compounds

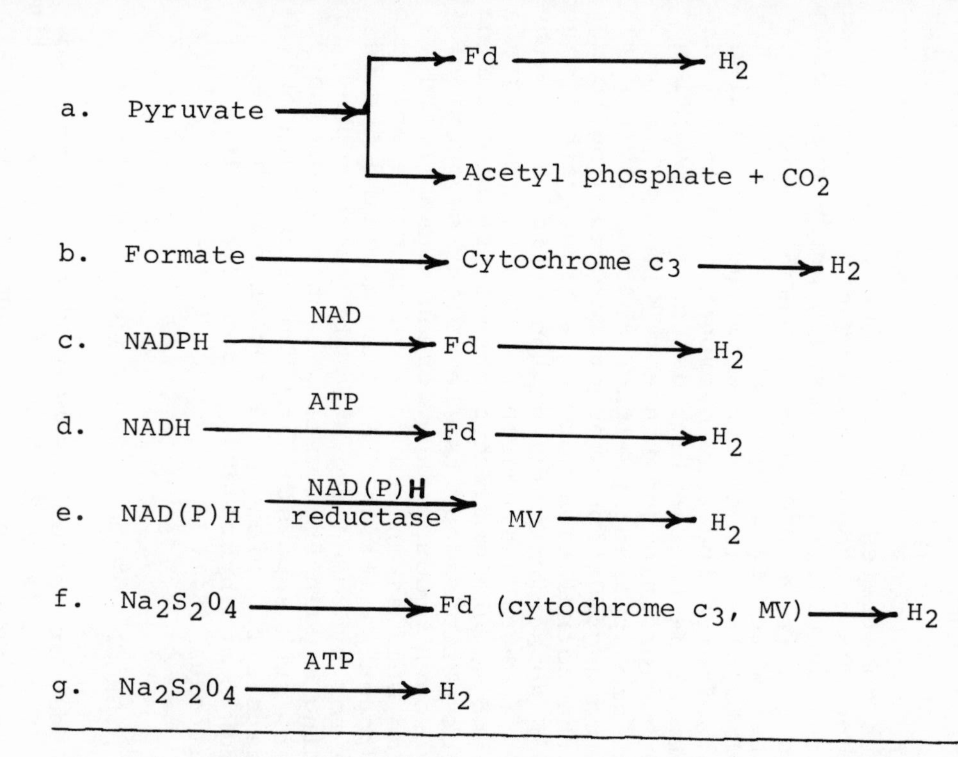

Fig. 1. Some schematic reactions for hydrogen production. Fd stands for ferrodoxin.

which are the products of carbon dioxide fixation. This is supported by the fact that in light and the presence of 1% carbon dioxide *A. cylindrica* cells are observed to evolve hydrogen much longer (28 hr) than in the absence of carbon dioxide (6-8 hr). Also, hydrogen production by cyanobacteria in light as well as by purple bacteria is due to the action of nitrogenase but not hydrogenase (14,15).

Hydrogen evolution by green algae in the dark is also connected with decomposition of organic substances, which may be exogenous as well as endogenous substrates (30,31). According to some data hydrogen photoevolution by green algae is accompanied by simultaneous oxygen production in the ratio of 2:1. Due to this a suggestion has been made that water is a source of the hydrogen evolved by algae in light. However, there also are data indicating that hydrogen photoevolution by green algae takes place due to decomposition of endogenous organic substances (32). Such substances are apparently either reduced NAD(P)H (33) or starch degradation products. According to our data, the correlation between starch synthesis and hydrogenase activity during the period of growth of *Chl. reinhardi* synchronic culture, its degradation, and simultaneous hydrogen production during the dark period supports the latter suggestion. And hydrogen production by green algae in the light as well as in the dark takes place only due to hydrogenase action. At the same time the possibility for hydrogen evolution by phototrophes, producing oxygen from water, cannot be excluded. However, such a possibility has been shown only in the tests with spinach chloroplasts with added ferredoxin and hydrogenase from *Clostridium pasteurianum* (34).

III. HYDROGEN PRODUCTION BY MODEL SYSTEMS.

Great attention has been shown recently to the different factors that influence the stability and effectiveness of hydrogen photoevolution by the systems containing chloroplasts, ferredoxin, and hydrogenase (35-37). The hydrogen evolution rate in such systems, as it has been shown in our investigations, depends on a number of factors, including the ferredoxin and chlorophyll concentrations and the nature of the hydrogenases. For example, *Cl. pasteurianum* hydrogenase catalyze hydrogen evolution with a high efficiency in the presence of ferredoxins from different sources. However, hydrogenase from *E. coli*, *R. rubrum*, *T. roseopersicina*, *Ch. vinosum*, and *Desulfovibrio ethylica* produce hydrogen in great amounts when cytochrome c_3 is added to such systems (Table 4). Thus, depending on the properties of the hydrogenases, either ferredoxin or low potential cytochrome c_3 can be used as an intermediate electron carrier in the system of water biophotolysis with chloroplasts.

The maximum rate of hydrogen evolution obtained in the system containing spinach chloroplasts, hydrogenase, and *S. maxima* ferredoxin was up to 94 μmoles/hr/mg of cholrophyll. However, this represents only part of the electron transport ability of the chloroplasts (22). Hydrogenase or ferredoxin autooxidation can be one of the factors effecting the low rate of hydrogen production. In the presence of a great abundance of hydrogenase and ferredoxin, illuminated chloroplasts are able to evolve hydrogen and oxygen simultaneously for up to two hours. Higher rates of hydrogen production and increased duration of evolution in the presence of catalase and ethanol indicate that hydrogen peroxide formed in the system has an inhibiting action on hydrogen production.

The stability of chloroplasts influences greatly the duration of hydrogen production. However, the attempts to use immobilized chloroplasts or hydrogenase plus ferredoxin in hydrogen production systems have given negative results (22). The causes can be due to a low concentration of immobilized ferredoxin, coupling of the ferredoxin electron mediating site to the sepharose support, or coupling of the ferredoxin electron accepting site with the hydrogenase.

TABLE 4

HYDROGEN PHOTOEVOLUTION FROM THE CHLOROPLAST SYSTEM WITH VARIOUS ELECTRON CARRIERS AND HYDROGENASES

Variable Components of System[a]	Hydrogen Rate (μmoles hydrogen/hr/mg chlorophyll)
Cl. pasteurianum H-ase + *S. maxima* Fd*	94.6
Cl. pasteurianum H-ase + *S. maxima* Fd	3.4
Cl. pasteurianum H-ase + *D. ethylica* Fd	2.5
Cl. pasteurianum H-ase (cytochrome c_3) + *D. ethylica*	0.2
Cl. pasteurianum H-ase (Fd + cytochrome c_3) + *D. ethylica*	3.2
D. ethylica H-ase + *S. maxima* Fd	0.4
D. ethylica H-ase + *D. ethylica* Fd	0.7
D. ethylica H-ase + *D. ethylica* (cytochrome c_3)	7.3
D. ethylica H-ase + *D. ethylica* (Fd + cytochrome c_3)	19.0

(a)

Reaction mixture contained 0.1 mg chlorophyll in spinach chloroplasts and the following amounts of the other items: *Cl. pasteurianum* hydrogenase 0.005 mg; *D. ethylica* hydrogenase 0.04 mg; *S. maxima* ferredoxin (Fd) 50 nmoles in * and 5 nmoles in others; *D. ethylica* Fd 5 nmoles; *D. ethylica* cytochrome c_3 5 nmoles.

IV. STABILITY AND SPECIFICITY FOR ELECTRON CARRIERS OF ENZYMES PARTICIPATING IN HYDROGEN PRODUCTION.

Nitrogen fixation phototrophic microorganisms produce two enzymes which participate in the evolution of molecular hydrogen (11,12,14), i.e. hydrogenase and nitrogenase. Nitrogenase catalyzes the light-dependent hydrogen evolution from exogenous or endogenous H-donors, while hydrogenase catalysis does not depend on the light or ATP reverse hydrogen oxidation-reduction.

One important approach towards increasing the hydrogen production efficiency consists of searching for hydrogenases stable to temperature and oxygen. The hydrogenases most stable to oxygen have been purified from the cells of *Chromatium vinosum* (39), *R. rubrum* (38), *T. roseopersicina* (40) and *Alcaligenes eutrophus* (41,42). However it is possible that these hydrogenases are reversibly bound with oxygen; and during the analysis of their activity in the presence of NADH or dithionite and methyl viologen the oxygen connected with the enzyme is removed and the enzymes become active (Table 5). However, such type of hydrogenase reduction is impossible apparently in the illuminated system of chloroplasts, where the enzyme is inactivated sufficiently by the oxygen evolved during photosynthesis. It also has been shown that *Al. eutrophus* Z-1 (43) is able to evolve hydrogen from NAD(P)H under air and even in the presence of 100% oxygen. However, the conditions for hydrogen evolution by this hydrogenase in the system with chloroplasts have not been determined yet.

The resistance of hydrogenases to heating is an important stability parameter. Hydrogenases from purple bacteria *T. roseopersicina*, *Rh. capsulata*, and cyanobacteria *Sp. platensis* (Table 5) are more stable in this respect (44). Such thermostable hydrogenases are likely to be of greater value for practical purposes, in particular for biochemical fuel elements or in systems of large-scale hydrogen production.

Table 5

HYDROGENASE STABILITY FROM DIFFERENT MICROORGANISMS

Microorganism	Half Life * (hr)	Useful Temperature (°C)	Optimum Temperature (°C)
Clostridium pasteurianum	0.5	30	20
Desulfovibrio vulgaris	72.0	50	38
Chromatium vinosum	-	70	-
Thiocapsa roseopersicina BBS	144.	80	75
Rhodospirillum rubrum	148.	70	55
Rhodopseudomonas capsulata	120.	75	70
Anabaena variabilis	28.0	60	50
Spirulina platensis	48.0	75	70
Chlorella vulgaris K	120.	65	60
Chlamydomonas reinhardi	0.1	40	30
Alcaligenes eutrophus Z-1	110.	40	30

*Enzyme Inactivation

REFERENCES

1. BISHOP, N. I., FRICK, M. & JONES, L. W. In "Biological Solar Energy Conversion" (A. Mitsui, Sh. Miyachi, A. San Pietro, and S. Tamura, eds.) Academic Press, New York 1977, p. 3.
2. KONDRATIEVA, E. N. & GOGOTOV, I. N. *Izvestija Akad. Nauk SSSR Biology Ser. 1*: 69, 1976.
3. KESSLER, E. In "Algal Physiological Biochemistry" (W. S. P. Steward, ed.) Blackwell, Oxford, 1974, p. 456.
4. HUNGANTE, R. E. *Arch. Microbiol. 59*: 158, 1967.
5. EFIMCEV, E. I., BOJCHENKO, V. A. & LITVIN, E. F. *Dokladi Akad, Nauk SSR 220*: 986, 1975.
6. WOOD, H.G. In "The Bacteria. A Treatise on Structure and Function" (I. C. Gunsalus and R. Y. Stanier, eds.) Academic Press, New York, 1961, p. 63.
7. BLARCHARD, G. C. & FOLEY, R. T. *J. Electrochem. Soc. 116*: 1232, 1971.
8. GRAY, C. T. & GEST, H. *Science 148*: 186, 1965.
9. LE GALL, J. & POSTGATE, J. R. *Adv. Microbiol. Physiol. 10*: 82, 1973.
10. HILLMER, P. & GEST, H. *J. Bacteriol. 129*: 732, 1977.
11. GOGOTOV, I. N. *Biochimie 60*: 267, 1978.
12. KONDRATIEVA, E. N., GOGOTOV, I. N. & GRUZINSKIJ, I. V. *Mikrobiologija*, in press.
13. BENEMANN, I. R. & WEARE, N. M. *Science 184*: 174, 1974.
14. BOTHE, H., DISTLER, E. & EISBRENNER, G. *Biochimie 60*: 277, 1978.
15. KOSIAK, A. V., GOGOTOV, I. N. & KULAKOVA, S. M. *Mikrobiologija 37*: 605, 1978.
16. GAFFRON, H. In "Plant Physiology" Academic Press, New York, 1960, p. 3.
17. BISHOP, N. I. *Ann. Rev. Plant Physiol. 17*: 185, 1966.
18. OSCHEPKOV, V. P. & KRASNOVSKIJ. A. A. *Izvestija Akad. Nauk SSSR Biology Ser. 1*: 87, 1976.
19. GEST, H. *Adv. Microbiol. Physiol. 7*: 243, 1972.

20. YOCH, D. C. & VALENTINE, R. C. *Ann. Rev. Microbiol* 26: 139, 1972.
21. TAUER, R. K., KAUFER, P., ZAHRINGER, M. & JUNGERMANN, K. *Eur. J. Biochem.* 42: 447, 1974.
22. RAO, K. K., GOGOTOV, I. N. & HALL, D. O. *Biochimie* 60: 291, 1978.
23. BELL, G. R., LEE, J.-P., PECK, H. D. & GALL, J. LE *Biochimie* 60: 315, 1978.
24. BENNET, R., RIGOPOULOS, N. & FULLER, R. C. *Proc. Nat. Acad. Sci. U.S.A.* 52: 762, 1964.
25. GOGOTOV, I. N. In "Abstracts of Symposium on Prokariotic Photosynthetic Organisms", Freiburg, 1973, p. 118.
26. ARNON, D. I. *Science* 149: 1460, 1965.
27. GOGOTOV, I. N. & LAURINAVICHENE, T. V. *Mikrobiologija* 44: 581, 1975.
28. LEACH, C. K. & CARR, N. G. *Biochim. Biophys. Acta* 245: 165, 1971.
29. OSCHEPKOV, V. P., NIKITINA, K. A., GUSEV, M. V & KRASNOVSKIJ, A. A. *Dokladi Akad. Nauk SSSR* 213: 739, 1973.
30. HEALEY, F. P. *Planta* 91: 290, 1970.
31. PERSANOV, V. M. & GOGOTOV, I. N. *Mikrobiologija* 47: 212, 1978.
32. STUART, T. S. In "Proceedings of the Workshop on Bioconversion of Solar Energy", Bethesda, Maryland, USA, 1973, p. 45.
33. ABELES, F. B. *Plant Physiol.* 39: 169, 1964.
34. BENEMANN, I. R., BERENSON, I. A., KAPLAN, N. O. & KAMEN, M. D. *Proc. Nat. Acad. Sci. U.S.A.* 70: 2317, 1973.
35. RAO, K. K., ROSA, L. & HALL, D. O. *Biochem. Biophys. Res. Comm.* 68: 21, 1976.
36. FRY, I., PAPAGEORGIOUS, G., TEL-OR, E. & PACKER, L. *Z. Naturforsch.* 32: 110, 1977.
37. PERSANOV, V. M., GOGOTOV, I. N., GINS, V. K. & MUCHIN, E. N. *Fiziologija Rastenij* 24: 699, 1977.
38. ADAMS, M. W. W. & HALL, D. O. *Biochem. Biophys. Res. Comm.* 77: 730, 1977.
39. GITLITZ, P. H. & KRASNA, A. I. *Biochemistry* 14: 2561, 1975.
40. GOGOTOV, I. N., ZORIN, N. A. & KONDRATIEVA, E. N. *Biokhimija* 41: 836, 1976.

41. SCHNEIDER, K. & SCHLEGEL, H. G. *Biochim. Biophys. Acta 452*: 66, 1976.
42. PINCHUKOVA, E. E., VARFOLOMEEV, S. D. & BEREZIN, I. V. *Dokladi Acad. Nauk SSSR 236*: 1253, 1977.
43. ZORIN, N. A., GOGOTOV, I. N. & KONDRATIEVA, E. N. *FEMS Lett.*, in press.
44. GOGOTOV, I. N. *Uspechi Mikrobiologii 14*: 3, 1979.

SPATIALLY STRUCTURED ENZYME SUPPORT ARRANGEMENTS IN ELECTROCHEMICAL SYSTEMS

L. B. Wingard Jr.

University of Pittsburgh Medical School
Pittsburgh, Pennsylvania, USA

Numerous ideas have been developed and others have been proposed for combining electrochemistry and immobilized enzymes to produce unique in vitro processes of possible use in energy transfer, analytical chemistry, and organic synthesis. Some recent and pertinent reviews have been published elsewhere (1-3). Early in vitro enzyme catalyzed energy transfer studies utilized microbial cells or isolated enzymes in solution or attached loosely to electron conducting supports; but the results of these studies showed rather poor electron transfer efficiencies (1,4,5). In analytical chemistry a wide variety of enzyme electrodes or immobilized enzyme assay techniques have been devised, often with polarographic or potentiometric readout (1,2). While in organic syntheses, the electrochemical regeneration of enzyme oxidation-reduction cofactors holds promise for the development of complex enzyme-catalyst reaction systems (6). In most of these studies little attention has been given to the relative spatial arrangement of the immobilized enzyme, the support matrix, and the electron conducting electrode material. However, some of the more recent studies have begun to explore the possibilities inherent in constructing specific relative spatial arrangements for the enzyme, the matrix, and the conducting electrode material. The purpose of this paper is to describe several immobilized enzyme systems where the spatial arrangement of the components may be an important factor in the overall electrochemical performance of the process.

I. SPATIAL CONSIDERATIONS IN ENZYME CATALYZED ENERGY TRANSFER

Two energy transfer examples where the relative spatial arrangement of the enzyme, matrix, and electrode material appear to be important are enzyme-catalyzed fuel cells and the electrochemical production of adenosine triphosphate (ATP). The fuel cell example is discussed first.

A typical fuel cell is shown schematically in Fig. 1, with A and C representing the anodic and cathodic half-cell reactions, respectively, and W_u representing the external work obtained from the electrons. Note that no potential difference is applied between the two half cells; instead, the driving force for the reactions comes from the net change in Gibbs free energy between the reactants and the products, as described in Fig. 2. The useful available free energy (E_a - E_c) is less than the reversible free energy due to irreversibilities, called overpotentials, for activation, ohmic, and concentration factors. The relation between these three types of overpotential and the current density is summarized in Fig. 3.

The activation overpotential is due to slow chemical reaction at the electrode surface or to slow charge transfer across the electrical double layer at the solid electrode surface. Compounds, such as methyl viologen and methylene blue, have low activation overpotentials (high electroactivity), while enzyme cofactors, such as nicotinamide adenine dinucleotide (NAD) and flavin adenine dinucleotide (FAD), have relatively high activation overpotentials under the experimental conditions reported in the literature. For enzyme catalyzed oxidation reduction reactions, it may be possible to reduce the cofactor activation overpotential through covalent attachment of the cofactor to the electrode surface so as to give a pathway for easy

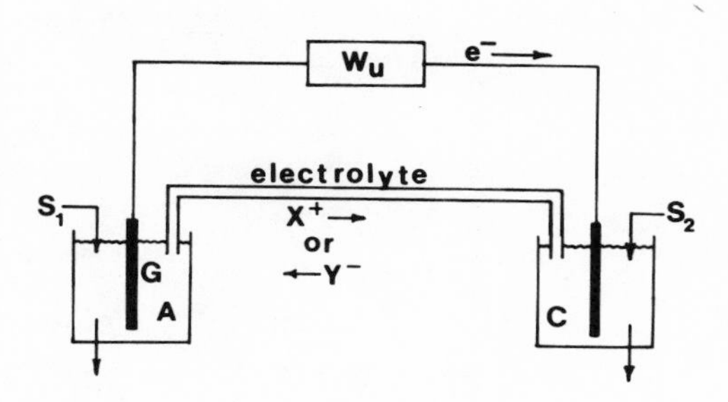

Fig. 1. Fuel cell schematic, showing anode A, cathode C, substrates S, glucose oxidase G on anode, electrolyte ions X or Y, and external work W_u.

transfer of electrons from the cofactor to the electron conducting surface of the electrode. This approach is discussed later for flavin cofactors.

The ohmic overpotential, due to the resistance of the electrolyte, does not appear subject to reduction by the use of enzymes. However, the concentration overpotential may be subject to reduction by the use of cofactor modified electrodes or by the use of special semi-conducting media between the enzyme-cofactor-substrate complex and the solid electrode surface; both of these approaches are discussed later.

Several possible spatial arrangements for the apoenzyme, the cofactor, and the electron conducting solid electrode surface are shown in Fig. 4. In Case 1 the enzyme protein is shown covalently coupled to the solid electrode surface; this approach has not given very high current densities for use as a glucose oxidase fuel cell anode (7). For Case 2 the cofactor is shown covalently coupled to the electrode surface; this approach appears promising and is discussed more later.

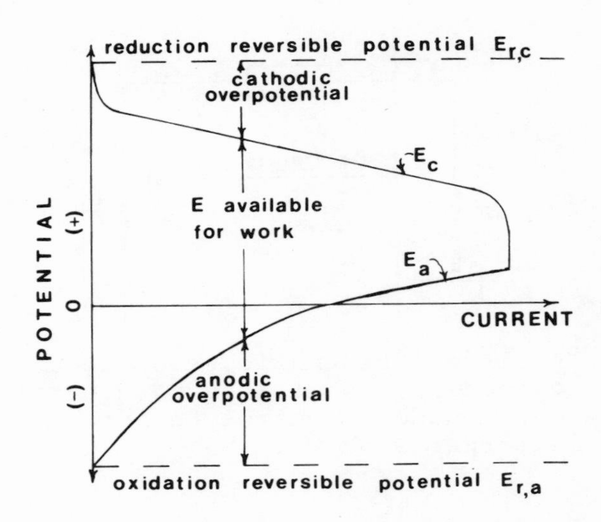

Fig. 2. Theoretical available free energy, expressed as the potential difference be tween the reversible cathodic (reduction) and anodic (oxidation) half cell reactions, for fuel cell operation. The overpotentials increase with current, which lowers the usable (E_a - E_c) available free energy.

Case 3 involves the use of an electron conducting spacer between the electrode surface and the cofactor and thus is similar to Case 2. In Cases 4 and 5 either the cofactor (Case 4) or the cofactor-apoenzyme (Case 5) must move from the bulk solution to the electrode surface. The latter two cases would be especially appropriate in situations where the cofactor was highly electroactive (low activation overpotential). Since NAD, FAD, and maybe other cofactors are not very electroactive, charge transfer agents, such as methyl viologen, have been suggested as mediators between the cofactor and the electrode surface. Oxidized mediator and reduced cofactor must react to produce reduced

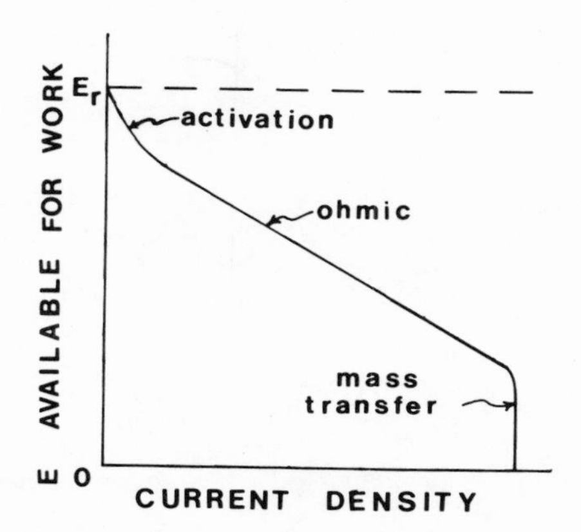

Fig. 3. Summary of how the useful potential difference (free energy) is used up as the current density increases in a fuel cell.

mediator, which then must undergo diffusional transport to the electrode surface for subsequent re-oxidation and diffusional transport back to the cofactor-apoenzyme complex. Another version of Cases 4 and 5 can involve the use of a semi-conductor medium between the reduced cofactor, or reduced mediator, and the electrode surface; however, one of the species still must undergo diffusional transport from the enzyme active site to the semi-conductor medium.

We have been using the glucose oxidase-FAD cofactor system as a model for the study of enzyme-cofactor fuel cells since the flavin molecules provide an opportunity for a novel approach to the construction of cofactor modified electrodes. Glucose oxidase catalyzes the oxidation of β-D-glucose to glucono-δ-lactone and the parallel reduction of FAD to $FADH_2$. The flavin cofactor is bound tightly to the apoenzyme (8); so there is little chance that appreciable quantities of

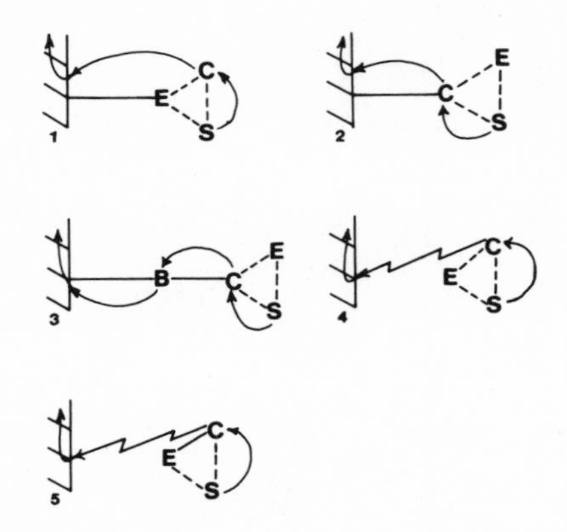

Fig. 4. Some possible spatial arrangements for apoenzyme (E), cofactor (C), substrate (S), bridging compound (B), and electron conducting electrode material (slant lines at left). Dashed lines represent weak binding of complex formation; solid lines between compounds represent covalent bonds; and zigzag lines represent diffusional transport. Other solid lines with arrows designate electron transfer.

$FADH_2$ would be free for diffusional transport to the electrode surface. In such a system it would be preferable to bind the flavin cofactor directly to the electrode surface so that only the substrate (glucose) and product (glucono-lactone) were required to undergo diffusional transport. The glucose oxidase apoenzyme should stay bound to the immobilized flavin cofactor; and the immobilized FAD-apoenzyme complex should retain enzyme activity. The retention of activity is expected since several natural flavin enzymes are known in which the flavin is covalently bound to the apoenzyme, through the position 8 methyl group of the isoalloxazine ring system (9,10).

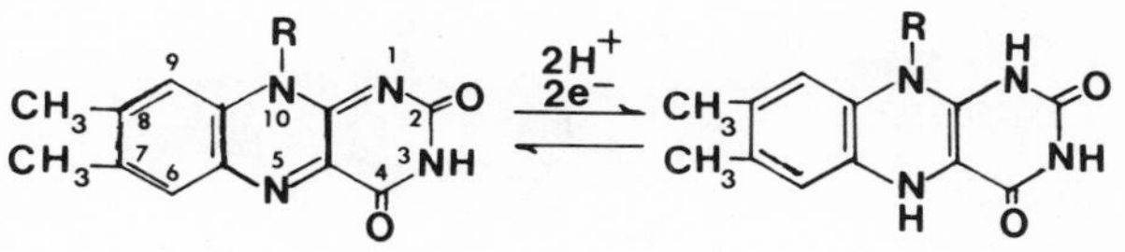

Fig. 5. Electron and proton receptor portion of flavin. R stands for the remainder of the riboflavin, FAD, or other flavin compound.

The number 8 position of the isoalloxazine ring provides a convenient location for attachment of flavin compounds to electrode surfaces. Actually, coupling through position 8 provides a unique pathway of overlapping π orbitals between the electron acceptor portion of the flavin molecole and the electrode surface. The electron acceptor portion of the flavin molecule is shown in Fig. 5. Since the phosphate linkage in FAD is rather labile, we elected to do our initial studies with riboflavin. In Fig. 5 the R consists of a ribosyl group for riboflavin.

The aim of the initial work was to couple riboflavin to a solid carbon electrode with the double bond linkage shown in Fig. 6. With this Fig. 6 linkage, the electrons accepted during reduction of the flavin should undergo enhanced delocalization, as compared to the non-coupled flavin. The expectation is that the electrons will delocalize more readily onto the external electrical circuit, leaving the flavin in the oxidized state. This assumes, of course, the presence of sufficient buffering capacity to take up the protons released during the oxidation of glucose and reduction of FAD.

Our method for synthesizing the linkage shown in Fig. 6 was based on converting the 8-methyl group of riboflavin to a triphenylphosphonium salt

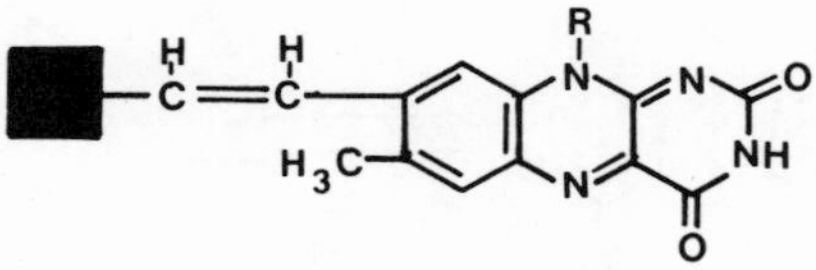

Fig. 6. Desired linkage of riboflavin to solid carbon electrode through position 8 to give overlapping π orbitals. Black square represents solid carbon electrode.

for coupling by a Wittig reaction to the surface of carbonyl derivatized glassy carbon. The overall reaction sequence is shown in Fig. 7 (11). Glassy carbon was derivatized by sequential treatment with nitric and sulfuric acids, thionyl chloride, and lithium tritertbutoxy aluminum hydride. Riboflavin was acetylated to protect the ribosyl hydroxyl groups during subsequent bromination. The active position 8 methyl group underwent selective bromination, followed by conversion to the 8-triphenylphosphonium bromide derivative of riboflavin. In strong base the phosphonium salt underwent rearrangement to the corresponding ylide, which was then coupled immediately to the derivatized glassy carbon via the Wittig olefin reaction (12).

The riboflavin, tetraacetylriboflavin, and bromotetraacetylriboflavin compounds were verified by comparison with literature NMR spectra at 60 and 250 MHz. The NMR spectra of the triphenylphosphonium riboflavin derivative was obtained; however, no spectra was found in the literature for this compound. Cyclic voltammetry measurements of the riboflavin-glassy carbon product showed peaks at -0.16 and -0.44 V, with reference to a standard hydrogen electrode at pH 7 and 50% reduction. The standard electrochemistry

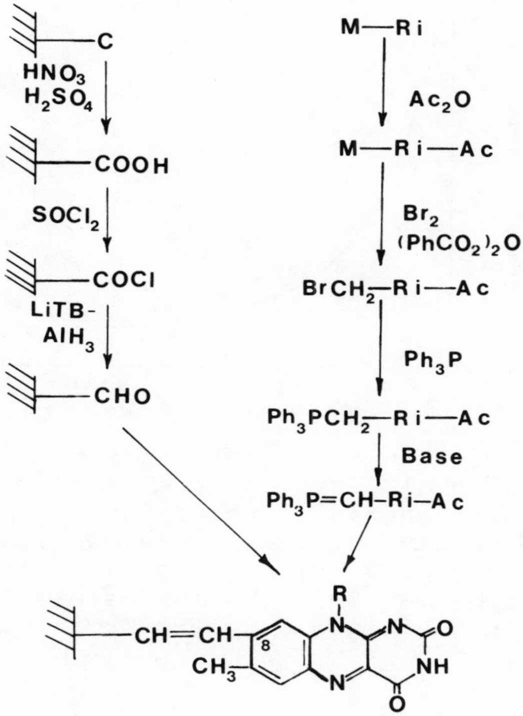

Fig. 7. Reaction sequence for carbonyl derivatization of glassy carbon surface (left side) and for phosphonium salt derivatization of riboflavin (right side) and subsequent Wittig coupling. M stands for the 8-methyl group and Ri for the remainder of riboflavin. Ac_2O represents acetic anhydride; other reagents are mentioned in the text.

notation for this redox state is designated E_o'. Literature E_o' values for riboflavin and some derivatives of riboflavin range from -0.14 to -0.22 V; however, a larger shift in E_o' might be expected

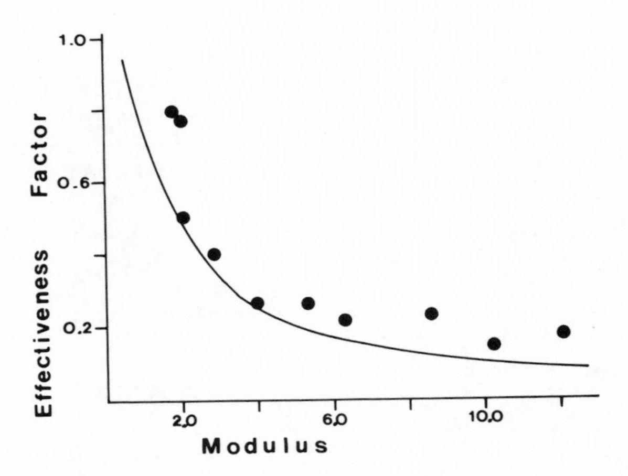

Fig. 8. Effectiveness factor for the yeast alcohol dehydrogenase catalyzed oxidation of ethanol and reduction of NAD as a function of the Thiele Modulus for NAD transport through the immobilized enzyme matrix. The enzyme was immobilized in a glutaraldehyde-albumin matrix.

for an extended π orbital system at position 8 (11). Additional work remains to be done to increase the concentration of bound riboflavin, to identify more completely the bound material, and to determine the activity of the bound cofactor in the presence of an apoenzyme for which riboflavin serves as the cofactor. Subsequent studies are planned for more complex forms of flavin cofactors, including FAD. However, the immobilized riboflavin serves as a model compound for the spatially oriented cofactor-apoenzyme-substrate approach of Fig. 4 Case 2.

In the alternative approach to the spatial arrangement of electrode components, where the cofactor or a charge transfer agent must move by diffusion from the enzyme active site to the electrode surface (Fig. 4 Cases 4 or 5), it is important to examine the degree of diffusional resistance for different enzyme-cofactor systems and different immobilization matrices. For example, we have measured the diffusional resistance for the transport of NAD through a matrix of yeast alcohol dehydrogenase (YADH) immobilized in a glutaraldehyde-albumin matrix (13). The diffusional parameters were expressed in terms of the Thiele Modulus, M, defined as:

$$M = L \left(E/(D_c\, k_c) \right)^{0.5} \qquad \text{(Eq. 1)}$$

where L is the thickness of the matrix, E the enzyme concentration in the matrix, D_c the effective diffusion coefficient of NAD within the matrix, and k_c a kinetic constant for the YADH catalyzed oxidation of ethanol. The influence of diffusional resistance on the velocity of the YADH catalyzed reaction was expressed in terms of the effectiveness factor. This latter variable is defined as the ratio of the measured velocity to the velocity in the absence of diffusional effects. The results, shown in Fig. 8, indicate that the reaction velicity was decreased markedly for even small values of the NAD modulus. Thus, the glutaraldehyde-albumin-YADH matrix used in this study would not be a good choice for an enzyme electrode system that required the diffusional transport of NAD through the matrix.

In order to minimize diffusional resistances for transport between the enzyme active center and the electrode surface, the small molecular weight compound methyl viologen has been used as a charge transfer agent. This has been studied with the enzyme hydrogenase, both immobilized and free in solution (3,14,15). In an alternative approach, the same authors have sought to reduce the length of the diffusional path through adsorption of the enzyme in a semi-conducting gel. This gel is a copolymer of poly-1-propargylpyridinium bromide (PPGP) and tetracyanoquinedimethane (TCNQ) (3). Once the reduced cofactor molecules reach the PPGP or TCNQ groups and undergo oxidation, the released electrons can travel along the conjugated PPGP chain or move by transport of the TCNQ groups.

The second energy transfer example, where the relative spatial arrangement of the enzyme, the matrix, and the electrode material appear to be important factors, is the enzyme-catalyzed electrochemical production of ATP (16). In vivo, the synthesis of ATP is presumed to occur by the ATPase catalyzed reaction of adenosine diphosphate (ADP) with orthophosphate (Pi) to form ATP and water. The energy to drive this synthesis comes from an electrochemical gradient across the enzyme membrane, according to the chemiosmotic theory of Mitchell (17,18). In this theory the electrochemical gradient is assumed to arise from the sum of a pH gradient and an electrical gradient across the enzyme-containing membrane. It occurred to us that this approach might be used in vitro to prepare ATP from ADP and Pi. A suggested experimental scheme, shown in Fig. 9, provides for both a pH and an electrical gradient across an ATPase membrane as well as for countercurrent transmembrane transport of hydrogen ions and electrons, with the hydrogen transported as hydrogen atoms as suggested by Mandel (19). This experimental system has not been tested yet; however, the scheme serves as an example of the concept of spatially structured immobilized enzyme systems for possible in vitro use in practical energy transfer.

II. SPATIAL CONSIDERATIONS FOR A GLUCOSE OXIDASE ANALYTICAL ELECTRODE

A wide variety of enzyme electrodes for the in vitro determination of glucose concentrations have been described in the literature (for references see 20). Essentially all of these electrodes are designed to operate without any special needs as to the relative spatial arrangements of the enzyme, the immobilization matrix, and the solid electrode surface. In our work on immobilized glucose oxidase, we observed that a potentiometric

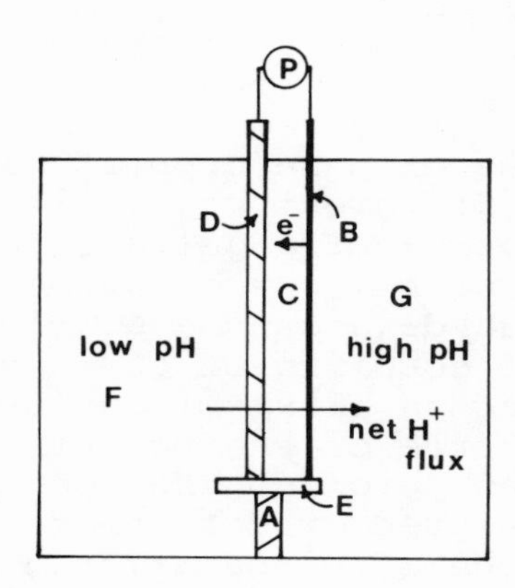

Fig. 9. Suggested experimental approach for the electrochemical synthesis of ATP from ADP and Pi in the presence of immobilized ATPase. Symbols: F, buffer at low pH; G, buffer at high pH, also contains ADP, Pi, and product ATP; P, potentiostat to apply voltage with essentially no current across membrane; C, semi-conductor gel and ATPase; B, porous carbon electrode containing gel and enzyme; D, palladium electrode, chosen because of high porosity to hydrogen atoms; A, membrane to pass negative ions; and E, an insulator.

response that varied with the glucose concentration was obtained when the enzyme was immobilized close to the surface of a platinum electrode. The potentiometric response was measured by placing the enzyme electrode and a reference Ag/AgCl electrode in the same beaker of buffered glucose solution and noting the steady state potential difference between the two electrodes (20). A very high impedance voltmeter was used in making the measurements, so that essentially no current was flowing between the two electrodes. The relationship between the potential difference and the logarithm of the glucose concentration was linear over the concentration range of 5 to 150 mg glucose/100 ml of solution.

The glucose oxidase support matrix was prepared in three different ways: 1) trapping the enzyme in polyacrylamide gel around a platinum screen, 2) crosslinking the enzyme and albumin with glutaraldehyde around a platinum-iridium wire, and 3) covalent coupling of the enzyme to a platinum screen using an alkylaminosilane and glutaraldehyde. The source of the potentiometric response was the hydrogen peroxide that was produced during the enzyme catalyzed oxidation of glucose. With all three methods of preparation of the immobilized enzyme, the matrix appeared to play a key role in establishing and maintaining a steady state concentration of hydrogen peroxide in the vicinity of the platinum surface (20-22). The value of the steady state peroxide concentration presumedly was different at each glucose concentration.

In the above system, hydrogen peroxide was produced continuously by the glucose oxidase catalyzed oxidation of glucose. Several pathways were available for the removal of the peroxide from the vicinity of the enzyme molecules within the matrix. Some of the hydrogen peroxide diffused into the bulk solution in the beaker; however, the peroxide level was very low since we were not able to detect any in the bulk solution. Additional hydrogen peroxide diffused to the platinum surface, where the platinum catalyzed decomposition of hydrogen peroxide would be expected to proceed quite rapidly. And finally, some of the hydrogen peroxide would have undergone thermal decomposition. The net result of these rate processes would be the attainment of a steady state hydrogen peroxide concentration within the enzyme support matrix. This in turn should establish a constant concentration gradient for the diffusion of the hydrogen peroxide to the platinum surface. At present, we suspect that the measured potentiometric response is created by the decomposition of hydrogen peroxide at the platinum surface; however, considerable research remains to be done to prove this. In the case of the enzyme covalently coupled to the platinum surface, the stagnant fluid layer adjacent to the platinum surface also may be a contributing factor in establishing the steady state level of hydrogen peroxide within the enzyme matrix vicinity.

The above results suggest so far that the spatial arrangement of the enzyme, the immobilization matrix, and the platinum support all are important factors in obtaining a potentiometric response that appears to be related to the concentration of glucose in a controlled manner. Our interest in studying this approach further lies in the possibility of adapting it for the continuous measurement of in vivo glucose levels in diabetic patients, without the need for removal of body fluids.

III. SUMMARY

The examples given in this paper are representative of the general concept that some unique uses for immobilized enzymes may be developed when the relative spatial arrangement of the immobilization environment and support are considered. This concept is in keeping with the ordered arrangements, for example of membrane bound enzyme sequences and electron transfer macromolecular complexes, found in vivo in subcellular particulate matter. A greater understanding of structured enzyme support systems certainly is needed before many novel uses can be expected; however, the glimpse of some of the possibilities serves as a challenge to pursue this matter with vigor.

ACKNOWLEDGMENTS

Portions of this work were supported by National Science Foundation grant ENG7516403 and

National Institute of Arthritis, Metabolism and Digestive Diseases grant 5R01AM19331.

REFERENCES

1. WINGARD JR., L. B. *Process Biochemistry 14:* 6, 1979.
2. GUILBAULT, G. G. & SADAR, M. H. *Accounts Chem. Res. 12:* 344, 1979.
3. BEREZIN, I. V. & VARFOLOMEEV, S. D. In "Enzyme Technology" (L. B. Wingard Jr., E. Katchalski-Katzir, and L. Goldstein, eds.) Academic Press, New York, 1979, p. 259.
4. WINGARD JR., L.B. *Hindustan Antibiotics Bull. 20:* 109, 1978.
5. WINGARD JR., L. B. & SHAW, C. H. In "Biotechnology of Electron Transfer" (E. K. Pye, ed.) Intl. Fed. Inst. Adv. Study, Solna, Sweden, in press.
6. AIZAWA, M., COUGHLIN, R. W. & CHARLES, M. *Biotechnol. Bioeng. 18:* 209, 1976.
7. LAHODA, E. J., LIU, C. C. & WINGARD JR., L. B. *Biotechnol. Bioeng. 17:* 413, 1975.
8. TSUGE, H. & MITSUDA, H. *J. Vitaminology 17:* 24, 1971.
9. SINGER, T. P. & EDMONDSON, D. E. *FEBS Lett. 42:* 1, 1974.
10. EDMONDSON, D. E. & SINGER, T. P. *FEBS Lett. 64:* 255, 1976.
11. WINGARD JR., L. B. & GURECKA JR., J. L., submitted.
12. TRIPPETT, S. *Quart. Rev. 17:* 406, 1963.
13. MILLIS, J. R. & WINGARD JR., L. B., submitted.
14. VARFOLOMEEV, S. D., YAROPOLOV, A. I., BEREZIN, I. V., TARASEVICH, M. R. & BOGDANOVSKAYA, V. A. *Bioelectrochem. Bioenergetics 4:* 314, 1977.
15. YAROPOLOV, A. I., Ph.D. Thesis, Moscow State University, 1978.
16. WINGARD JR., L. B., YAO, S. J. & LIU, C. C. *J. Mol. Cat.*, in press.
17. MITCHELL, P. *Biol. Rev. Cambridge Phil. Soc. 41:* 445, 1966.

18. MITCHELL, P. *FEBS Lett.* *78*: 1, 1977.
19. MANDEL, L. J. In "Modern Aspects of Electrochemistry," No. 8 (J. O'M. Bockris and B. E. Conway, eds.) Plenum Press, New York, 1972, p. 239.
20. LIU, C. C., WINGARD JR., L. B., WOLFSON JR., S. K., YAO, S. J., DRASH, A. L. & SCHILLER, J. G. *Bioelectrochem. Bioenergetics* *6*: 19, 1979.
21. WINGARD JR., L. B., ELLIS, D., YAO, S. J., SCHILLER, J. G., LIU, C. C., WOLFSON JR., S. K. & DRASH, A. L. *J. Solid Phase Biochem.*, in press.
22. WINGARD JR., L. B., SCHILLER, J. G., WOLFSON JR., S. K., LIU, C. C., DRASH, A. L. & YAO, S. J. *J. Biomed. Matls. Res.* *13*: 245, 1979.

APPLICATION OF IMMOBILIZED ENZYME SYSTEMS IN NONSILVER PHOTOGRAPHY

I. V. Berezin, N. F. Kazanskaya, and K. Martinek

Moscow State University
Moscow, USSR

All photographic processes are based on the ability of a material to change its properties, especially transparency and color, under the action of light. These changes which arise from photochemical conversions of light sensitive compounds, result in the formation of a photographic image. There are very many light sensitive compounds; but only a few have found application in photography. Examples are 1) silver salts that are reduced to silver metal by light, 2) complex salts of iron which are converted to divalent iron by light and in turn can reduce salts of other metals, 3) bichromates of alkaline metals that can tan a gelatin carrier under the action of light, 4) diazo compounds that undergo degradation under the action of light; 5) dyes that change their color under the action of light, and 6) monomers or low molecular weight soluble polymers that under the action of light are changed into insoluble high molecular weight compounds (1-3). As things are today, light signals can be recorded and converted into visible images in any region of the spectrum.

At the present time there are many light sensitive photographic processes that provide both semitone and total contrast images; but none of them are as good as the processes involving silver halogenides. As part of the search for new silverless photographic processes, we believe that enzymatic systems are of interest (10,11). Enzymes possess a catalytic activity that by many orders of magnitude exceed that of ordinary chemical catalysts (12,13). In addition enzymes have a unique macromolecular structure that is capable of being regulated, i.e. of changing its catalytic activity sharply under the action of even minute alterations in the environment (14). The problem of creating new silverless photographic processes with the use of biological catalysts involves two stages: learning to control enzymatic reactions with the help of light and using this to prepare a light sensitive material. Both of these stages are considered in this paper.

I. PRESENT PHOTOGRAPHIC PROCESSES

Classical diazo type materials possess high resolution capabilities, with the best giving an image of 1500 lines/mm (4). Moreover, the simplicity for development of the image and new possibilities for increased spectral sensitivity have expanded the range of application of diazo type materials (5,6). However, the light sensitivity of the classical diazo type materials has not been high. Their quantum yield is close to unity, which by Stark-Enstein's law of chemical equivalence means that each absorbed quantum of light converts no more than one molecule of light sensitive compound.

Systems in which absorption of a primary light quantum initiates a conjugated dark process, that ensures amplification of the primary signal, are free from the Stark-Einstein limitation. In principle, dark processes that follow the primary photochemical reaction are not always chemical (2,3,7). For example, degradation of diazo compounds under the action of light gives birth to bubbles of nitrogen, which on being heated dilate and being entrapped in a polymeric film, render the film nontransparent. However, as a result of such a physical process of amplification, the light sensitivity increases insignificantly and the resolution decreases drastically.

Materials in which light initiates a primary photochemical reaction that leads to either a catalytic or chain reaction possess a much higher sensitivity. In this case, the effective quantum yield can exceed unity by a significant degree. For example, the excellent light sensitivity of silver photographic materials is due primarily to the fact that atoms of silver, under the action of light, serve as catalytic centers around which crystalline grains form during a dark process of chemical or physical development. One such grain of silver can consist of up to 10^9 atoms, giving a rather high upper limit for the quantum yield of processes involving silver halogenides. Unfortunately, the resolution of silver materials is inversely proportional to their light sensitivity. On the other hand, the possibility of sensitizing silver halogenides by dyes has allowed one to extend the full range of registered wave lengths of silver materials up to 700 nm. Both these advantages have made silver materials extensively used.

However, silver is becoming more and more scarce, especially in the last decade. Therefore the search for new silverless photographic materials is quite necessary (3). New light initiated catalytic processes have been developed, based mostly on palladium salts (8); and metal complexes of alterable valency have been offered (9). Systems where the secondary dark process for light amplification sensitivity is a chain reaction also have come into use in recent years. This is the line along which the diazo type process has been developing. Recently the direction of photodegradation of diazo compounds has been shifted towards a free radical chain reaction with an effective quantum yield of about 10 (4).

Much greater success has been achieved in photopolymerization processes used for formation of relief images on the surface of a support (4,5).

For example, the Du Pont company photopolymer material for production of matrices for off-set printing reportedly gives a quantum yield of 10^6. The resolution is sufficiently high to allow submicron size relief to be made. Unfortunately, both the processes of photopolymerization and those with photoinduced initiation of metal ions as catalysts are very sensitive to oxygen and other contaminations, which interrupt the catalytic reaction and the propogation of polymeric chains. That is why a very high purity of the reagents at all stages is essential for these processes (5).

II. HOW AN ARTIFICIAL ENZYMATIC SYSTEM CAN RECEIVE AND AMPLIFY A LIGHT SIGNAL

The enzymatic approach (15-17) can be analyzed by use of a system that consists of a catalytically inactive derivative of an enzyme, E_{inact}, and a substrate of the enzyme. The inactive enzyme can be converted into an active catalyst, E, under the action of light. The system, if illuminated by light of a certain wavelength, will give rise to the following reactions:

Photochemical: $$E_{inact} \xrightarrow{\text{light}} E \qquad \text{(Eq. 1)}$$

Dark reaction: $$E + S \longrightarrow E + P \qquad \text{(Eq. 2)}$$

If Eq. 2 obeys the Michaelis equation, and the concentration of the substrate is sufficiently high to saturate the enzyme, the effective quantum yield of product P can be described by the following expression:

$$\Phi = \rho k_{cat} t \qquad \text{(Eq. 3)}$$

where k_{cat} is the catalytic rate constant for Eq. 2, as determined from the maximum reaction rate and equal to the turnover of the enzyme per unit of time, ρ is the quantum yield for Eq. 1, and t is the duration of the Eq. 2 reaction. With the given values of k_{cat} and ρ and with a sufficiently high concentration of substrate, the effective quantum yield can be extremely high, as it is determined only by the duration of the dark reaction.

Amplification of a light signal may be visualized in the following way: if a colored product, P, is formed as a result of Eq. 2, then just one quantum of light gives rise to such a great number of molecules of the colored substance that the latter can be seen with the naked eye.

Eq. 3 is a theoretical description of the operation of light sensitive enzymatic systems. To realize such performance experimentally, one must have enzyme derivatives, which are capable of undergoing alteration in activity under the action of light. In principle, one can use an enzyme involved in the process of vision, rhodopsin (18). But in practice this is hardly feasible, because natural light sensitive systems have a very fine and complex organization and are very unstable if withdrawn from their natural environment (19). Hence there arises the problem of creating artificial light sensitive enzymatic systems.

We have listed six general principles for creating artificial light sensitive enzymatic systems. The first principle is based on the use of light sensitive molecules that act as effectors. Erlanger and coworkers (20,21) in analyzing the photoregulatory processes occurring *in vivo*, arrived at the conclusion that all such processes have a common underlying principle, i.e. small light sensitive molecules play the role of effectors (inhibitors, activators) towards biologically active macromolecules, such as enzymes, which as a rule, are insensitive to light. Illumination induces a photochemical modification of the effector molecule, which in turn alters the character or degree of its interaction with the biomacromolecule. In applying this principle to creating artificial photochemical systems, one is able to draw on the many possibilities from present day photochemistry (22-24). In other words, one must screen light sensitive low molecular weight substances to find ones that will interact with enzymes.

The second and third principles for creating artificial light sensitive enzymatic systems are based on creating a system where light induces a change in the reactivity of the substrate or the enzyme, respectively. As a fourth principle, the action of light may be used to alter some property, such as pH, of the medium in which the enzymatic reaction takes place. The final two principles make use of light induced changes to the structure of a support membrane in which an enzyme has been immobilized (principle 5) or changes to the structure of a support so as to create new centers for subsequent immobilization of an enzyme (principle 6) (25).

The following sections contain a critical analysis of existing artificial light sensitive enzymatic systems described in the literature, in which the level of catalytic activity can be changed under the action of light, and in keeping with one of the above six principles. Of course, from a practical viewpoint the chemical amplification of weak signals can only be achieved where light initiates or gives birth to a catalytic activity. This is true of systems where the activity after illumination increases, for example, by three or four orders of magnitude. In other words, practicable systems are ones where a mixture of enzyme, substrate, low molecular weight effector, and medium gives no catalytic activity in the absence of light but gives a high level of catalytic activity after exposure to light. If under the action of light the change in enzymatic activity does not exceed one or two orders of magnitude, the action of the light may be called only regulatory. The difference between these regulatory and amplification systems is only quantitative; but we shall deal with each separately.

A. Systems With a Light Sensitive Effector

Light regulation of the rates of inactivation or reactivation of enzymes can be based on enzyme interaction with irreversible photochromic inhibitors. For example, diphenylcarbamyl chloride (Fig. 1, I) is known to be an irreversible inhibitor of α-chymotrypsin (26-28) and of acetylcholinesterase (29). Its photochromic azoderivative (Fig. 1, II), N-p-phenylazophenyl-N-phenylcarbamyl chloride, can exist in the trans and cis configurations. The equilibrium between the two forms may be shifted by illuminating II with light of a certain wavelength (20):

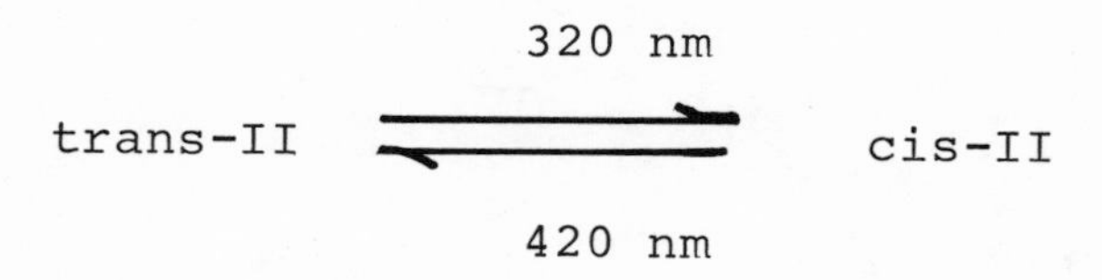

It turns out that cis-II inactivates α-chymotrypsin five times as fast as trans-II. Approximately the same difference in the rates of inactivation by the geometrical isomers of II also has been reported for acetylcholinesterase (30,31). Consequently, by illuminating the system consisting of the enzyme plus II by light of a longer or shorter wavelength, one may vary the ratio of the inhibitor geometrical isomers and thus regulate the rate of inactivation of the enzyme and catalytic activity of the system. In another example Goeldner and Hirth (32) reported the activation of acetylcholinesterase as a result of the light initiated conversion, at 300 nm, of a product resulting from the enzymatic hydrolysis of methyl(acetoxymethyl)nitrosamine. However, the nature of the product and of the irreversible inhibitor generated by the irradiation have not been elucidated. In a third example o-azidobenzoylphenylalanine ester (Fig. 1, III) can be regarded as a phenylalanine derivative analogue of α-chymotrypsin substrates (34). Therefore Stevens et al. (33) suggested that III could be bound to the active center of this enzyme. A biradical (Fig. 1, IV) is formed by the action of light on

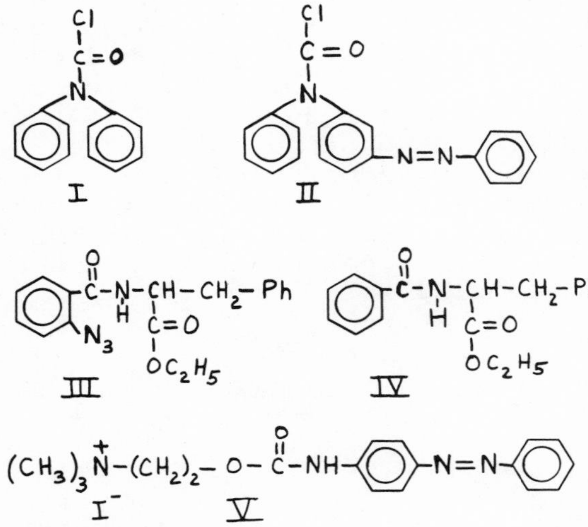

Fig. 1. Compounds relating to light sensitive effector molecules. I, diphenylcarbamyl chloride; II, N-p-phenylazophenyl-N-phenylcarbamyl chloride; III, o-azidobenzoylphenylalanine ester; IV, product formed by action of light on III; V, N-p-phenylazophenylcarbamylcholine iodide.

III, with the release of N_2. Thus, IV could react with a favorably localized nucleophile in the protein and thus inactivate the enzyme. But, as far as we know, no such investigation has been carried out.

The enzymatic activity also can be regulated by use of light sensitive reversible inhibitors. The cis, trans photostereoisomerization of azocompounds was used by Erlanger and coworkers (30,35) for the reversible regulation of the catalytic activity of acetylcholinesterase. N-p-phenylazophenylcarbamylcholine iodine (Fig. 1, V), a competitive inhibitor of this enzyme, exists in the cis and trans forms. The two forms can undergo photochemical interconversion as follows:

$$\text{trans-V} \underset{\text{darkness}}{\overset{\text{light (348 nm)}}{\rightleftharpoons}} \text{cis-V}$$

The values of the inhibition constants for the trans and cis isomers are 1.6 μM and 3.6 μM, respectively (35). This difference is due to the specificity of the structure of the active center of acetylcholinesterase (36). Using the difference in the binding ability of the stereoisomers with respect to the active center, one can, by switching on and off the light, reversibly change the degree of inhibition of the enzyme and hence the level of the catalytic activity of the enzymatic system.

In an example of the reversible regulation of the catalytic activity of α-chymotrypsin, we used (37) a different reaction. Proflavin (Fig. 2, VI), a competitive inhibitor of the enzyme (38, 39), under the action of visible light and in the presence of an organic reducer is capable of being photoreduced to form 3,6-diaminoacridane (Fig. 2, VII); in the dark VII reoxidizes to proflavin (40). We have demonstrated (37) that photoreduction of proflavin entails disappearance of its inhibiting ability towards α-chymotrypsin; probably due to distortion of the planar structure of the molecule so that the C_9 atom becomes tetrahedral. Consequently, when the system consisting of α-chymotrypsin plus proflavin plus a reducer like ascorbic acid is illuminated, the level of the catalytic activity increases, since the inhibitor becomes inactivated; in the dark the level of catalytic activity decreases.

The potential difference across a biological membrane can be regulated with the help of photochromic effectors. This is in keeping with the observation (41-43) that the potential difference across certain biomembranes, which contain acetylcholinesterase, decreases in the presence of agents which interact specifically with the enzyme. Such agents include acetylcholine and carbamylcholine. These observations were used (44,45) in creating a system for transmembrane potential photoregulation. As noted above, Erlanger and colleagues worked with irreversible (20,30,31) and reversible (30,35) photochromic inhibitors of acetylcholinesterase. Under the action of light one could change the ratio of the trans and cis isomers of the inhibitor; and by using their different reactivity towards the enzyme, the degree of inhibition of acetylcholinesterase could be varied.

Therefore, in systems with light sensitive effectors, one can really regulate the level of the catalytic activity over a wide range. However, for creating chemical amplifiers of weak signals this approach is far from sufficient. For amplifier purpose it is important that the light initiate the catalytic activity. This may be done if an effector, such as a cofactor or an activator, is created without which the enzyme would be unable to catalyze the reaction.

The initiation of the catalytic activity of heme containing enzymes by light is an example where an active species is created as a result of the primary light process and subsequent reactions. The majority of the heme containing oxidizing enzymes have in the active center a ferrous ion. For example, peroxidase is an enzyme whose molecule contains, in addition to a protein moiety, a hemin complex of ferrous ion with protoporphyrin. Neither apoperoxidase nor the complex of apoperoxidase with protoporphyrin possess enzymatic activity; and even the catalytic activity of hemin is very low. In our laboratory the principle of initiation of the activity of such enzymes with light was elaborated (46). One takes a mixture of apoperoxidase, photoporphyrin, and a ferrous salt such as potassium ferrioxalate that does not interact with protoporphyrin. Such a system should not possess enzymatic activity. When the complex salt is illuminated, the ferrous ion is liberated (47, 48) and can form a complex with protoporphyrin. The resulting ferroprotoporphyrin reacts with oxygen to create hemin, which in the presence of the apoenzyme gives peroxidase. Thus, an active enzyme can be created as a result of a primary photochemical process and subsequent dark reactions.

B. Systems with a Light Sensitive Substrate

This section is begun by describing several systems for light regulation of the rates of reaction of enzymes with synthetic substrates capable of photoisomerization. The first system involves esters of β-arylacrylic acids, since these may be regarded as structural analogues of specific substrates of α-chymotrypsin (34). The second order rate constants for the ester hydrolysis by α-chymotrypsin are much higher for trans isomers than for cis isomers, i.e. 5×10^3 times for p-nitrophenyl p-nitrocinnamate (Fig. 2, VIII) (49) and 10^3 times for p-nitrophenyl indolylacrylate (Fig. 2, IX) (50). Consequently, by illuminating the mixture of the enzyme and the cis isomer of the substrate, one may greatly accelerate the enzymatic process. A second system uses amides and esters of N-α-benzoly-L-arginine, which are classical low molecular weight substrates of trypsin (34). The reactivity of the stereoisomers of the azoderivatives, such as the methyl ester (Fig. 2, X) and hydroxyamide, of N-α-p-phenylazophenyl-L-arginine towards trypsin differ by several times (51). The cis and trans isomers interconvert under the action of light of different wavelength. This means that with the help of light, one may switch the rates of the enzymatic reaction.

The third system uses stereoisomerization around the C=C bond (52) to regulate with light the rate of α-chymotryptic hydrolysis of N-cinnamoyl-L-tyrosine ethyl ester (Fig. 3, XI). Compound XI may be regarded as an analogue of the classical specific substrate of α-chymotrypsin,

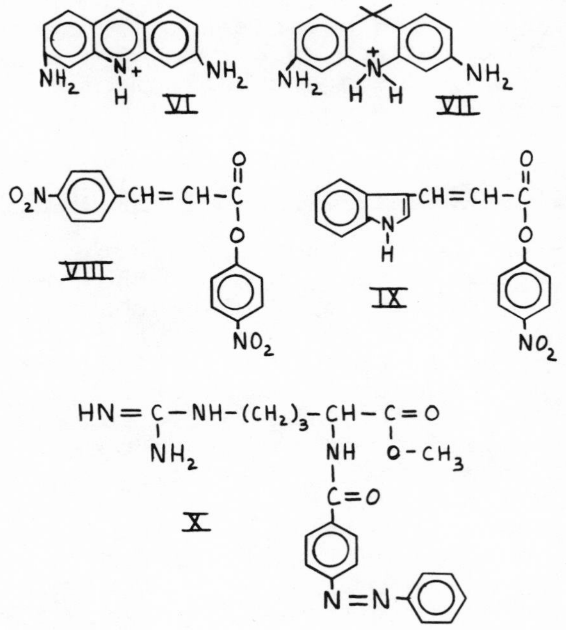

Fig. 2. Additional compounds relating to light sensitive effector molecules plus other compounds that act as light sensitive substrates. VI, proflavin; VII, 3,6-diaminoacridane; VIII, p-nitrophenyl+p-nitrocinnamate; IX, p-nitrophenyl indolylacrylate; X, N-α-p-phenylazophenyl-L-arginine methyl ester.

N-acetyl-L-tyrosine ethyl ester (34). The final system is based on the observation that the rates of hydrolysis of compounds of type XII, catalyzed by butyrylcholinesterase, differ by several times for the cis and trans isomers (53). This means that, by using photostereoisomerization of a substrate of type XII, one can regulate the rate of the cholinesterase reaction. Brestkin et al. (52)

believe that the rate limiting step of the enzymatic hydrolysis process is the desorption of the product alcohol from the active center. This is quite likely, as it is known that low molecular weight products can be desorbed from enzymes rather slowly (54, 55).

Another approach to the use of light sensitive substrates may be to examine the reaction of enzymes with photoexcited substrates. Molecules in their photoexcited states have entirely different chemical properties than do molecules in the ground states (56). Hence one may expect that, in principle, the reaction of an enzyme with a photoexcited substrate will have a different rate than the same reaction with a nonexcited substrate. This may have a bearing on the surprising results obtained by Comorosan and coworkers (57-59), who showed that low molecular weight substrates illuminated with low doses of UV light reacted with enzymes much more quickly than did nonexcited substrates. The effect strongly depended on the conditions of illumination; but no mechanism was suggested to explain the phenomenon.

As has been demonstrated above, with the help of light sensitive substrates one can not only regulate the rate but even initiate an enzymatic reaction. However, this is not sufficient for creating chemical amplifiers of weak signals; to accomplish this, one must not only initiate the reaction by activating the substrate but also must give rise to catalytic activity in the system. This may be attained by using quasi-substrates, as discussed below.

The concept of quasi-substrates is based on the use of compounds which can be convected at the enzyme active site to fairly stable intermediates. Since the intermediates are stable, they remain at the active site for extended periods of time and thus form a catalytically inactive enzyme. With the active site modified by the quasi-substrate, one may attempt to increase the enzyme reactivity by removing the intermediate with the help of light and thus regenerate the free active site. Our laboratory was the first to demonstrate the possibility of inducing catalytic activity in artificial enzymatic systems using light (15-17). It is known from the literature (34) that acylation of a serine hydroxyl in the active center of α-chymotrypsin completely inactivates this biocatalyst. Deacylation of the acylenzyme induces total regeneration of the catalytic activity. With this in mind, the active center of α-chymotrypsin was acylated with cis-cinnamoyl imidazole (Fig. 3, XIII). The trans stereoisomer of XIII is a well known titrant of α-chymotrypsin (60). Cis-cinnamoyl-α-chymotrypsin (Fig. 3, XIV) formed as a result of the acylation of chymotrypsin is a rather stable acylenzyme, possessing no catalytic activity. Even at the optimum pH, the rate constant of deacylation is 1×10^{-6} min^{-1} and the half-time of deacylation exceeds 24 hr. This time can be further increased by shifting away from the optimum pH. But when illuminated with UV light the cis acylenzyme is converted to the trans stereoisomer that deacylates rather quickly, with a half-time of conversion of about 1 min. Thereby the native enzyme is restored.

The reactivity of the cis and trans stereoisomers of acylenzymes differ by a factor of about 1,500. This difference can be increased still further by introducing a nitro group in the para position of the aromatic ring of the cinnamoyl radical (17). In this case trans acylenzyme deacylates 2.5×10^4 times as rapidly as the respective cis acylenzyme. The reasons for the trans specificity of α-chymotrypsin have been discussed earlier (50, 61).

The cis and trans isomers of another proteolytic enzyme, trypsin, show a somewhat less striking difference in reactivity (62-64).

Another approach was employed by Westheimer and coworkers (65,66) who suggested a photochemical method for studying the amino acid environment of the active center of an enzyme. As a result of

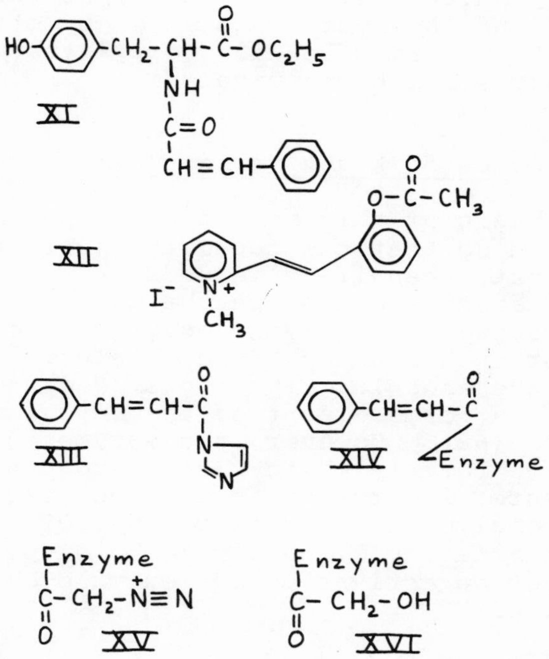

Fig. 3. Additional compounds pertaining to light sensitive substrates. XI, N-cinnamoyl-L-tryosine ethyl ester; XII, a cholinesterase substrate; XIII, cis cinnamoyl imidazole; XIV product of reacting XIII with α-chymotrypsin; XV, diazoacetyl-α-chymotrypsin; XVI, hydroxyacetyl-α-chymotrypsin.

the interaction of α-chymotrypsin or trypsin with p-nitrophenyl diazoacetate, which is an analogue of the classical nonspecific protease substrate p-nitrophenyl acetate (67), there forms a diazoacetyl enzyme (Fig. 3, XV) which possess no catalytic activity. This diazoacetyl-α-chymotrypsin hardly ever deacylates within 48 hr at pH 6.5. Illumination of XV with UV light results in the cleavage of a nitrogen from the diazoacetyl residue with a carbene radical being formed. This radical hydrolyses; and the hydroxyacetyl-α-chymotrypsin (Fig. 3, XVI) formed readily deacylates with the active enzyme being regenerated.

C. The Action of Light on Enzymes

The basis of this approach is that an enzyme is modified so that the catalytic activity is completely lost; then light is used to remove the modifying agent and restore the catalytic activity. Papain, a proteolytic enzyme, is a handy object for such studies. The catalytic function of papain depends largely on the SH-group of cysteine, which forms a part of the active center of the enzyme (68). However, the enzyme is usually in an inactive form in which the SH-group of the active center is present as a mixed disulphide with an additional cysteine residue (69,70). For the catalytic activity of papain to be displayed, the enzyme must be activated, for example by breaking the S-S bond.

In principle, S-S bonds both in peptides and in proteins may be cleaved not only chemically but also photochemically (71-73). This was used by Dose and Risi (74) for photochemical activation of papain; they illuminated papain-S-S-cysteine with UV light and formed the native enzyme plus cysteine. This approach was modified in our work (75) where a disulphide dimer of the enzyme, papain-S-S-papain was obtained as a result of oxidation of active papain catalyzed by copper ions. If this dimer was illuminated by UV light, the S-S bond broke with two molecules of the catalytically active enzyme being formed. The quantum yield of photoactivation in both cases was 0.1-0.2.

Reactivation of enzymes also can be achieved by light stimulated cleavage of reduced pyridoxal 5'-phosphate-enzyme complexes. This photochemical reaction was observed (76) with aspartase transcarbamoylase, D-serine dehydratase, and tryptophanase. This approach was based on a Shiff base adduct being formed between pyridoxyl 5'-phosphate and a lysyl residue in the active site of the enzyme and then reduced with sodium borohydride to give an inactive derivative. This derivative was found to be light sensitive, with light irradiation entailing an almost quantitative regeneration of the active enzyme. Thereby total cleavage of the covalent modification occurred; and the original lysyl residue was regenerated from the reduced Shiff base. However, the pathways leading to the destruction of the pyridoxyl group are complicated. Unfortunately, no attempt has been made to elucidate either the photochemical mechanism or the nature of the final products.

Light regulation of the catalytic activity of enzymes also can be done with the help of photochronic effectors attached to the enzyme away from the active center. Use is made of photochromic polymers (77,78), which when illuminated change their structure (79-84). In these studies azodyes, for example compounds like II or V, were attached covalently or noncovalently to synthetic polymers or proteins. Although the trans configuration of these azodyes is the more stable form thermodynamically, on illumination there occurs trans to cis stereoisomerization. Such photochemical reactions cause a marked alteration in the structure of the polymer (79-81) and can be followed by changes in the viscosity of the polymer solution. Enzymes possess biological activity which is very sensitive to the conformation of the protein globule (14). From this point of view, it would be interesting to modify enzymes with photochromic reagents and to study the dependence of the catalytic activity on the light regulated structure of the biocatalyst.

This idea was first realized experimentally in our laboratory (52). α-Chymotrypsin was modified by coupling its tyrosine residues with p-diazonium salts of some derivatives of azobenzene. The resulting colored protein contained groups that were capable of cis to trans photostereoisomerization (Fig. 4, XVII).

Recently, Suzuki and coworkers (85,86) employed for the modification of α-amylase a compound of the spiropyran series (Fig. 4, XVIII), that is widely used for preparing photochromic polymers (77,78). Compound XVIII when illuminated with UV light underwent isomerization (Fig. 4, XVIIIa), which could be reversed by illuminating the solution with visible light or by letting the solution sit in the dark (22,87). The anhydride of spiropyran XVIII was prepared (85) and then used to modify the amino groups of α-amylase. The modified enzyme when illuminated with near UV light, lost some of its catalytic activity; in the dark the enzyme activity was restored. The mechanism of regulation in these systems is based on the fact that under the action of light the molecular structure of the modifier changes.

Enzyme molecules usually contain some chromophores of their own, for example aromatic amino acid residues, capable of effective light absorption. In some cases such absorption of light may induce activation of the biocatalyst. Thus, UV light illumination of an aldolase solution causes a reversible increase in enzymatic activity (88, 89). A similar effect, but with aldolase illuminated in the presence of a substrate, was described elsewhere (90) An interesting explanation was offered by suggesting that some of the light absorbed by the tryptophan residues was converted without radiation into heat, which may have caused distortion in the structure of the solvatational shell of the macromolecule. This in turn might facilitate the allosteric conformational transition of the enzyme subunits to give a conformation of aldolase with a high catalytic activity. The data of Montagnoli and coworkers (91,92) agree with this hypothesis. They differed from other works (88-90) in that primary light acceptors were not the chromophores of the protein but residues of diazotized aminobenzoic acid covalently bound to

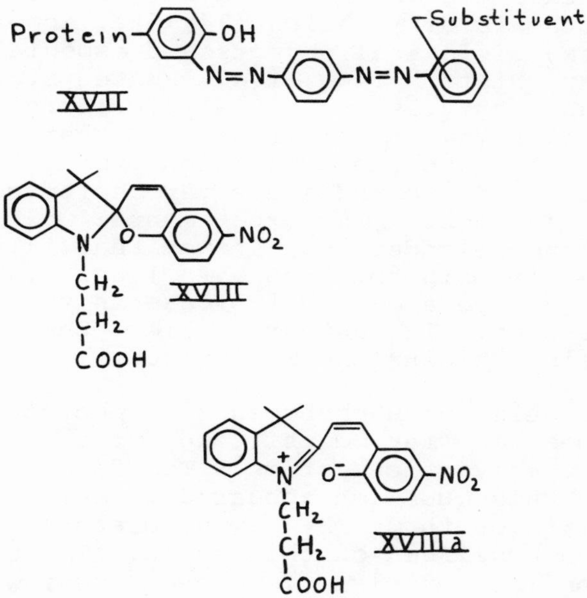

Fig. 4. Compounds used in photochanges of enzymes or the enzymatic reaction medium, XVII, chymotrypsin tryosine residues coupled to diazonium salts of azobenzene derivative; XVIII, spiropyran type compound; XVIIIa, product from UV illumination of XVIII.

the SH- groups of aldolase. Such modification should, obviously, not be a great influence on the photoactivation properties of the enzyme, as follows from the mechanism of Volotovski (90). This proved to be the case, in that illumination of modified aldolase caused a change in the conformation of the enzyme as noted by alteration of the absorption spectra and electrophoretic mobility (92). The kinetics of the enzymatic cleavage of fructose-1,6-disphosphate also were changed (91,

93). Thus, the mechanism of Volotovski (90) may hold for different enzymes.

Another mechanism can be suggested, based on the hypothesis of Frohlich (94,95). According to this idea, biological objects and especially proteins may exist in long lived excited states. In the case of enzymes, the catalytic activity of these states may be different from that for the ground state; and the transition from the ground state into the excited state may be induced, for example, by light. The experimental work of Kollias and Melander (96) showed that illumination of a chymotrypsin solution with laser light of 457.9 nm caused a marked increase in the activity of the enzyme. The authors believe that Frohlich's hypothesis explains the phenomenon.

The well known phenomenon of photoinactivation of enzymes also may be utilized for creating light sensitive enzymatic systems. The first works dealing with this question appeared as early as the end of the 19th century. By now hundreds of papers have been published on this topic. Therefore we shall confine ourselves to a short review (97-102). UV light induced inactivation of enzymes is caused by photodestruction of certain amino acid residues of the protein globule, especially tryptophan and cysteine. The quantum yield of photoinactivation in this case is 0.001 to 0.1 and depends on the nature of the enzyme, its conformation, and the environment. Some flavin enzymes can be inactivated by visible (blue) light. Here the flavin cofactor plays the role of a sensitizer for the degradation of the apoenzyme (103). Also well known is the photodynamic effect, i.e. photooxidation of protein molecules sensitized by dyes (102-105). The mechanism of the photodynamic effect is that of transfer of molecular oxygen, dissolved in water, to the protein. An intermediate compound is formed that consists of a complex of a triplet excited molecule of the dye with oxygen. The photodynamic action is initiated with protein solutions illuminated and in the presence of dyes like eosine or methylene blue.

The quantum yield of enzyme photoinactivation is usually 0.001 to 0.1.

D. Light Induced Alteration of Enzymatic Reaction Medium

Almost all enzymatic reactions are rather sensitive to the pH of the solution; the rates of reaction change sharply as the pH is moved away from the pH optimum (12,106). This prompted the idea of regulating enzymatic reactions by photoinduced alteration of the pH of the medium (107). For this purpose we have used photochemical reactions that produce strong acids and bases (107). For example, a solution of o-nitrobenzaldehyde, when illuminated at 436 nm gives rise to phototautomerization; o-nitrobenzaldehyde converts to o-nitrosobenzoid acid with a quantum yield of 0.5 (22). In aqueous solution the resulting acid dissociates to form hydroxonium ion, which results in a lowering of the solution pH. Ammonia complexes of transition metal ions, when illuminated with visible light, undergo photoaquatation at quantum yields of 0.01 to 1.0 (108,109). The ammonia, released as a result of the reaction, will react with water to cause alkalinization of the solution. Oxalates of transition metal ions when illuminated with visible light also undergo a photochemical redox reaction with a quantum yield of about 1.0 (109). For example ferric oxalate is converted to ferrous oxalate plus oxalate and carbon dioxide (47,48); the carbon dioxide thereby reacts with water to produce a more acidified medium.

If we take an enzyme and substrate at a pH value where the rate of enzymatic reaction is negligibly small and add for example o-nitrobenzaldehyde described above, then the pH of the solution will change upon illumination. If the change is in the direction of the optimum pH for the enzyme, then the rate of the enzymatic reaction will increase. We verified this principle (107) with photoinduced reactions catalyzed by α-chymotrypsin, using nitrobenzaldehyde, (52) and by trypsin, using transition metal complexes (110).

E. Light Induced Alteration of a Membrane in Which an Enzyme is Immobilized

The idea is to entrap a photochromic effector into a membrane in which an enzyme has been immobilized. If the structure of the effector is altered under the action of light, the structure of the membrane itself may also be changed. The mechanism of such an indirect action on the immobilized enzyme will be very complicated. The microenvironment of the active center in which the enzymatic reaction proceeds may become different. Also, the distribution of the substrate between the bulk phase and the carrier, as well as within the carrier, may change to give an altered steady state concentration of substrate in the vicinity of the active center. And finally, the conformation of the protein active center may change. The systems described in the literature differ not only in the nature of the components used but first and foremost in the fact that both the enzyme and the effector can be incorporated into the membrane, either mechanically or covalently. But whatever the mode of immobilization, the light induced change in the catalytic activity is not great.

Chymotrypsin and azobenzene have been incorporated mechanically into a membrane of hexadecane, hexanol, and potassium oleate formed in a water and oil microemulsion (111). The azobenzene did not interact with the enzyme; yet a light induced change in the structure of this photochromic effector, due to cis-trans photostereoisomerization, altered the catalytic activity of the entrapped chymotrypsin.

In the work of Suzuki et al. (112,113) the anhydride of spiropyran XVIII was used to modify collagen amino groups. The modified collagen was formed into films containing entrapped urease or lactate dehydrogenase. Illumination of the films induced an alteration in the structure of the modified polymer, which led to a decrease in the catalytic activity of the entrapped enzyme. After storage in the dark, the photochemical reaction was reversed; and the activity of the enzymes went back to the initial level. The same results were obtained by Nakamoto et al. (114) with trypsin covalently immobilized on agarose gel modified with polyethyleneimine and the spiropyran compound XVIII. It also should be possible to use light to regulate protein-protein interactions such as complex formation between trypsin and pancreatic inhibitor incorporated into a gel membrane labelled with spiropyran compound XVIII (115).

F. Systems with a Support in Which Light Produces Centers for Enzyme Immobilization

In almost all the examples described above, the sensitivity to light is associated with the specificity of an enzymatic system. This means that each of the mechanisms depended strongly on the structure of the active center or on the character of how the center interacted with low molecular weight substrates or how it responded to alterations of the protein globule or the reaction medium. Hence, the method worked out for one enzyme may not hold for another system.

Recently, we suggested an entirely new approach (11). The idea is presented in Fig. 5. The functional group R on the support does not interact with the enzyme. Under the action of light, Group R undergoes chemical conversion to form a functional group X capable of interacting with the enzyme chemically or sorptionally. Thus, the surface of the support, only after illumination by light, is ready to be attached to the enzyme. The attached enzyme may be visualized if the system is designed to use a substrate that gives a colored product.

The evident advantage of this approach (11) is its applicability to any enzyme. Since rather

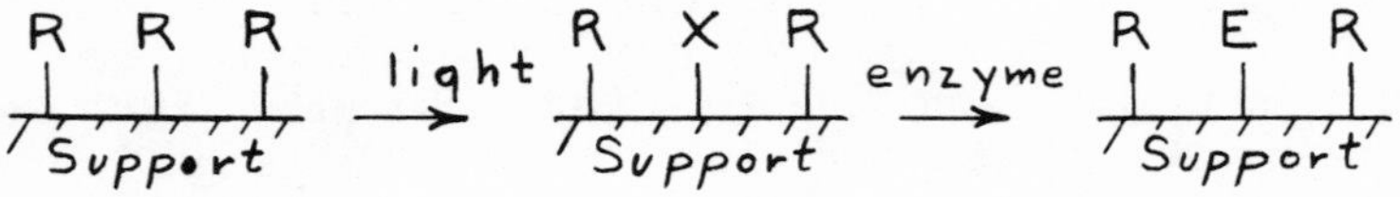

R R R light R X R enzyme R E R
Support Support Support

Fig. 5. Representation of scheme for light initiated production of sites for enzyme immobilization. Functional group R does not interact to bind the enzyme until photoconverted to group Z.

comprehensive reviews of methods of immobilization of enzymes on supports have been elaborated (116-118), we shall dwell here only on the photochemical methods which help endow the support with the ability of attaching the enzyme. However, the methods described in the literature have by no means been designed for creating chemical amplifiers of light signals. This latter idea is discussed elsewhere (11).

Guire (119,120) gave a description of the immobilization of asparaginase with the help of 1-fluoro-2-nitro-4-azidobenzene as the photochemical reagent (121). This compound was attached to alkylaminoglass as a result of substitution of fluoride by the nucleophilic amino groups of the support. The resulting modified support (Fig. 6, XIX), after exposure to light to cleave off nitrogen, yielded a rather reactive nitrene radical. If the enzyme was present during the illumination, the enzyme became covalently attached to the support via interaction with the nitrene radical.

Irradiation at 315 nm of another modified support (Fig. 6, XX), in the presence of acetoyl or acetyl benzoate sensitizers, enabled attachment of protein at the alpha carbon position of the glycine residue to give the structure shown in Fig. 6, XXI.

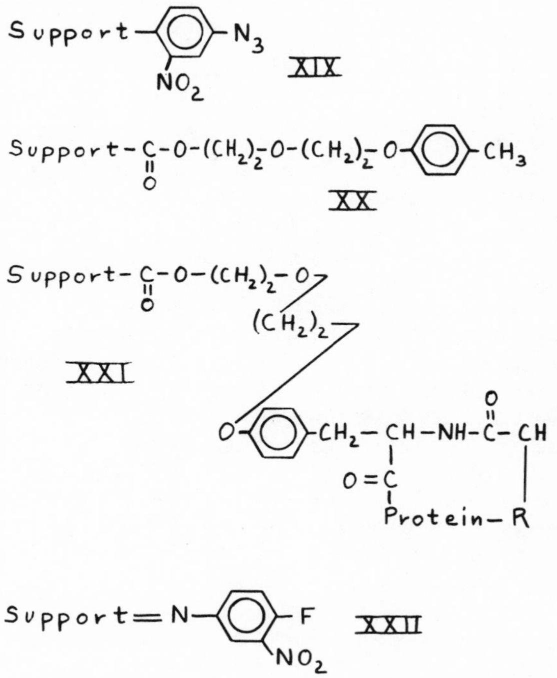

Support–$C_6H_3(NO_2)$–N_3 XIX

Support–C(=O)–O–$(CH_2)_2$–O–$(CH_2)_2$–O–C_6H_4–CH_3 XX

Support–C(=O)–O–$(CH_2)_2$–O–$(CH_2)_2$–O–C_6H_4–CH_2–CH(–C(=O)–Protein–R)–NH–C(=O)–CH XXI

Support=N–$C_6H_3(NO_2)$–F XXII

Fig. 6. Compounds associated with light activated support modification. XIX, product of 1-fluoro-2-nitro-4-azidobenzene coupled to alkylaminoglass; XX, another modified support; XXI, product of XX plus light, a sensitizer, and protein containing a lysine residue; XXII, product of reaction of a nitrene radical with the support.

In the above procedures, the enzyme as well as the support are irradiated. However, an approach suggested in our laboratory (11) is free from this disadvantage. An optically transparent support containing o-nitrobenzaldehyde groups underwent photochemical reaction (16) to give carboxy groups. Subsequent amine blocking of the remaining aldehyde groups, at nonilluminated sites, followed by carbodiimide activation of the carboxy groups, provided a suitable site for attachment of enzymes.

In another version, the photochemical immobilization of enzymes with 1-fluoro-2-nitro-4-azidobenzene may be performed either by attachment to a support, using nucleophilic substitution to an aromatic nucleus to give XIX, or by illuminating a mixture of 1-fluoro-2-nitro-4-azidobenzene with a support containing no nucleophilic groups. In the latter case, after illumination the nitrene radical becomes covalently attached to the support with XXII (Fig. 6) being formed. If then a solution of enzyme is added, the amino groups of the protein will interact with the nitrofluorophenyl residue of XXII; and the biocatalyst will be covalently immobilized. To enhance the rate of the latter reaction, 1-fluoro-2-nitro-4-azidobenzene may be substituted by more reactive derivatives (123), a whole series of which was synthesized by Guire (120,124).

Photoimmobilization of enzymes also can be carried out with the use of gel formation. For example, polyacrylamide gel, a classical gel for enzyme immobilization (125), and other polyacrylate gels may be prepared by illuminating the solutions of the respective monomers by visible light in the presence of riboflavin (126) or by ultraviolet light (127,128). It was suggested (129,130) that microparticles of enzyme containing gel could be made this way for use in chemical reactors. However, as was demonstrated by us (131), a thin layer of monomer solution applied to a plate and then illuminated will form a gel attached to the plate. However, any sites that were not illuminated are easily washed off with water. To the gel relief, thus formed, an enzyme may be attached covalently (131). An important feature of this method is that the light induced polymerization processes are characterized by high quantum yields, by far exceeding unity. This means that one quantum of light can in principle ensure immobilization of more than one molecule of an enzyme, thereby signifying possibilities for light signal amplification (131).

III. SILVERLESS PHOTOGRAPHIC PROCESSES BASED ON ENZYMES

It is commonly believed that the high cost of enzymes should limit their application, especially in traditional processes. This is not so, since their cost is constantly decreasing as the enzymes of animal and plant origin give way to microbial enzymes and the high catalytic activity of enzymes allows their use in low concentrations.

As to light sensitivity, the effective quantum yield of a light initiated enzymatic reaction can be high. In terms of Eq. 3, the quantum yield is estimated first by assuming that the quantum yield of a primary photochemical act that initiates a catalytic activity is close to unity. This is the case, for example, in the cis to trans photostereoisomerization of the nitrocinnamoyl label in the active center of chymotrypsin. In their natural environment the most common enzymes operate with a rate constant, k_{cat}, of 10^2 to 10^3 sec^{-1}; but there are biocatalysts with a k_{cat} value of 10^4 to 10^6 sec^{-1} (Fig. 7). Consequently, for a k_{cat} of 10^5 sec^{-1} and a time for the dark enzymatic reaction of the order of 10 min, the effective quantum yield will be 10^9. This value is close to the sensitivity of the silver photographic materials described earlier. The light sensitivity of enzyme systems can be increased further by using special approaches discussed later.

A high resolution also might be expected in the case of enzymatic photography since histochemical methods, based mainly on enzymatic reactions that stain the tissue microsections, allow one to discern objects 10^4 Å in size (132).

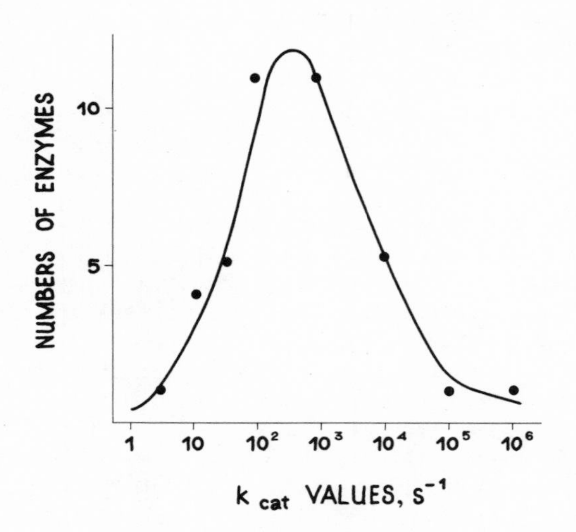

Fig. 7. A statistical analysis of the k_{cat} values for 50 native enzymes in their native environments (13).

A. Photographic Methods Based on Light Induced Enzyme Inactivation

Several modes of enzymatic production of photoimages have been patented. Almost all of them are based on the inactivation of enzymes by light. The active enzyme retained in the nonexposed sites reacts with a relevant substrate and forms a visible image. For such images essentially any enzyme could be used.

For development of the image with the help of a dye forming substrate, a mixture of enzyme, gelatine, and sensitizers is applied to a solid support and dried (133,134). Then the material is exposed, thereby inducing the inactivation of the enzyme. The quatum yield of inactivation usually does not exceed 0.1; therefore, the time of exposure needs to be as long as an hour (133). The image is visualized by submerging the material into a bath containing the proper substrate. In principle, even a semitone image can be obtained with this procedure. The greatest difficulty is to attain total inactivation and make it possible to get rid of background pigmentation. One of the patents that utilizes enzymes to catalyze the decomposition of hydrogen peroxide allows a vesicular image to be obtained by retention of the bubbles of oxygen in the support under certain conditions (133).

Other patents describe photographic processes in which the substrate is a relief forming support. Such patents describe the use of proteolytic enzymes that are capable of splitting protein supports (135, 136). For example, gelatine is mixed with a dye capable of inducing, when illuminated, sensitized inactivation of the enzyme. The mixture is supplemented with a proteolytic enzyme and placed on a plate. At the exposed sites, the enzyme undergoes photosensitized inactivation; and in the nonilluminated sites the active enzyme cleaves the gelatine so that it can be washed off the plate. The remaining gelatine is stained to give a positive image. The processes of enzyme photoinactivation have basic limitations. First, while working with the above described materials, one should take into consideration the possiblility that the gelatine may undergo splitting during storage, since the active enzyme is present in the light sensitive layer. Second, the gelatine support always contains admixtures attaching the photosensitive layer and the support; so that the action of light usually results in the tanning of this layer at the exposed sites and thus reducing the ability of the galatine to split. Third, for the action of the light to become appreciable, the major portion of the enzyme must be inactivated. This requires a great number of light quanta, so that such systems cannot be too sensitive. More promising are the approaches based on the photoinitiation of enzymatic activity, where the extremely low initial activity of these systems presents a situation where even a small number of

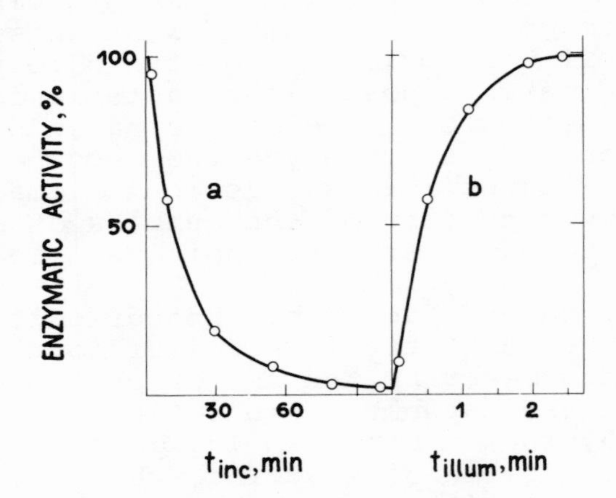

Fig. 8. Decrease in the enzymatic activity of α-chymotrypsin solution depending on the incubation time (t_{inc}) of the enzyme with the p-nitrophenyl ester of cis 4-nitrocinnamic acid, Curve a. The cis 4-nitrocinnamoyl-α-chymotrypsin formed was illuminated with UV light at 313 nm, giving the change in enzymatic activity in Curve b, shown as a function of the illumination time (t_{illum}) (17).

newly formed molecules of catalyst will be noticeable, thereby giving a high sensitivity to light.

B. Photographic Processes Based on Photoinitiation of Enzymatic Activity

The literature describes a number of ways for light initiation of catalytic activity in artificial enzymatic systems. However, the greatest success so far has been achieved with the proteolytic enzymes trypsin and chymotrypsin with the active centers modified by photochromic quasi-substrates. These systems can be used for amplification of weak light signals. Let us take, for example, cis cinnamoyl-α-chymotrypsin, which is catalytically inactive (Fig. 8a) and under the experimental conditions does not deacylate (15, 16). If such a system is illuminated with UV light, the acylenzyme will undergo cis to trans photoisomerization; and the resulting trans acylenzyme will hydrolyze quickly. As a result of the illumination, the catalytic activity will regenerate (Fig. 8b); and hence in the presence of the substrate, a product will be formed (Fig. 9).

In connection with creating photographic materials, at least two major problems arise: the need for a fully inactive photosensitive precursor and diffusional problems with the light sensitive

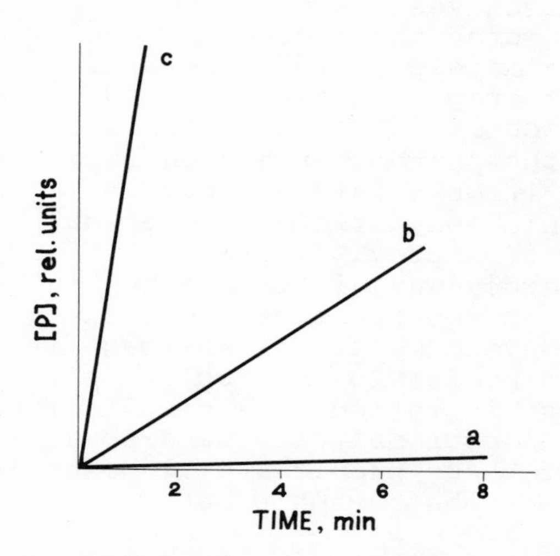

Fig. 9. The product versus time curves for the hydrolysis of N-acetyl-tyrosine ethyl ester under the action of chymotrypsin reactivated as a result of illumination of cis cinnamoyl-chymotrypsin. Time of illumination: Curve a, 0 sec; Curve b, 3 sec; Curve c, reaction with unmodified enzyme (15).

material. A fully inactive stable photosensitive precursor of an enzyme needs to be synthesized. In this laboratory a series of cinnamic acid derivatives of proteolytic enzymes (Table 1) has been produced. These light sensitive acylenzymes are activated by light at 280 to 420 nm and can serve as the basis for technical photographic materials (137-139). Table 2 lists the rate constants of deacylation for some light sensitive acylenzymes. It is obvious that the half-time of conversion of the more stable stereoisomer is several hours at the optimum pH. However, in a synthesized preparation there are always small admixtures of the starting enzyme that should be inactivated by treatment with fast reacting irreversible inhibitors. For trypsin, consecutive treatment with diisopropylfluorophosphate and p-nitrophenylguanidinebenzoate were carried out (110,140). As a result, the preparations of light sensitive enzyme derivatives possessed no catalytic activity and thus ensured the absence of background activity (approximately 0.1%) at nonilluminated sites of the photographic material during development. The diffusion of the light sensitive enzyme component in the photomaterial can be excluded by immobilizing it on a suitable carrier, such as entrapment in a polymeric gel, chemical binding to a support, or microcapsulation. In dry material there is hardly any diffusion; but in humid material the enzyme can move along the gel in the course of development. If an image is developed with the help of a low molecular weight substrate, the dye that is formed as a result of development also should occur only at the exposed site. The solution of these problems will be analyzed below in describing photographic materials.

The use of light sensitive material with a relief forming support as substrate is an approach of interest (138). Any light sensitive precursor of a proteolytic enzyme in a concentration of about 10^{-6}M is mixed with a solution of gelatine, the concentration of which can vary between 4 to 6% w/w. The solution also contains appropriate tanners used in silver photography; the type of

TABLE 1

WAVELENGTH OF LIGHT TO INDUCE STEREOISOMERIZATION OF A CHROMOPHORE IN THE ACTIVE SITE OF AN ENZYME[a]

Light Sensitive Derivative of Proteolytic Enzyme	Wavelength (nm)
cis cinnamoyltrypsin or cis cinnamoyl-α-chymotrypsin	280-320
trans Trimethylaminocinnamolytrypsin	300-320
cis p-nitrocinnamoyl-α-chymotrypsin	330-350
4-nitro-3-methoxy cis cinnamoyl-α-chymotrypsin	350-400
p-dimethylamino cis cinnamoyl-α-chymotrypsin	370-420

[a] (139)

TABLE 2

RATE CONSTANTS FOR DEACYLATION FOR STEREOISOMERIC ACYLENZYMES[a]

Acylenzyme	Ratio of Rate Constants for Trans/Cis Stereoisomers	Rate Constant for More Stable Acylenzyme (sec^{-1})	Ref.
cinnamoyl-α-chymotrypsin	1.5×10^3	6×10^{-6}	(15,16)
p-nitrochinnamoyl-α-chymotrypsin	2.5×10^4	1.7×10^{-5}	(17)
cinnamoyltrypsin	460	2.6×10^{-5}	(63)
p-N,N,N-trimethyl-amino-cinnamoyltrypsin	0.1	6.3×10^{-5}	(62)

[a] pH 8.0, 25°C

tanner depending on the gelatine. This is applied to a support suitable for binding with the gelatine. The pH value of the solution should not exceed 4, as deacylation of the enzyme derivative should be avoided. The temperature of the solution should be about 30°C. Like in silver photography, the layer is first gelatinized and then dryed in air. The material when kept dry is stable; the photographic properties of some samples have not deteriorated after storage for a year in the dark. Depending on the light sensitive component, the material is exposed to light of a suitable wavelength through a negative. As a result of isomerization in the film, the stable isomer of the acylenzyme is converted to the labile isomer. The quantum yield of this process in a dry film is approximately three times as low as in solution for an exposure time of 0.1 to 1 sec. In the gelatine film there appears a latent image, like in a dry sample prepared on the basis of an acidified solution. This latent image can be stored for months. The exposed material is placed in a buffer solution corresponding to the optimum pH for the enzyme action. Fast deacylation of the enzyme is followed by hydrolysis of the gelatine. In this process the gelatine has a double function; it is a carrier of the light sensitive component and also a substrate, the conversion of which results in the visualization of the latent image. The concentration of the enzyme in the dry gelatine layer is high, about 10^{-4}M. The process of cleavage proceeds at a very high rate concurrently with the swelling of the gelatine; the peptides formed thereby go into solution. That is why the edges of the image are not damaged by hydrolysis. Nevertheless, attention should be paid to the increase in concentration of the enzyme in the developing solution. This enzyme comes into the solution after splitting of the gelatine and can interact with unexposed sites; hence the development should be done in a flow system. The solution can be recirculated after thermal treatment to inactivate the enzyme in the solution.

Depending on the degree of cleavage of the gelatine, a relief image can be obtained by treating the partially cleaved retained gelatine with a dye, for example 3% methylene blue (141). The relief formed after complete washing away of the peptides can be used, after additional treatment with tanning agents, as a printing form. If a gelatine layer has been applied on a metal surface, it can be transformed into a support that survives thousands of operations. The resolution can be as high as 20 lines/mm.

The photographic process described above is very sensitive to the development and other processes of relief image formation. For example, the resolution decreases as the concentration of the enzyme in the exposed material increases, because the enzyme that goes to the solution after splitting of gelatine can interact with gelatine at the unexposed sites. In addition, the rates of lateral diffusion, even if very low, can increase with the concentration of enzyme. On the other hand, low concentrations of the enzyme require longer development times, which is not very good for the gelatine support. These two limitations can be coped with (142) if the light sensitive precursor of the enzyme is microcapsulated (143). The walls of the microcapsules should be prepared from polymers that are light permeable in the desired wavelength region. For example, films could be made of cellulose with microscopic water bubbles containing a light sensitive component. In this case low molecular weight substrates have to be used for developing the latent image, as the shell of the microcapsule is impermeable for high molecular weight substances. In this process a suspension of microcapsules in a certain support is applied on any surface then dried. The material is exposed to light and subsequently developed in a buffer solution containing a substrate. One of the most difficult problems here is to find a substrate, the product of hydrolysis of which would be an insoluble dye. We have suggested (139) the use of indoxyl esters of acetylated amino acids. Hydrolysis of these esters in air results in the formation of indigoid dyes that are totally insoluble. The degree of coloration of the microcapsules on the support will depend on the time of exposure. This arises because it is the exposure time that determines the concentration of the enzyme formed under the action of light; and for equal development times different quantities of a stained insoluble reaction product are formed on the illuminated sites of the photomaterial. The resolution of the image will depend only on the size of the capsule (5-25 microns).

It is noteworthy that in this method there are serious diffusional hindrances at the solution-capsule wall boundary (144), which will of course increase the time of development. A photographic material in which the light sensitive component is covalently immobilized on the surface of a support (131, 142) should be free from these limitations. But the resolution here will be lower because of diffusion of the dark reaction product into the unexposed regions.

In the process described by Lee (145), the dye that makes the image visible is formed at the expense of a secondary, also enzymatic, dark reaction. The first dark reaction is the splitting of polytyrosine under the action of α-chymotrypsin. The tyrosine formed is oxidized by tyrosinase via indoquinone into melanine, which is an insoluble dye. The stage of visualization of the light signal is hindered by the slowness of the polytyrosine hydrolysis and slow formation of melanine. But of most importance, the visualization is hindered by the background processes of spontaneous pigmentation. Nevertheless, the method of Lee can be helpful in choosing the components for the dark reactions.

IV. CONCLUSIONS

It must be stated that the silverless processes involving the use of enzymes are not perfect; but it is beyond doubt that they have a future.

The first important reason for this is that the amplification coefficient or effective quantum yield in the enzymatic systems may be rather high. In addition there are two more ways of increasing it.

The first way to increase the amplification coefficient is to make activation of the respective zymogen the first of the dark reactions. Zymogens are catalytically inactive precursors of enzymes, capable of being activated. For example, trypsinogen possesses no enzymatic activity but can be converted into the active enzyme under the action of trypsin (146). This is an autocatalytic process, and in principle, one molecule of trypsin can activate all of the zymogen. This means that trypsinogen is sensitive to the lowest concentrations of trypsin (140). Hence, the idea of creating a light sensitive enzymatic system consisting of a catalytically inactive trypsin derivative, trypsinogen, and a substrate of trypsin is very promising. If this system is illuminated, the inactive derivative of trypsin gives rise to the active enzyme which then can activate the zymogen; and the product of this activation will hydrolyze the substrate. Thereby the linear dependence of the effective quantum yield on the time of the dark reaction becomes exponential (64,110), as shown in Fig. 10.

The second way to increase the amplification coefficient is based on the concept that photochemical formation of the support makes it possible to have, in response to the absorption of one quantum of light, more than one molecule of an immobilized enzyme (131). For this purpose photopolymerizable reproduction systems, developed previously (4,5), may prove feasible.

The second important reason for believing that light sensitive enzyme systems have a promising future is that their special sensitivity may be selected by varying the molecular structure of the light sensitive low molecular weight components, such as an effector of the enzymatic

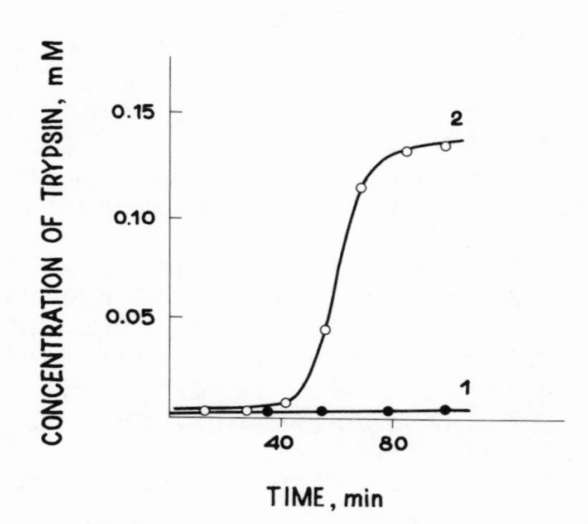

Fig. 10. The kinetics of accumulation of trypsin in a solution containing cis cinnamoyl-trypsin (3×10^{-6}M) and trypsinogen (1.5×10^{-4}M). Curve 1: without illumination; Curve 2: after illumination of the solution for 4 min. Conditions: pH 8.0, Ca^{+2} 5×10^{-3}M, 25°C. (63).

reaction, a protein modifier, or the substrate (120,124,146). Sensitizable photochemical reactions are most appropriate objects for that. In this case, to vary the spectral sensitivity of the system, it is sufficient just to change the molecular structure of a sensitizer. The most promising for this purpose is photopolymerization creating a support capable of immobilization of enzymes (131). Photopolymerization processes do not restrict very much the structure of the sensitizer molecule (127) Here the experience gained during the development of photopolymerizable reproduction (4,5) can prove very useful.

REFERENCES

1. MEES, K. C. E. & JAMES, T. H. "Theory of the Photographic Process," 3rd edition, MacMillan, New York, 1966.
2. LYALIKOV, K. S. "The Theory of the Photographic Process," Iskusstvo Press, Moscow, 1960.
3. COX, R. I. "Non-Silver Photographic Processes," Academic Press, London, 1975.
4. IURRE, T. A., SHABUROV, V. V. & ELTSOV, A. V. *Shurn. Vses. Khim.* (*J. All-Union Chem. Soc.*, (Russ.) *19*: 412-423, 1974.
5. WALKER, P., WEBERS, V. J., THOMMES, G. A. *Journ. Photographic Sci. 18*: 150, 1970.
6. DELZENNE, G. A. *Chem. Weekblad*, December 637, 1976.
7. ROSE, A. "Vision, Human and Electronic," Plenum Press, New York, 1973.
8. CALLABY, D. R. & BROTTO, M. *J. Photographic Sci. 18*: 8, 1970.
9. JONKER, H., BIEK, H. K. U., JANSSEN, C. & KLOSTERMANN, F. T. *Photograph. Sci. Eng. 13*: 33, 1969.
10. BEREZIN, I. V., VARFOLOMEYEV, S. D., KAZANSKAYA, N. F. & MARTINEK, K. "Fotoreguljacis Metabolizma i Morfogeneza Rastenij" "(Photoregulation of Metabolism and Morphogenese of Plants"), Nauka, Moscow, 1975, p. 48.
11. MARTINEK, K. & BEREZIN, I. V. *Photochem. Photobiol.* (in press).
12. LAIDLER, K. J. & BUNTING, P. S. "The Chemical Kinetics of Enzyme Action," Clarendon Press, Oxford, 1973.
13. BEREZIN, I. V. & MARTINEK, K. "Osnovv Fizicheskoi Khimii Fermentativnogo Kataliza" ("Physicochemical Fundamentals of Enzyme Catalysis"), Vysshaya Shloka Press, Moscow, 1977.

14. LUMRY, R. & BILTONEN, R. In "Structure and Stability of Biological Macromolecules" S. N. Timasheff and G. D. Fasman, eds.), Marcel Dekker, New York, 1969, p. 65.
15. BEREZIN, I. V., VARFOLOMEYEV, S. D. & MARTINEK, K. *Dokl. Akad, Nauk SSSR* (Russ.) *193*: 932, 1970.
16. MARTINEK, K., VARFOLOMEYEV, S. D. & BEREZIN, I. V. *Eur. J. Biochem. 19*: 242, 1971.
17. VARFOLOMEYEV, S. D., KLIBANOV, A. M., MARTINEK, K. & BEREZIN, I. V. *FEBS Lett. 15*: 118, 1971.
18. WALD, G. *Nature 219*: 800, 1968.
19. OSTROVSKI, M. A. In "Ion-Transport across Membranes" D. C. Tosteson and Yu. A. Ovchinnikov, eds.), Raven Press, New York, 1977.
20. KAUFMAN, H., VRATSANOS, S. M. & ERLANGER, B. F. *Science 162*: 1487, 1968.
21. ERLANGER, B. F. *Ann. Rev. Biochem. 45*: 267, 1976.
22. TERENIN, A. N. "Photonics of Dye Molecules," Nauka Press, Leningrad, 1967.
23. CALVERT, J. G. & PITTS, J. N. JR. "Photochemistry", John Wiley and Sons, New York, 1966.
24. KAN, R. O. "Organic Photochemistry," Mcgraw-Hill, New York, 1966.
25. BEREZIN, I. V., KLIBANOV, A. M., SAMOKHIN, G. P., GOLDMACHER, V. S. & MARTINEK, K. In "Biomedical Applications of Immobilized Enzymes and Proteins," vol. 2 (T. M. S. Chang, ed.), Plenum Press, New York, 1977.
26. ERLANGER, B. F. & COHEN, W. *J. Amer. Chem. Soc. 85*: 348, 1963.
27. ERLANGER, B. F., CASTLEMAN, H. & COOPER, A. G. *J. Amer. Chem. Soc. 85*: 1872, 1963.
28. ERLANGER, B. F., COOPER, A. G. & COHEN, W. *Biochemistry 5*: 190, 1966.
29. METZGER, H. P. & WILSON, I. B. *Biochemistry 3*: 926, 1964.
30. BIETH, J., VRATSANOS, S. M., WASSERMANN, N. & ERLANGER, B. F. *Proc. Nat. Acad. Sci. USA 64*: 1103, 1969.

31. BIETH, J., VRASTSANOS, S. M., WASSERMANN, N., COOPER, A. G. & ERLANGER, B. F. *Biochemistry* *12*: 3023, 1973.
32. GOELDNER, M. P. & HIRTH, C. G. *FEBS Lett.* *82*: 151, 1977.
33. STEVENS, M. F. G., MAIR, A. C. & REISCH, J. *Photochem. Photobiol.* *13*: 441, 1971.
34. CUNNINGHAM, L. W. *Comprehensive Biochem.* *16*: 85, 1965.
35. BIETH, J., WASSERMANN, N., VRATSANOS, S. M. & ERLANGER, B. F. *Proc. Nat. Acad. Sci. USA* *66*: 850, 1970.
36. GALLEY, K. T., DE SORGO, M. & PRINS, W. *Biochem. Biophys. Res. Commun.* *50*: 300, 1973.
37. BEREZIN, I. V., VARFOLOMEYEV, S. D., KLIBANOV, A. M. & MARTINEK, K. *FEBS Lett.* *39*: 329, 1974.
38. GLAZER, A. N. *Proc. Nat. Acad. Sci. USA* *54*: 171, 1965.
39. BERNHARD, S. A., LEE, B. F. & TASHJIAN, Z. H. *J. Mol. Biol.* *18*: 405, 1966.
40. MILLICH, F. & OSTER, G. J. *Amer. Chem. Soc.* *81*: 1357, 1959.
41. SCHOFFENIELS, E. & NACHMANSOHN, D. *Biochim. Biophys. Acta* *26*: 1, 1957.
42. SCHOFFENIELS, E. *Biochim. Biophys. Acta* *26*: 585, 1957.
43. HIGMAN, H. B., PODLESKI, T. R. & BARTELS, E. *Biochim. Biophys. Acta* *79*: 138, 1964.
44. DEAL, W. J., ERLANGER, B. F. & NACHMANSOHN, D. *Proc. Nat. Acad. Sci. USA* *61*: 1230, 1969.
45. BARTELS, E., WASSERMANN, N. & ERLANGER, B. F. *Proc. Nat. Acad. Sci. USA* *68*: 1820, 1971.
46. BEREZIN, I. V., VARFOLOMEYEV, S. D., SAVITSKI, A. P. & UGAROVA, N. N. *Dokl. Akad. Nauk SSSR* *222*: 380, 1975.
47. PARKER, C. A. *Proc. Roy. Soc.* *A220*: 104, 1953.
48. HATCHARD, C. G. & PARKER, C. A. *Proc. Roy. Soc.* *A235*: 518, 1956.
49. BEREZIN, I. V., VARFOLOMEYEV, S. D. & MARTINEK, K. *FEBS Lett.* *8*: 173, 170.
50. MARTINEK, K., VARFOLOMEYEV, S. D., PREOBRAZHENSKAYA, M. N., SAVEL'EVA, L.A. & BEREZIN, I. V. *Biokhim.* (Russ.) *37*: 614, 1972.

51. WAINBERG, M. A. & ERLANGER, B. F. *Biochemistry 10*: 3816, 1971.
52. VARFOLOMEYEV, S. D., Ph.D. Thesis, Lomonosov State University, Moscow, 1971.
53. BRESTKIN, A. P., ZHUKOVSKI, YU. G., MURASHKINA, S. K., SAMOKISH, V. A., STRELETS, B. H. & TRAKHNOVA, G. M. *Dokl. Akad. Nauk SSSR 232*: 1438, 1977.
54. TAYLOR, P. W., KING, R. W. & BURGEN, A. S. V. *Biochemistry 9*: 2638, 1970.
55. MAGUIRE, P. J., HIJAZI, N. H. & LAIDLER, K. J. *Biochim. Biophys. Acta 341*: 1, 1974.
56. BURSTEIN, E. A. "Luminescence of Protein Chromophores," Chap. 5, Viniti Press, Moscow, 1976.
57. COMOROSAN, S. *Nature 227*: 64, 1970.
58. COMOROSAN, S., VIERU, S. & SANDRU, D. *Int. J. Radiat. Biol. 17*: 105, 1970.
59. COMOROSAN, S., SANDRU, D. & ALEXANDRESKU, E. *Enzymologia 38*: 317, 1970.
60. SCHONBAUM, G. B., ZORNER, B. & BENDER, M. L. *J. Biol. Chem. 236*: 2930, 1961.
61. VARFOLOMEYEV, S. D., KLIBANOV, A. M., MARTINEK, K. & BEREZIN, I. V. *Dokl. Akad. Nauk SSSR 203*: 616, 1972.
62. AISINA, R. B., VASIL'EVA, T. E., KAZANSKAYA, N. F., TIKHODEEVA, A. S. & BEREZIN, I. V. *Biokhimiya* (Russ.) *38*: 601, 1973.
63. BEREZIN, I. V., AISINA, R. B., BRONNIKOV, G. E. & KAZANSKAYA, N. F. *Bioorg. Khim.* (Russ.) *1*: 402, 1975.
64. BEREZIN, I. V., AISINA, R. B., VARFOLOMEYEV, S. D. & KAZANSKAYA, N. F. *Dokl. Akad, Nauk SSSR 219*: 1255, 1975.
65. SINGH, A., THORNTON, E. R. & WESTHEIMER, F. H. *J. Biol. Chem. 237*: 3006, 1962.
66. VAUGHAN, R. J. & WESTHEIMER, F. H. *J. Amer. Chem. Soc. 91*: 217, 1969.
67. SPENCER, T. & STURTEVANT, J. M. *J. Amer. Chem. Soc. 81*: 1874, 1959.
68. TORCHINSKI, YU. M. "Sulfhydryl and Disulfide Groups of Proteins," Plenum Press, New York, 1974.
69. KLEIN, I. B. & KIRSCH, J. F. *Biochem. Biophys. Res. Commun. 34*: 575, 1969.

70. KLEIN, I. B. & KIRSCH, J. F. *J. Biol. Chem.* *244*: 5928, 1969.
71. EAGER, J. E. & SAVIGE, W. E. *Photochem. Photobiol.* *2*: 25, 1963.
72. RISI, S., DOSE, K., RATHINASAMY, T. & AUGENSTEIN, L. G. *Photochem. Photobiol.* *6*: 423, 1967.
73. ASQUITH, R. & HIRST, L. *Biochim. Biophys. Acta* *184*: 345, 1969.
74. DOSE, K. & RISI, S. *Photochem. Photobiol.* *15*: 43, 1972.
75. KAZANSKAYA, N. F. & NIKOL'SKAYA, I. I., VESTNIK MGU *Ser. Khim.* (Russ.) *16*: 49, 1975.
76. RITCHEY, J. M., GIBBONS, T. & SCHACHMAN, H. K. *Biochemistry* *16*: 4584, 1977.
77. SMETS, G. *Pure Appl. Chem.* *30*: 1, 1972.
78. ERMAKOVA, E. D., ARSENOV, V. D., TCHERKASHIN, M. I. & KISILITSA, P. P. *Uspekhi Khim.* (Russ.) *46*: 292, 1977.
79. LOVRIEN, R. & WADDINGTON, J. C. B. *J. Amer. Chem. Soc.* *86*: 2315, 1964.
80. LOVRIEN, R. *Proc. Nat. Acad. Sci. USA* *57*: 236, 1967.
81. LOVRIEN, R. *J. Amer. Chem. Soc.* *96*: 244, 1974.
82. VAN DER VEEN, G. & PRINS, W. *Nature* *230*: 70, 1971.
83. VAN DER VEEN, G. & PRINS, W. *Photochem. Photobiol.* *19*: 191, 1974.
84. VAN DER VEEN, G., HOGUET, R. & PRINS, W. *Photochem. Photobiol.* *19*: 197, 1974.
85. NAMBA, K. & SUZUKI, S. *Chem. Lett.* *9*: 947, 1975.
86. AIZAWA, M., NAMBA, K. & SUZUKI, S. *Arch Biochem. Biophys.* *180*: 41, 1977.
87. INOUE, E., KOKADO, H., SHIMIZU, I. & KOBAYASHI, H. *Bull. Chem. Soc. Japan* *45*: 1951, 1972.
88. KOLOMIYCHENKO, M. A. *Ukrainsk. Biokhim. Zh.* (Russ.) *28*: 164, 1956.
89. KOLOMIYCHENKO, M. A. *Ukrainsk. Biokhim. Zh.* (Russ.) *29*: 361, 1957.
90. VOLOTOVSKI, I. D., VOSKRESENSKAYA, L. G. & KONEV, S. V. *Biofizika* (Russ.) *17*: 971, 1972.

91. MONTAGNOLI, G. *Acta Vitamin. Enzymol.* (Milano) *28*: 268, 1974.
92. MONTAGNOLI, G., MONTI, S., NANNICINI, L. & FELICIOLI, R. *Photochem. Photobiol.* *23*: 29, 1976.
93. MONTAGNOLI, G., MONTI, S., NANNICINI, L., GIOVANNITTI, M. P. & RISTORI, M. G. *Photochem. Photobiol.* *27*: 43, 1978.
94. FROHLICH, H. *Int. J. Quant. Chem.* *2*: 641, 1968.
95. FROHLICH, H. *Nature* *228*: 1093, 1970.
96. KOLLIAS, N. & MELANDER, W. R. *Phys. Lett.* *57A*: 102, 1976.
97. MCLAREN, A. D. *Adv. Enzymol.* *9*: 75, 1949.
98. AUGENSTEIN, L. G. *Adv. Enzymol.* *24*: 359, 1962.
99. MCLAREN, A. D. & SHUGAR, D. "Photochemistry of Proteins and Nucleic Acids," Oxford University Press, Oxford, 1964.
100. VLADIMIROV, Y. A. "Photochemistry and Luminescence of Proteins," Nauka Press, Moscow, 1965.
101. MCLAREN, A. D. *Enzymologia* *37*: 273, 1969.
102. KONEV, S. V. & VOLOTOVSKI, I. D. "Introduction in Molecular Photobiology," Nauka i Tekhnika Press, Minsk, 1971.
103. SCHMID, G. H. *Hoppe-Seyler Z. Physiol. Chem.* *351*: 575, 1970.
104. SPIKES, J. D. In "Photophysiology," vol. 3 (A. C. Giese, ed.) Academic Press, New York, 1968, p. 33.
105. SPIKES, J. D. & STRAIGHT, R. *Ann. Rev. Phys. Chem.* *18*: 409, 1967.
106. DIXON, M. & WEBB, E. C. "Enzymes," Academic Press, New York, 1964.
107. VARFOLOMEYEV, S. D., MARTINEK, K. & BEREZIN, I. V. "Proc. Lab. Bioorg. Chem. Moscow State Univ. MGU Press, Moscow, 1970, p. 289.
108. ADAMSON, A. W. & SPORER, A. H. *J. Amer. Chem. Soc.* *80*: 3865, 1958.
109. BASOLO, F. & PEARSON, R. G. "Mechanisms of Inorganic Reactions. A Study of Metal Complexes in Solution," John Wiley and Sons, New York, 1967, Chap. 8.

110. BEREZIN, I. V., KAZANSKAYA, N. F. & AISINA, R. B. *Dokl. Akad. Nauk SSSR 207*: 1383, 1972.
111. BALASUBRAMANIA, D., SUBRAMANI, S. & KUMAR, C. *Nature 254*: 252, 1975.
112. KARUBE, I., NAKAMOTO, Y., NAMBA, K. & SUZUKI, S. *Biochim. Biophys. Acta 429*: 975, 1976.
113. NAKAMOTO, Y., KARUBE, I., TARAWAKI, S. & SUZUKI, S. *J. Solid Phase Biochem. 1*: 143, 1976.
114. NAKAMOTO, Y., NISHIDA, M., KARUBE, I. & SUZUKI, S. *Biotechnol. Bioeng. 19*: 1115, 1977.
115. KARUBE, I., SUZUKI, S., NAKAMOTO, Y. & NISHIDA, M. *Biotechnol. Bioeng. 19*: 1549, 1977.
116. ZABORSKY, O. R. "Immobilized Enzymes," Chem. Rubber Co. Press, Cleveland, 1973.
117. MOSBACH, K. *Meth. Enzymol. 44*: 149, 1976.
118. CHANG, T. M. S., "Biomedical Applications of Immobilized Enzymes and Proteins," vol. 1, Plenum Press, New York 1977.
119. YAQUB, M. & GUIRE, P. *J. Biomed. Mater. Res. 8*: 291, 1974.
120. GUIRE, P. *Meth. Enzymol. 44*: 280, 1976.
121. FLEET, G. W. J., PORTER, R. R. & KNOWLES, J. R. *Nature 224*: 511, 1969.
122. KRAMER, D. M., LEHMANN, K., PENNEWISS, H. & PLAINER, H. "Proc. Conf. on Protides of the Biological Fluids, Brugge, Belgium, May 1975," Pergamon Press, 1975, p. 505.
123. WILSON, D. F., MIYATA, Y., ERECINSKA, M. & VANDERKOOI, J. M. *Arch. Biochem. Biophys. 171*: 104, 1975.
124. GUIRE, P. In "Enzyme Engineering," vol. 3 (E. K. Pye and H. H. Weetall, eds.) Plenum Press, New York, 1978.
125. BERNFELD, P. & WAN. J. *Science 142*: 678, 1963.
126. MAURER, H. R. "Disk-Elektrophorese," Walter de Gruyter and Co., Berlin, 1968, Chap. 1.
127. ODIAN, G. "Principles of Polymerization," 1970, Chap. 3.

128. FUKUI, S. & TANAKA, A. *FEBS Lett.* *66*: 179, 1976.
129. JOHANSSON, A.-C. & MOSBACH, K. *Biochim. Biophys. Acta* *370*: 339, 1974.
130. EKMAN, B. & SJOHOLM, J. *Nature* *257*: 825, 1975.
131. SAMOKHIN, G. P., KLIBANOV, A. M. & MARTINEK, K. *Vestnik MGU Ser. Khim.* (Russ.) *19*: 433, 1978.
132. BURSTONE, M. S. "Enzyme Histochemistry and its Application in the Study of Neoplasms," Academic Press, New York, 1962.
133. USA Pat. No. 3,694,207; 1972.
134. USA Pat. No. 3,515,551; 1970.
135. France Pat. No. 1,492,872; 1967.
136. USA Pat. No. 3,649,207; 1970.
137. USSR Pat. No. 439,780; 1971.
138. USSR Pat. No. 595,693; 1973.
139. USSR Pat. No. 584,280; 1975.
140. AISINA, R. B., KAZANSKAYA, N. F. & BEREZIN, I. V. *Biokhimiya* (Russ.) *39*: 577, 1974.
141. LILLIE, R. D. "Histopathologic Technic and Practical Histochemistry," McGraw-Hill Book Company, New York, 1965.
142. LUKASHEVA, E. V., AISINA, R. B., KAZANSKAYA, N. F. & BEREZIN, I. V. *Biochimiya* *42*: 465, 1977.
143. CHANG. T. M. S. "Artificial Cells," Charles C. Thomas Publisher, Springfield, Illinois, USA, 1971.
144. LUKASHEVA, E. V., AISINA, R. B., GRACHOVA, I. I., KAZANSKAYA, N. F. & BEREZIN, I. V. *Biochimiya* *42*: 2013, 1977.
145. LEE, Y. Y., MELINA, O. & TEBBETT, L. *Biotechnol. Bioeng.* (in press).
146. NEURATH, H. & WALSH, K. A., *Proc. Nat. Acad. Sci. USA* *73*: 3825, 1976.

IMMOBILIZED ENZYMES: A BREAKTHROUGH IN FINE CHEMICALS PROCESSING

E. Cernia

Laboratories for Biochemical Process
Assoreni
Monterotondo (Roma), Italy

There are several reactions in which enzymes have been applied for the production of fine chemicals. The main thrust of this paper is to discuss these processes, examining in particular the advantages that enzymes have to offer.

The synthesis of certain penicillins involves the hydrolysis of a side chain, with the enzyme penicillin acylase a convenient catalyst for this reaction. Bacterial penicillin acylase hydrolyzes preferentially benzyl penicillin, while the enzyme found in actinomycetes and molds hydrolyzes phenoxy penicillin. Penicillin acylase also catalyzes the reverse reaction, i.e. the synthesis of a penicillin from a side chain molecule and 6-amino-penicillanic acid (6-APA). While the hydrolytic reaction is favored at alkaline pH, the synthesis takes place at acidic pH (1). The same enzyme also catalyzes the hydrolysis of cephalosporins obtained by the chemical ring expansion of penicillins (2). The commercial importance of 6-APA as an intermediate for semi-synthetic penicillins has concentrated many efforts on both the chemical and the enzymatic hydrolysis of penicillins and cephalosporins (3) In many of the efforts the penicillin acylase has been immobilized on solid supports; however this aspect has been covered in many other reports and is not discussed here.

The procedures for the chemical and enzymic hydrolysis of penicillins are illustrated schematically in Fig. 1.

From this figure, it appears manifest at once the greater complexity of the chemical route in contrast with the enzymic one. The contrast is even more clear if we examine the chemicals and the utilities necessary to carry out the same reaction by the two methods (Tables 1 and 2). This example clearly demonstrates the advantages of using the enzyme over the chemical method: simplicity, low energy consumption, no pollution, etc. However, even with the disadvantages, the chemical hydrolysis of penicillins is still widely used. This has been possible because a good deal of effort has been done to optimize the chemical method, unlike the enzymic process. In fact, most of the works in the literature deal with the properties of penicillin acylase as a biochemical compound. Rather few of the literature works view the enzyme with the same chemical reasoning used to consider the reactivity of normal organic reagents.

II. OTHER REACTIONS

Little success has been encountered with the enzymic acylation of 6-APA and 7-amino-desacetoxy cephalosporanic acid (7-ADCA), mainly because of the poor substrate specificity of *E. coli* penicillin acylase. This can be seen better by examining the enzymic synthesis of ampicillin. This synthesis can be done with a reaction mixture containing the acylase enzyme, 6-APA, and an energy rich derivative of the side chain, i.e. phenylglycine methyl ester. The enzyme catalyzes not only the synthetic reaction but also the hydrolysis of the synthesized ampicillin. At the end of the incubation, there are four compounds in the reaction mixture: unconverted 6-APA, unconverted phenylglycine ester, phenylglycine, and ampicillin. The difficulties in the separation of such a mixture have prevented the practical use of this synthesis approach. Toyo Jozo, a Japanese company, has

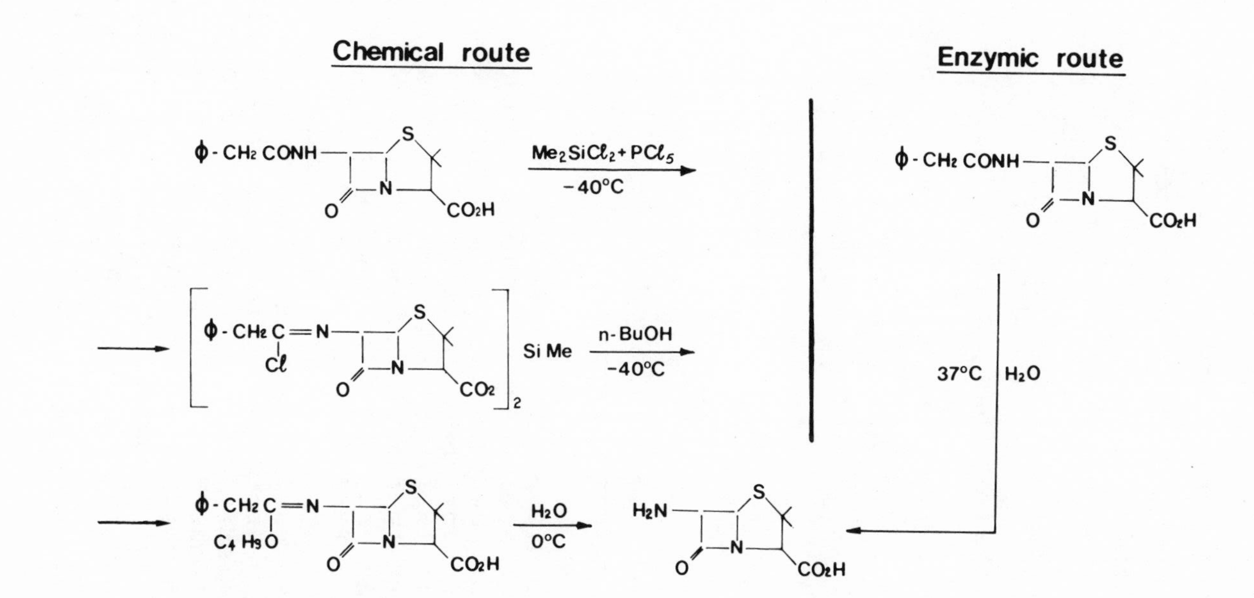

Fig. 1. Chemical and enzymic routes for the production of 6-amino-penicillanic acid (6-APA) (8).

TABLE 1

CHEMICALS FOR THE PRODUCTION OF 6-APA (8)

Chemical Route	Enzymic Route
Dimethyl Aniline	Enzyme
Dimethyl Dichlorosilane	Phosphate Buffer
Phosphorous Pentachloride	Sodium Hydroxide
Ammonia	Hydrochloric Acid
Methylene Chloride	Methanol
Butanol	
Acetone	
Methanol	

developed a different type of acylase produced by *Pseudomonase melanogenum*. This new acylase is specific for phenylglycine acyl donors; and it can overcome some of the disadvantages of the *E. Coli* acylase. It is likely that the *Pseudomonas* enzyme will make the enzymic synthesis competitive with the chemical one, provided that an efficient immobilization method will be found.

An interesting example of enzymic removal of a protecting group can be found in the cephalosporin chemistry. An extracellular esterase from *B. subtilis* has been developed that selectively removes the acetate ester at the C-3 position. The enzymic reaction takes place under very mild conditions and at high yields. On the contrary, the chemical removal of acetate occurs with very low yield because of migration of the double bond from

TABLE 2

UTILITIES CONSUMPTION FOR THE PRODUCTION OF 1 KG OF 6-APA (8)

	Chemical Route	Enzymic Route
Process Water (m^3)	0.19	0.11
Cooling Water (m^3)	1.25	4.2
Low Pressure Steam (Kg)	23.1	73.7
Electric Power (Kwh)	13	6.1
Liquid Nitrogen (Kg)	17.5	-

C-3 to C-2 and lactone formation between the hydroxy methyl and the adjacent carboxyl groups.

Another type of organic synthesis where enzymes have great potential is in the preparation of optically active molecules. Amino acids are a good example. They are used as food supplements and in the pharmaceutical industry; and only the L-isomers are biologically active. The chemical methods lead to racemic mixtures whose resolution is normally quite expensive and tedious. Some of the immobilized enzymes, used in aminoacid synthesis, that are of particular interest are tryptophanase and β-tyrosinase. Both enzymes catalyse α-β-elimination and β-replacement reactions: tryptophanase with tryptophan, serine, cysteine, S-methylcysteine and other substrates; β-tryosinase with tyrosine, phenylalanine, and L-dopa (4-5).

The reversible α-β-elimination can be used to synthesize L-tryptophan, L-tyrosine, L-phenylalanine and L-Dopa (6). Tryptophan synthetase, one of the few enzymes composed of two dissimilar subunits, catalyzes among a number of other reactions the condensation of indole with L-serine. The condensation reaction takes place at normal temperatures and slightly alkaline pH. The same reaction can be performed by chemical methods in the presence of acetic anhydride, under reflux; however the yield is no higher than 50%, the quality of the product is quite poor, and a racemic mixture is obtained.

On examining these examples, another problem related to the use of enzymes in organic synthesis may arise, i.e. the requirement of many enzymes or a coenzyme. The enzymes mentioned before require pyridoxal-5-phosphate to form the catalytically active species. A good deal of work has been done recently for obtaining active coenzyme derivatives in which the coenzymes are attached to high molecular weight compounds. In this manner the coenzymes can be regenerated and reused many times (7). Although the enzymic synthesis of optically active aminoacids remains at present time the undisputed route for industrial use, it must be recorded that recent progress in the fields of both chemical catalysis and stereochemistry may reduce the gap between the enzymic and chemical approaches. The use of certain asymmetrically modified rhodium complexes as catalysts for the hydrogenation of suitable olefinic substrates now offers a promising route for the synthesis of a number of amino acids, e.g. L-dopa, L-alanine, or L-tryptophan. Chemical catalysts owe to enzymes the concept of multiple substrate-catalyst interaction as a primary requirement for their high stereoselectivity. But, chemical catalysts are by far less complex in terms of structure and size. Transition metal catalysts are not as sensitive as enzymes to reaction parameters, such as temperature and solvents, which can be manipulated to help achieve the goal of maximum stereoselectivity.

Table 3 shows another group of fine chemicals produced by immobilized enzymes or cells. Here again are examples of poor substrate specificity. Fumarase catalyzes the stereospecific hydration of a variety of fumaric acid derivatives to malic acids. In fact, a number of α,β-unsaturated dicarboxylic acids can be hydrated. Other hydrating enzymes, such as aconitase, are active on different substrates. Hydantoinase accepts a large number of hydantoins as substrates and catalyzes the stereospecific opening of the hydantoin ring. Finally, the enzyme hydroxy-nitrilolyase catalyses the stereospecific addition of hydrogen cyanide to several aldehydes. These examples suggest that intensive research on the substrate specificity of enzymes might help to uncover other reactions useful in organic synthesis.

Where enzymes cannot be replaced by chemical reagents is in the transformation or synthesis of complex molecules. Steroid transformations provide several examples where immobilized enzymes or cells have been used to carry out certain steps. For example $\Delta^{1,2}$-dehydrogenation, 11α- and 11β-hydroxylation, and 20β-dehydrogenation have been investigated using entrapped cells. Steroid transformations also present the problem of substrate solubility. Although the use of organic solvents in enzymic catalysis needs further investigation, there are many enzymes that can tolerate significant amounts of organic solvents mixed with water. In the steroid field, 15% ethanol has been used fruitfully for the transformation of hydrocortisone to prednisolone; and acetone has been employed to solubilize cholesterol. Certain enzymes, such as lipases, appear to work mainly in organic solvents or in two-phase systems.

The enzymic synthesis of coenzyme A is another example of the synthesis of a complex molecule through sequential reactions. This sequence is illustrated in Fig. 2. A similar process has been carried out in Japan using *Brevibacterium ammoniagenes* cells entrapped in gel or covalently coupled to a support. This coenzyme A synthesis

TABLE 3

ADDITIONAL FINE CHEMICALS PRODUCED USING ENTRAPPED ENZYMES OR CELLS

Substrate	Product(s)	Enzyme
Histidine	Urocanic Acid	Histidine ammonia lyase (a)
Fumaric Acid	Malic Acid	Fumarase (a)
Sorbose	Sorbosone	Sorbose Dehydrogenase
Methionine	S-Adenosyl Methionine	S-Adenosyl Methionine Transferase (b)
D,L-Hydantoins	D-Carbamoyl Amino Acids	Hydantoinase (b)
-Aldehyde	-Hydroxy Acid	Hydroxynitrilo Lyase (c)

(a) Tanabe Co., (b) Snamprogetti; (c) University of Marbury

is typical of a so-called "one-pot" overall reaction. All of the enzymes involved in this synthesis can work together without interfering with each other. On the contrary there are few examples of "one-pot" syntheses in classical organic chemistry, because the conventional chemical reagents often react with each other rather than with the compound to be transformed. Moreover the conditions under which one of the reactions takes place may not be suitable for the other steps. The application of multi-step "one-pot" enzyme catalyzed syntheses has a large potential; but a good deal of work must be done in exploring the possibilities.

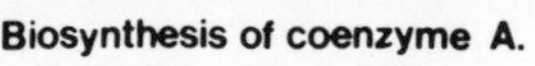

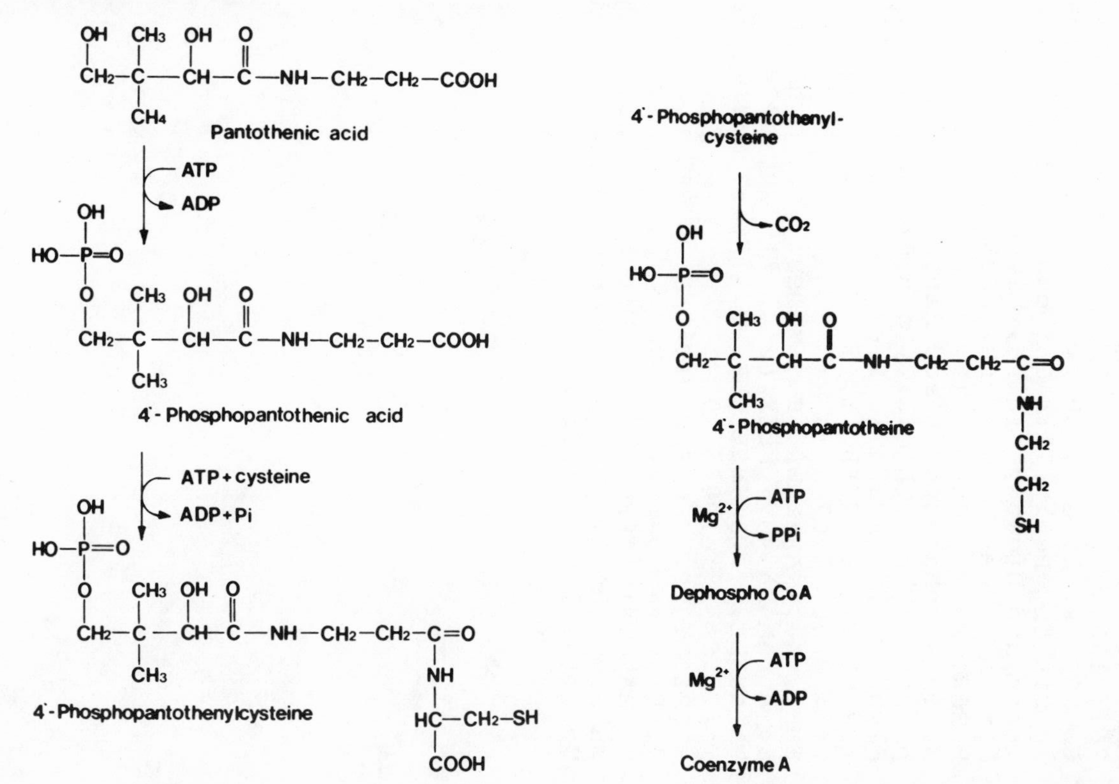

Fig. 2. Biosynthesis of coenzyme A by immobilized microbial cells (8).

REFERENCES

1. HAMILTON-MILLER, J. M. T. *Bacteriol. Rev. 30:* 761, 1966.
2. COOPER, R. D. G. & SPRY, D. O. In "Cephalosporin and Penicillin Compounds: Their Chemistry and Biology" (E. H. Flynn, ed.) Academic Press, New York, 1973.
3. HUBER, F. M., CHAUVETTE, R. R. & JACKSON, B. G. In "Cephalosporin and Penicillin Compounds: Their Chemistry and Biology" (E. H. Flynn, ed.) Academic Press, New York, 1973.
4. MORINO, Y. & SNELL, E. E. *J. Biol. Chem. 242:* 2800, 1967.
5. ENEI, A., NAKAZAWA, H., MATSUI, A., OKUMURA, S. & YAMADA, H. *FEBS Lett. 21:* 39, 1972.
6. WATANABE, T. & SNELL, E. E. *Proc. Natl. Acad. Sci. USA 69:* 1086, 1972.
7. ZAPPELLI, P., ROSSODIVITA, A. & RE, L. *Eur. J. Biochem. 54:* 475, 1975.
8. W. MARCONI, "Proceedings First European Congress on Biotechnology 1978," Dechema Monographien No. 1693-1703, Vol. 82, p. 119.

PROBLEMS OF EFFICIENCY AND OPTIMIZATION IN ENZYME ENGINEERING

A. Köstner and E. Siimer

Laboratory of Enzyme Technology
Tallinn Technical University
Tallinn, USSR

In spite of rapid research developments in the field of enzyme immobilization and the application of immobilized enzymes, only a very limited number of real industrial applications have been reported. We consider this gap between prospects and practical application as being caused in part by inadequate attention to the optimization and scale up of laboratory processes. We have come to the conclusion that the theoretical basis for the strategy of optimization in enzyme technology must be elaborated by analyzing combined kinetic and economic approaches. Such approaches have been discussed by a few authors (1,2) and have been developed in our earlier papers (3,4). Here we propose a general kinetic and economic equation for the analysis of a process in enzyme engineering.

From this equation, important guidelines for the optimization of immobilized enzyme processes and the operation of enzyme reactors may be derived, even when we can only approximate the economic characteristics. As examples of our theoretical approaches, penicillin amidase and β-galactosidase will be considered. The necessary kinetic and economic indices for these enzymes are taken from the literature (5,6) and from our own experimental data.

I. GENERAL CRITERION OF EFFICIENCY

Let us consider a general process in which a substrate is converted to a desired product with the aid of an immobilized enzyme; and the reactor is reloaded with fresh catalyst when necessary. As a general measure for the economic efficiency we propose a dimensionless criterion, E_p, which is defined as the ratio of the costs of the product C_p during one reloading cycle of the reactor to the costs for the operation C_E including the cost of substrate and enzyme.

$$E_p = \frac{C_p}{C_E} = \frac{P_p A_o Y(\bar{X},t)}{P_S A_o W(\bar{X},t) + A_o P_E/\gamma + C_I + C_R \tau} \quad \text{(Eq. 1)}$$

where C_I is the expense of the carrier, the immobilization, and the reloading of the reactor; C_R is the operating expense of the reactor per hour; P_p is the price/mole of product; P_S is the price/mole of substrate; P_E is the price/activity unit of the initial native enzyme preparation; A_o is the total initial standard activity of the immobilized enzyme preparation used in one reloading cycle of the reactor in moles per hour; γ is the yield of activity for immobilization; t is time in hr; τ is duration of the reloading cycle in hr; $\bar{X}$ describes the operating conditions as a vector; $Y(\bar{X},t)$ is a function of product formation with dimensions of time; and $W(\bar{X},t)$ is a function of substrate consumption, with dimensions of time.

The product formation function Y is regarded as proportional to the total enzyme activity under the operating conditions. For the case of changing operating conditions and enzyme activity decay takes the form of Eq. 2, where the multiple integral is taken between the vectors of initial and final operating conditions, $\bar{X}_o$ and $\bar{X}_F$, respectively, and time from zero to the recharging time τ.

$$Y(\bar{X},t) = \int_{\bar{X}_o}^{\bar{X}_F} \int_0^T f(\bar{X})\, \varphi(\bar{X},t)\, d\bar{X}\, dt \qquad \text{(Eq. 2)}$$

Usually, the operating conditions are kept constant; and the only changing component of the vector $\bar{X}$ is the degree of substrate conversion, X. To simplify Eq. 2, the rate of decay of activity is assumed to be independent of the conversion. Defining the function of inactivation under the operating conditions as $\rho_A(t)$ and the conversion function under the same conditions as $f_A(X)$, Eq. 2 may be transformed into Eq. 3:

$$Y(\bar{X},t) = K \int_{X_o}^{X_F} f_A(X)\, dX \int_0^T \varphi_A(t)\, dt \qquad \text{(Eq. 3)}$$

where K is the ratio of the initial activity under operating conditions to that under standard conditions; X_o is the initial degree of substrate conversion; and X_F is the final degree of substrate conversion.

Assuming that the total initial enzyme activity under operating conditions is proportional to that under standard conditions, the activity of the immobilized enzyme preparation used in one reloading cycle, A_1, may be expressed as follows:

$$A_1 = K\ A_o \qquad \text{(Eq. 4)}$$

To define function $W(\bar{X},t)$ the consumption of the substrate is regarded as being directly proportional to the formation of the product; and unreacted substrate is taken as being valueless. The molar yield of the product may differ from the stoichiometric yield due to side reactions and incomplete recovery and certainly depends on the operating conditions. Assuming, as mentioned above, that the operating conditions are kept constant and that the only changing component of vector $\bar{X}$ is the degree of conversion, X, the molar yield of the product may be ragarded as being a function of X. In this way, function $W(\bar{X},t)$ is defined as

$$W(\bar{X},t) = \frac{K}{X_F} \int_{X_o}^{X_F} \frac{f_A(X)}{\eta(X)} dx \Big/ \int_0^T \varphi_A(t)dt \qquad \text{(Eq. 5)}$$

where $\eta(X)$ is the dependence of the molar yield of the product on the degree of conversion. Usually the yield function, $\eta(X)$, is not defined and the yield of the product may be regarded as being constant at η. In this case Eq. 5 reduces to

$$W(\bar{X},t) = \frac{K}{X_F \eta} \int_{X_o}^{X_F} f_A(X)dX \Big/ \int_0^T \varphi_A(t)dt \qquad \text{(Eq. 6)}$$

From comparison of Eqs. 3 and 6 it is evident that

$$W(\bar{X},t) = \frac{Y(\bar{X},t)}{X_F\ \eta} \qquad \text{(Eq. 7)}$$

In order to express the cost of the immobilized enzyme preparation C_C in an expanded form, Eq. 4 is substituted into the appropriate part of Eq. 1 to give Eq. 8, which can aid in defining the optimum conditions for the process of enzyme immobilization.

$$C_C = \frac{A_o P_E}{\gamma} + C_I = \frac{A_I P_E}{\gamma K} + C_I \qquad \text{(Eq. 8)}$$

When analyzing the operating conditions, the cost of the immobilized enzyme preparation for one reloading may be taken as an arbitrary constant, C_C.

If we now express the integral of $f_A(X)dt$ in Eqs. 3 and 6, which gives the average relative effective rate as a function of X, in terms of $\psi_A(X)$, we obtain

$$\psi_A(X) = \int_{X_o}^{X_F} f_A(X)dX \qquad \text{(Eq. 9)}$$

The form of the function $\psi_A(X)$ depends on the basic kinetic equation and on the type of reactor. Later we will define $\psi_A(X)$ for two typical cases. For specific values of X_F, the function $\psi_A(X)$ may be replaced by a coefficient ψ_A.

Thus, for the optimization of the operating conditions, the process efficiency index may be represented in the following form:

$$E_P = \frac{P_P \, K \, A_o \psi_A(X) \int_0^{\tau} \varphi_A(t)dt}{\frac{P_S K A_o}{X \, \eta} \psi_A(X) \int_0^{\tau} \varphi_A(t)dt + C_C + C_R \, \tau} \qquad \text{(Eq. 10)}$$

The decay in activity under fixed conditions depends on the reloading time τ according to Eq. 11.

$$Q_A(\tau) = \int_0^{\tau} \varphi_A(t)dt \qquad \text{(Eq. 11)}$$

Thus, if the total initial activity according to Eq. 4 is introduced and Eq. 10 rearranged, we get Eq. 12, where K_p is the coefficient for the rise in the molar price due to the enzymatic reaction.

$$E_p = \frac{K_p \, X_F \, \eta \, \psi_A(X) \, Q_A(\tau)}{\psi_A(X) \, Q_A(\tau) + X_F \, K_E} \qquad \text{(Eq. 12)}$$

$$K_p = \frac{P_p}{P_S} \qquad \text{(Eq. 13)}$$

K_E is a coefficient derived from the operating costs and is expressed as

$$K_E = \frac{\eta\,(C_C + C_R T)}{P_S\;A_I} \qquad \text{(Eq. 14)}$$

In Eq. 12 the terms K_E and $Q_A(T)$ have the dimension of time, while K_p, X_F, η and $\psi_A(X)$ are dimensionless.

The first inference from Eqs. 12 and 14 is that the reactor must be loaded with the maximum possible amount of immobilized enzyme preparation, i.e. the maximum amount of activity A_1. Since the treatment of Eqs. 10 and 12 offers marked difficulties, some simplified forms of these basic equations and some aspects of the optimization of the process are discussed.

II. THE DURATION OF THE RELOADING CYCLE

When dealing with the catalyst exhaustion and the reactor reloading, we can assume that the conditions are fixed; therefore $\bar{X}$ becomes constant. This means that continuous reactors must be operated at a constant final degree of substrate conversion X_F. This would be the degree of conversion at the reactor outlet for a plug flow reactor, at the end of the run for a batch reactor, and at steady state for a continuous stirred tank reactor. Subsequently, the function $\psi_A(X)$ in Eqs. 11 and 12 may be replaced by a constant, $\psi_A(X_F)$.

Thus, the average operational activity of the immobilized enzyme preparation A_R will be proportional to the initial activity under the operating conditions A_1:

$$A_R = \psi_A(X_F)\;A_I \qquad \text{(Eq. 15)}$$

where ψ_A (X_F) is the coefficient of proportionality at a fixed degree of conversion.

$$\varphi_A = \varphi_A(X_F) \qquad \text{(Eq. 16)}$$

If X_F and η have constant values, the real molar consumption of the substrate for obtaining one mole of the product, K_y, may be expressed as follows

$$K_y = \frac{1}{X_F \eta} \qquad \text{(Eq. 17)}$$

Consequently, in this case the cost of the substrate, C_S, the first term of the denominator in Eq. 10, will be proportional to the operational activity, A_R, and the real molar consumption of the substrate, K_y:

$$C_S = P_S \ A_R \ K_y \ Q_A(\tau) \qquad \text{(Eq. 18)}$$

By analogy, the cost of the product, C_P, the numerator in Eq. 10, may be given as follows:

$$C_p = P_p \ A_R \ Q_A(\tau) \qquad \text{(Eq. 19)}$$

Assuming, that the immobilized enzyme inactivation reaction follows first order kinetics, there is an equation available to describe function $\varphi_A(\tau)$:

$$\varphi_A(t) = e^{-kt} \qquad \text{(Eq. 20)}$$

From Eq. 11 we get

$$Q_A(\tau) = \frac{1}{k}(1 - e^{-k\tau}) \qquad \text{(Eq. 21)}$$

where k is the rate constant of inactivation in reciprocal hr.

The maximum amount of product can be obtained only with total exhaustion of the enzyme activity; for this an infinite time would be necessary. However, the index of efficiency, E_p, can be optimized with regard to the coefficient of activity for exhaustion $Q_E(\tau)$, defined as

$$Q_E(\tau) = k\, Q_A(\tau) = 1 - e^{-k\tau} \qquad \text{(Eq. 22)}$$

Using the value of Q_E from Eq. 22, the values of C_S and C_p from Eqs. 18 and 19, respectively, and the value of K_y from Eq. 17, the general Eqs. 10 and 12 are transformed into the following forms:

$$E_p = \frac{P_p \; A_R \; Q_E(\tau)}{P_s \; A_R \; K_y \; Q_E(\tau) + k(C_C + C_R \tau)} \qquad \text{(Eq. 23)}$$

$$E_p = \frac{K_p}{K_y} \; \frac{Q_E(\tau)}{Q_E(\tau) + kX_F \; K_E / \varphi_A} \qquad \text{(Eq. 24)}$$

The dependence of the characteristics of the reactor on $Q_E(\tau)$ is presented in Fig. 1. One may see that the maximum yield (Curve 1) leads to the minimum average activity (Curve 3). Subsequently, a reasonable activity limit of the catalyst must be chosen. By differentiation of Eq. 23 with respect to time, it appears that the maximum efficiency is obtained at a certain reloading time τ, which can be calculated from Eq. 25:

$$e^{k\tau} = 1 + k\tau + K_B \qquad \text{(Eq. 25)}$$

Here, the dimensionless criterion K_B is calculated from the economic characteristics by Eq. 26:

$$K_B = \frac{k \; C_C}{C_R} \qquad \text{(Eq. 26)}$$

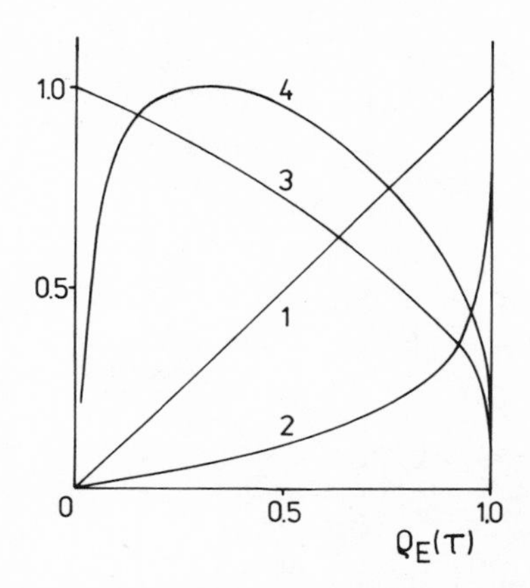

Fig. 1. Simulated operational characteristics of a β-galactosidase reactor in terms of the coefficient of activity for exhaustion $Q_E(\tau)$. Curve 1: total amount of product; Curve 2: duration of reloading cycle $10\tau/\tau_{1/2}$; Curve 3: relative average activity; Curve 4: relative efficiency of the process, E_p.

Thus K_B is proportional to the ratio of the expenditure for catalyst to that for the operation of the reactor. We note now that the optimum reloading time for the reactor is determined solely by the criterion of the cost structure. These functions are given graphically in Fig. 2. Hence, from the analysis of Eqs. 23 and 24 an important conclusion for the exploitation of immobilized enzyme reactors is derived, namely that the optimum reloading time increases with the criterion K_B.

III. EFFICIENCY CRITERION FOR IMMOBILIZATION

The above equations suggest that E_p should increase with a decrease in both k and C_C. Assuming

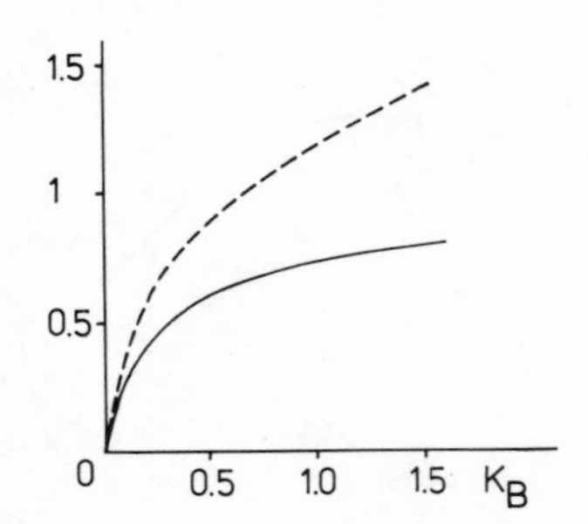

Fig. 2. Simulated optimum coefficient of activity for exhaustion Q_E (τ) (solid line) and for dimensionless reloading time k_τ (dashed line) as functions of the economic coefficient K_B.

the initial catalyst activity A_y and the activity yield γ to be independent of each other, we reach a trivial conclusion that the stability always must be maximized and the costs for immobilization minimized. However, these characteristics usually exhibit strong interdependency, and high activities are obtained due only to the decrease in the yield coefficient γ . Our experimental data (Fig. 3) show that linearity between the reciprocals $1/A_1$ and $1/\gamma$ can be obtained, so that the hyperbole type function, shown in Eq. 27, may be proposed to describe the activity versus yield relationship.

$$A_I = \frac{A_E \; A_M}{A_C + A_E} \qquad \text{(Eq. 27)}$$

In this equation A_E is the activity of the native enzyme; while A_M and A_C are empirical constants with the dimensions of activity. Using Eq. 26 in Eq. 24, we conclude that there must exist a maximum efficiency which depends on the amount of the

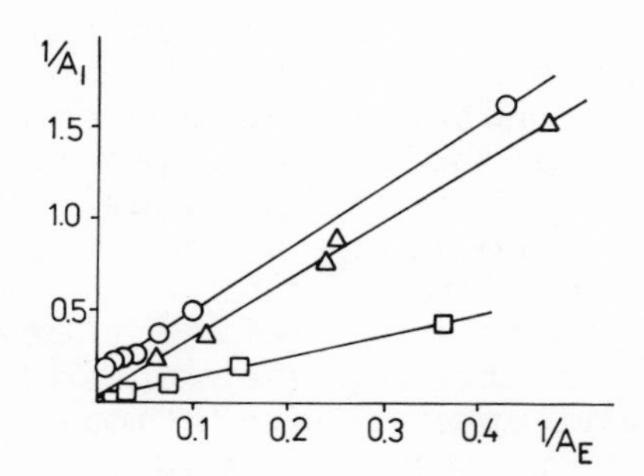

Fig. 3. Examples to show the relationship between the activity of the native enzyme A_E and the activity of the immobilized enzyme preparation A_1. Squares are for penicillin amidase entrapment in modified polyacrylamide gel, triangles are for β-fructosidase entrapment in modified polyacrylamide gel, and circles are for β-galactosidase covalently bound to a siliceous carrier.

native enzyme used. The corresponding optimum amount of enzyme can be calculated from the combination of Eqs. 8 and 27. In this way, we have reached another important conclusion about the necessity to optimize the immobilization process with regard to activity.

The performance of the immobilized enzyme preparation depends on several interacting variables in the immobilization process, so that it is advisable to optimize this process by use of planned multifactorial experiments and a quantitative criterion of efficiency. The exact values of the efficiency may be calculated from Eq. 23 only when all of the appropriate economical indices are available. During the initial development of a process, the economics usually have not yet been defined; and approximate optimization procedures must be used. With no information about the economics, the immobilization efficiency E_1 may be estimated as equal to $A_o\ \tau_{1/2}\gamma$, where $\tau_{1/2}$ is the half-life of the immobilized enzyme preparation under the operating conditions. When additional information about the process under consideration is available, this general criterion may be modified. For example, using an expensive enzyme and assuming that the economics of the process will depend mainly on this factor, it seems more reasonable to define the efficiency as $E_1 = \gamma\tau_{1/2}$. In contrast, when the costs due to immobilization and operation are more significant, the efficiency E_1 is nearly equal to $A_o\ \tau_{1/2}$.

IV. OPERATING CONDITIONS

The operating temperature, one of the most important variables, has been dealt with by several authors (7,8). Here we will concentrate our attention on the substrate conversion, which also plays a substantial role in the total efficiency.

With this in mind we can modify Eq. 12 to omit activation, and we can assume that the separation of the desired product or subsequent application does not depend on the degree of substrate conversion. Let us also consider the immobilization conditions and reloading time as being independently optimized. Assuming also that the final degree of substrate conversion is kept constant, which is always possible by regulation of the space time in the reactor, and is equal to X_F, Eq. 12 may be given in the following form:

$$E_p = \frac{K_p\ X_F\ \eta}{1 + \dfrac{k\ X_F\ K_E}{\varphi_A(X_F)}} \qquad \text{(Eq. 28)}$$

The substrate conversion, expressed as the integral of $f_A(X)dX$, is determined by the kinetic equation of the enzymatic process, and by the type of reactor. Here we recommend the use of a general kinetic equation, (Eq. 29) described elsewhere (9), for the general description of various two step single substrate enzymatic reactions.

$$-\frac{d[S]}{dt} = \frac{k_{cat}[E]_o(1-X)}{a + bX + cX^2} \quad \text{(Eq. 29)}$$

In Eq. 29, [S] and $[E]_o$ are the substrate and initial enzyme concentrations, respectively; k_{cat} is the catalytic rate constant; and a,b, and c are combinations of basic kinetic constants. In the case of an incompletely described reaction scheme at constant initial substrate concentration, a, b, and c may be considered as empirical coefficients. We have demonstrated that this equation is valid for the description of the enzymatic hydrolysis of benzylpenicillin, lactose, and saccharose (9).

If an enzymatic reaction produces one mole of product from one mole of substrate, Eq. 29 may be given also as follows:

$$\frac{d[P]}{dt} = \frac{[S]_o\,dX}{dt} = \frac{V_m(1-X)}{a + bX + cX^2} \quad \text{(Eq. 30)}$$

where [P] is the concentration of the product and V_m is the maximum velocity equal to k_{cat} $[E]_o$.

We have shown earlier (9) that the effective relative rate $\psi_A(X)$ may be expressed for different types of reactors. If we assume there is no product in the initial solution of substrate, then X_F will be a constant for a continuous stirred tank reactor at steady state; and the rate of reaction can be described by Eq. 29. In this case, the effective relative rate at X_F may be expressed as follows:

$$\psi_{A(1)}(X_F) = \frac{1 - X_F}{a + bX_F + cX_F^2} \quad \text{(Eq. 31)}$$

For reactors where the conversion changes with time, i.e. a batch reactor, or with length, i.e. a plug flow reactor, and with initial conditions of $\tau = 0$ and $X_o = 0$, the corresponding function has the following form:

$$\varphi_{A(2)}(X_F)=\frac{-X_F}{(a+b+c)\ln(1-X_F)+(b+c)X_F+CX_F^2/2} \quad \text{(Eq. 32)}$$

which characterizes the average rate of reaction in reactors of this type. As coefficients a, b, and c depend on the initial substrate concentration, the comparison of the efficiency of the two types of reactors must be carried out at the same values of $[S]_o$ and of a, b, and c.

Using Eqs. 31 and 32 in Eq. 28, the dependence of the efficiency on the final conversion X_F may be calculated for the two types of reactors; and the optimum X_F can be estimated. As examples we con - sider two enzymes, penicillin amidase and β-galactosidase. The first one is used for the treat ment of an expensive substrate, benzvlpenicillin, while the other is proposed for the treatment of cheap cheese whey waste. The values of the kinetic and economic coefficients have been calculated or estimated from our experiments or taken from the literature (5,6) and are given in Table 1. The results of the calculations are given in Fig. 4. From these results we concluded that an optimum reaction extent always exists. Under the same initial conditions, the continuous stirred tank reactor is always less effective and exhibits maximum efficiency at a lower conversion. The optimum conversion is again determined by the ratio of the economic indices. The criterion K_E (Eq. 28) mainly determines the optimum degree of substrate conversion. This criterion expresses the ratio of the total exploitation costs to the cost of the substrate. For arbitrary kinetic coefficients, the optimum X_F as a function of K_E is given in Fig. 5.

The relationship in Fig. 5 demonstrates that a lower conversion can be expected in the case of high K_E values and constant k. When processing a cheap substrate, a lower conversion could be tolerated, as it would result in a higher operational

TABLE 1

COEFFICIENTS FOR IMMOBILIZED ENZYME EXAMPLES

Coefficients	Penicillin Amidase	β-Galactosidase
Kinetic[a]		
a	1.60	1.22
b	14.67	6.53
c	-13.78	-4.58
k (hr^{-1})	2×10^{-3}	2×10^{-3}
η	0.8	1.0
Economic		
K_p	3.0	4.0
K_E (hr)	5.0	350
k K_E	0.01	0.7

(a) Calculated for an initial substrate concentration of 0.15 M; a, b, and c were calculated for immobilized penicillin amidase from the data of Warburton et al. (6) and for immobilized β-galactosidase from our data (10).

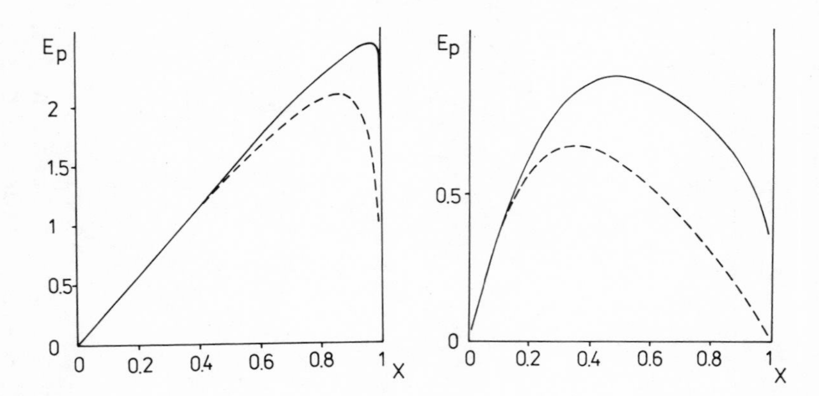

Fig. 4. Simulated efficiency of the process, E_p, as a function of the degree of conversion, X_F (Eq. 28) for reactors with changing conversion (solid line) and for a continuous stirred tank reactor at steady state (dashed line). Left side figure is for 6-aminopenicillanic acid production by penicillin amidase; right side figure is for acid whey treatment by β-galactosidase. For constants used in the simulations are given in Table 1.

activity of the catalyst. To illustrate this, we refer to the production of 6-aminopenicillanic acid; in this process it is necessary to obtain a high degree of substrate conversion in order to reduce the consumption of expensive benzylpenicillin. However, when treating cheap cheese whey with immobilized β-galactosidase, high conversion rates are not as important.

It must be pointed out that optimization of the conversion is especially important in the case of product inhibition. This type of inhibition seems to be rather usual, at least for all the hydrolases that we have investigated.

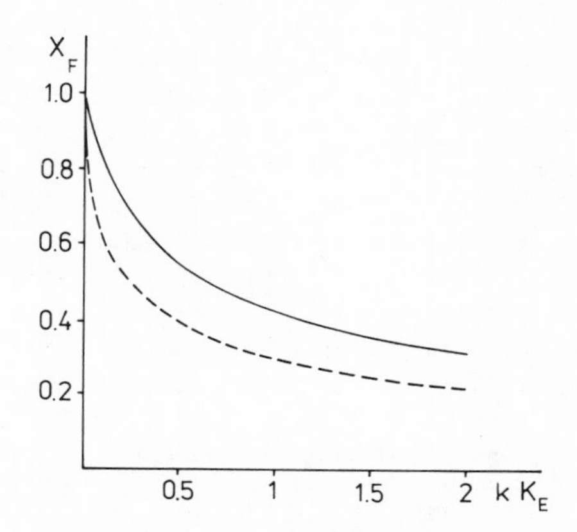

Fig. 5. Simulated optimum degree of substrate conversion X_F as a function of the product of the economic criterion K_E and the inactivation rate constant k, calculated using the kinetic coefficients for β-galactosidase in Table 1. Solid line is for reactors with changing conversion; and the dashed line is for a continuous stirred tank reactor.

V. APPLICATIONS AND CONCLUSIONS

The above theoretical approaches have been applied to the production of 6-aminopenicillanic acid (unpublished results). In cooperation with other institutes, the technology for the immobilization of penicillin amidase in a modified polyacrylamide gel has been developed and introduced on an industrial scale. In the USSR the entire production of 6-aminopenicillanic acid is based on the use of immobilized penicillin amidase; and the resulting industrial technology has proved to have high economic efficiency.

The technological process for the treatment of acid whey with silochrome bound β-galactosidase has been successfully extended to 50 liter pilot reactors (unpublished results).

In conclusion we have suggested a general efficiency criterion for the optimization of processes in enzyme technology. This approach is based on a combined kinetic-economic quantitative description and provides the following tentative guidelines: 1) the efficiency of the immobilization process is generally proportional to the multiple of activity, activity yield, and half-life; 2) the optimum duration of the reloading cycle of the reactor is determined by the ratio of the catalyst cost to the operational cost; and 3) the optimum degree of substrate conversion is determined by the ratio of the total exploitation costs, including the immobilized enzyme preparation, to the substrate cost.

In this work many assumptions have been made. Space limitations have prevented us from giving a clearer picture of the detailed development of the main equations and the experimental verification of these assumptions. In addition, the true test of the usefulness of this work will come when we can present a detailed analysis of a specific immobilized enzyme process, using actual experimental data and actual economic values obtained from plant scale or larger pilot scale operation.

ACKNOWLEDGMENT

The authors appreciate the skilled collaboration of Mrs. K. Pappel in the course of the kinetic experiments and Mrs. M. Semper for the economic calculations.

NOMENCLATURE

A	total activity of immobilized enzyme preparation used in one reactor reloading cycle, moles/hr
A_o	initial standard activity (maximum activity)
A_C	empirical constant defined in Eq. 27, moles/hr
A_E	activity of the native enzyme
A_M	empirical constant defined in Eq. 27, moles/hr
A_R	average operational activity at operating conditions
A_1	initial activity at operating conditions
a	kinetic coefficient defined in Eq. 29
b	kinetic coefficient defined in Eq. 29
C_C	cost of loaded immobilized enzyme preparation defined in Eq. 8
C_E	whole exploitation expense during one reloading cycle, including the costs of substrate and enzyme, and the expenses for the operation of the reactor
C_1	expenses for carrier, immobilization, and reloading of reactor
C_p	cost of the product over one reloading cycle
C_R	expenses for the operation of the reactor per hr
C_S	cost of consumed substrate per Eq. 18
c	kinetic coefficient per Eq. 29
$[E]_o$	initial concentration of enzyme
E_1	criterion for immobilization efficiency
E_p	dimensionless efficiency index for a process, per Eq. 1
$f(\bar{X})$	function determining the reaction rate as influenced by operating conditions
$f_A(X)$	same as previous symbol but considering only the degree of substrate conversion

K	ratio of the initial enzyme activity under operating conditions to that under standard conditions, A_1/A_o
K_B	economic coefficient per Eq. 26
K_E	economic coefficient per Eq. 14, hr
K_p	coefficient for molar price rise, P_p/P_S
K_y	coefficient for substrate consumption, per Eq. 17
k	rate constant for enzyme inactivation at operating conditions, 1/hr
k_{cat}	catalytic rate constant
$[P]$	concentration of product
P_E	price of one activity unit, with standard activity expressed as moles/hr, of the initial native enzyme preparation
P_p	price/mole of product
P_S	price/mole of substrate
$[S]$	concentration of substrate
$[S]_o$	initial concentration of substrate
t	time, hr
V_m	maximum activity of enzyme
$W(\bar{X},t)$	function of substrate consumption, per Eq. 5, hr
X	degree of substrate conversion
X_o	initial degree of substrate conversion
X_F	final degree of substrate conversion
$\bar{X}$	operating conditions presented as a vector
$\bar{X}_o$	initial operating conditions
$\bar{X}_F$	final operating conditions
$Y(\bar{X},t)$	function of the product formation, per Eq. 2, hr
γ	yield of activity from the immobilization process
η	molar yield of product
$\eta(X)$	function determining the molar yield of product
$Q_A(\tau)$	function for activity decay, per Eq. 11
$Q_E(\tau)$	coefficient of activity exhaustion, per Eq. 22
$\tau_{1/2}$	half-life of the catalyst, hr

τ	duration of the reactor reloading cycle, hr
$\rho(\bar{X},t)$	function for activity decay
$\rho_A(t)$	the same as previous symbol but at fixed conditions
ΨA	average effective relative rate at fixed conditions
$\Psi_A(X)$	function for average effective relative rate, per Eq. 9
$\Psi_A(X_F)$	same as previous symbol but at final degree of conversion

REFERENCES

1. SWANSON, S. J., EMERY, A. & LIM, A. C. *J. Solid-Phase Biochem.* *1:* 119, 1976.
2. LAMBA, H. S. & PUDUKOVIC, M. P. *Adv. Chem. Ser.* *133:* 106, 1974.
3. KOSTNER, A. I. Abstr. No. 3827 12-th FEBS Meeting, Drezden, 1978.
4. KOSTNER, A. I. *Trudy Tallinsk. Polytechn. Inst.* (Russ.) *424:* 9, 1977.
5. PITCHER, W. H., FORD, J. R. & WEETALL, H. H. *Meth. Enzymol.* *44:* 792, 1976.
6. WARBURTON, D., DUNNILL, P. & LILLY, M. D. *Biotechnol. Bioeng.* *15:* 13, 1973.
7. HO, L. Y. & HUMPHREY, A. E. *Biotechnol. Bioeng.* *12:* 291, 1970.
8. MARSH, D. R. & TSAO, G. T. *J. Solid-Phase Biochem.* *1:* 67, 1976.
9. SIIMER, E. H. *Biotechnol. Bioeng.* (in press).
10. SIIMER, E. & PAPPEL, K. *Biotechnol. Bioeng.* (in press).

SECTION V
ENZYMES IN FOOD AND NUTRITION

NOVEL ENZYMATIC PRODUCTION OF L-MALIC ACID AS AN ALTERNATIVE ACIDULANT TO CITRIC ACID

F. Giacobbe,* A. Iasonna,* W. Marconi,**
F. Morisi** and G. Prosperi**

Biochem Design S.p.A.* and
Assoreni Microbiological Laboratory**
Rome* and Monterotondo,** Italy

During the production of phthalic anhydride, the crude fumaric acid, which is obtained as a by-product, is often purified to obtain a food grade product which is then sold as a food acidulant. However, because of the low solubility of fumaric acid, its sales volume is limited. If crude fumaric acid could be converted to L-malic acid, then a much larger market would be created for this product. This would allow savings in the cost of acidulants for the formulation of processed foods and beverages. Although processors are reluctant to change ingredient formulations, they are most happy to use the lower priced items when fabricating new products. However in such products as infant foods, where DL-malic acid was recently banned by the Food and Drug Administration in the United States,[1] there is an immediate market for the L form of malic acid. A process for conversion of fumaric acid to L-malic acid has been developed on a laboratory scale at the Microbiological Laboratories of Snamprogetti, with the assistance of Biochem Design; this process is discussed in this paper.

I. DESCRIPTION OF THE PROCESS.

Crude fumaric acid is suspended in water and neutralized to pH 7.2. The soluble fumarate is mixed with the unconverted fumarate before being sent to the hydration reactor, where about 60% of the fumarate salt is enzymatically converted to L-malate using a fiber entrapped fumarase. The L-malic acid is precipitated as the calcium salt and is then recovered by filtration. The unconverted fumarate is recycled to the reactor to improve the conversion of the fumaric acid to L-malic acid. The calcium L-malate is suspended in water and solubilized with sulphuric acid, which then precipitates calcium sulphate. The calcium sulphate is removed by filtration and sent for disposal, following water washing. The L-malic solution is concentrated; and the L-malic acid is recovered by crystallization. The L-malic acid crystals are then separated by centrifugation and dried. The mother liquor is recycled to the sulphuric acid solubilization stage.

The fumarase is obtained by growing *Pseudomonas* sp. on a medium (2) containing:

Corn steep liquor	50 g/l
Glucose	10 g/l
Fumaric acid	5 g/l

and the enzyme is entrapped in the fiber using the procedure described by D. Dinelli (3). The fumarase purification steps necessary before entrappment, are shown in Fig. 1. The yield of the enzyme sufficiently pure for entrappment, is about 80% of the enzyme present in the microbial cells. In the wet-spinning immobilization process, the purified enzyme is dissolved in a buffer solution containing stabilizing agents. This is then mixed with cellulose triacetate, dissolved in a solvent that is not miscible with water, to obtain an emulsion. The mixture is filtered and forced through spinneret holes into a coagulation bath to obtain the fibers with the enzyme trapped within the cellulose triacetate matrix.

The pH activity profile of the entrapped enzyme was studied. The results are given in Fig. 2. Working with a substrate concentration of

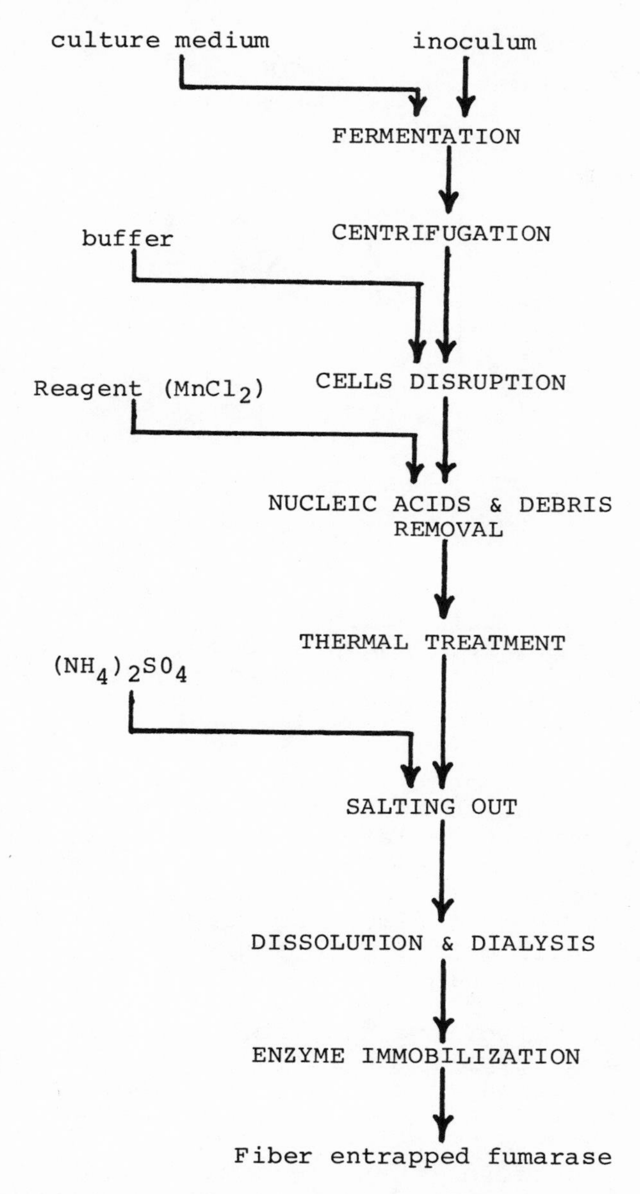

Fig. 1. Flow diagram for purification of fumarase. The overall recovery of fumarase is 80%, with an activity after dialysis of 4.2 μmoles/min/mg protein.

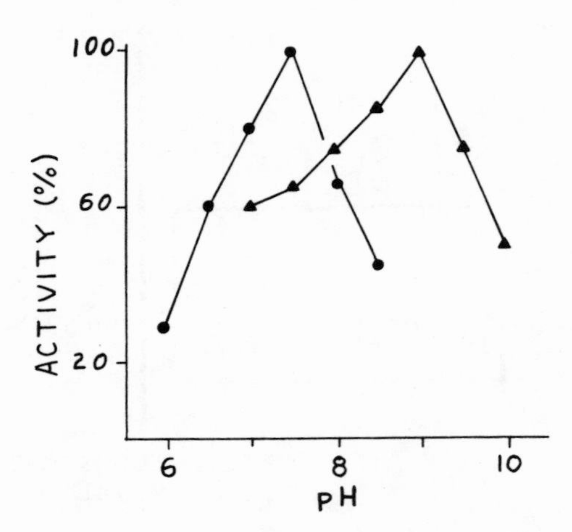

Fig. 2. Activity of fumarase entrapped in cellulose triacetate fibers for different pH values. Fumarate to malate (circles); malate to fumarate (triangles); 25°C and 0.42 M fumarate.

0.43 M, it was found that the optimal pH values were 9.0 for the malate to fumarate reaction and 7.5 for the fumarate to malate reaction. Fig. 3 illustrates the effects of substrate concentration on the activity of fiber entrapped fumarase, showing that full activity is achieved at substrate concentrations higher than 1 M. The effect of temperature on the activity of the entrapped enzyme was also investigated. The Arrhenius diagram for a temperature range from 5°C to 30°C is shown in Fig. 4. From this plot the activation energy of 9.06 Kcal/mole was calculated.

It was found that the overall reactor rate equations could be described as follows for the conversion of fumarate to L-malate in a batch reactor (Eq. 1) and a continuous flow reactor (Eq. 2):

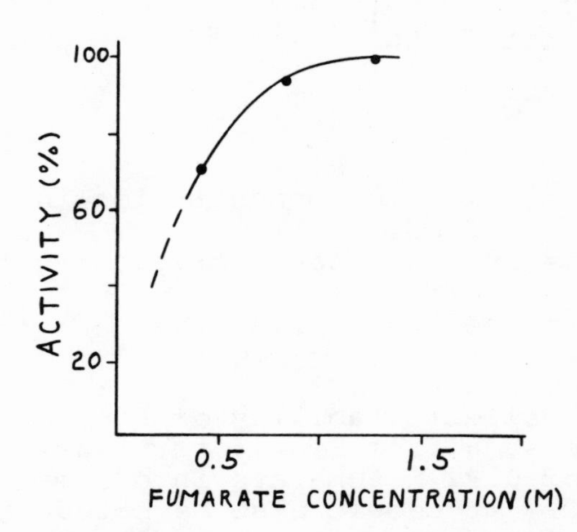

Fig. 3. Activity of fiber entrapped fumarase versus substrate concentration.

$$\ln\left[\frac{F_i - F_{eq}}{F_o - F_{eq}}\right] = K \frac{E}{V} t \qquad \text{(Eq. 1)}$$

where:

F_i = initial concentration of fumarate, M
F_o = concentration of fumarate at time t, M
F_{eq} = fumarate concentration at equilibrium, M
t = reaction time, hr
V = reaction volume, liter
E = amount of catalyst, g
K = kinetic constant depending on fiber activity and on initial concentration of fumarate

$$\ln \left[\frac{F_{in} - F_{eq}}{F_{out} - F_{eq}} \right] = K \frac{E}{Q} \qquad \text{(Eq. 2)}$$

where:

F_{in} = fumarate concentration in the reactor feed, M
F_{out} = fumarate concentration in the reactor effluent, M
Q = feed rate, liter/hr

The excellent stability of the entrapped enzyme can be seen from Fig. 5 for continuous transformation of 0.42 M fumarate into L-malate at pH 7.2 at 25°C. It should also be noted that the entrapped enzyme is very resistant to bacterial contaminations.

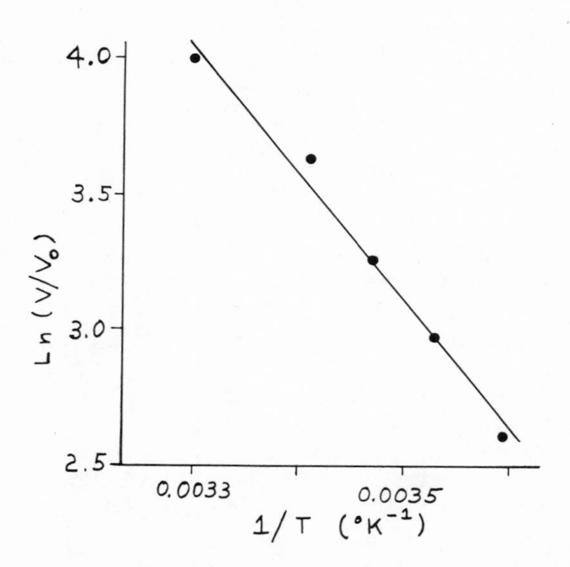

Fig. 4. Arrhenius plot for the reaction fumarate to L-malate. 0.45 M fumarate.

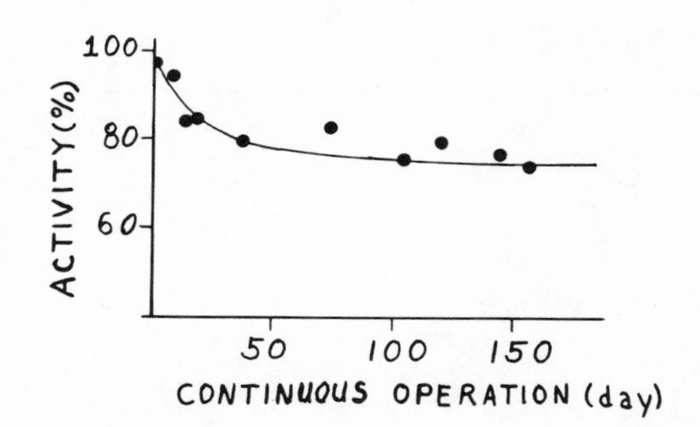

Fig. 5. Activity of fiber entrapped fumarase versus time for the fumarate to L-malate reaction.

II. ECONOMIC EVALUATION OF THE PROCESS.

Using the results obtained from the bench scale operation, it is possible to extrapolate the potential of this process for use on an industrial scale. A process design, based on laboratory data, has been made to estimate the capital and operating costs for a plant of 7,000 tons/yr capacity. The results are summarized in Table 1, which shows an estimated investment cost. Table 2 shows the estimated manufacturing cost, which is described further in a preliminary cost model of the process in Fig. 6. The cost of the fumaric acid, which is the major component of the manufacturing cost, depends on internal and external factors, such as:

1. Cost of the conversion of byproduct maleic acid, obtained in the production of phthalic anhydride, to crude fumaric acid.

2. Market demands for food grade fumaric acid.

TABLE 1

ESTIMATED INVESTMENT COST(a)

Item	Cost ($1,000)
Process Equipment	1027
Piping	371
Instruments	85
Electricals	55
Civil Works & Structure	404
Labor	558
Total	2500

(a) Basis: plant capacity of 7000 tons/yr

The effect of variations in the costs of the crude fumaric acid feedstocks, chemicals and catalyst, utilities, investment, plant capacity, and sales volume on the manufacturing cost is shown in Fig. 7. It can be seen from both Figs. 6 and 7, that the cost of the fumaric acid is the most important component of the manufacturing cost. The L-malic acid manufacturing cost varies from \$483/ton to \$764/ton as the cost of fumaric acid changes from \$100/ton to \$400/ton. It should also be noted that in the calculation of the manufacturing cost of the L-malic acid in Table 2, the cost of the fumaric acid has been assumed to be \$271/ton. At this price the estimated manufacturing cost of the L-malic acid is then equal to that of citric acid without incurring any of the

TABLE 2

ESTIMATED MANUFACTURING COST[a]

Item	Cost ($/ton)
Crude Fumaric Acid[b]	253.6
Catalyst and Chemicals	104.8
Utilities	115.4
Other Direct Costs	59.1
Total Direct Manufacturing Cost	532.9
Depreciation	88.2
Other Overheads	23.3
Total Indirect Manufacturing Cost	111.5
Total Manufacturing Cost	644.4

(a) Basis; plant capacity 7000 tons/yr

(b) Assumed cost of $271/ton

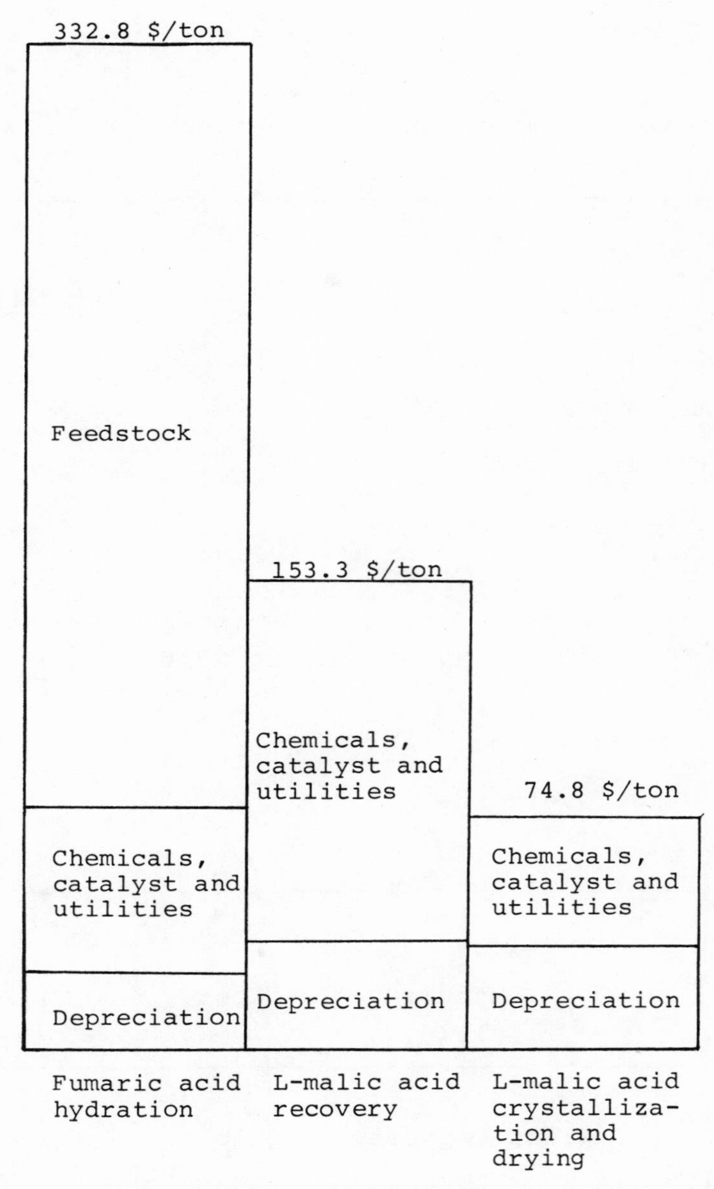

Fig. 6. Preliminary cost model of 7000 tons/yr L-malic acid plant.

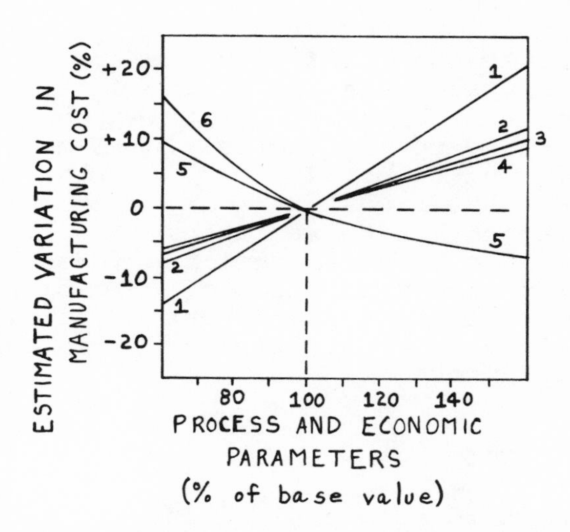

Fig. 7. Estimated variation in the manufacturing cost for different process and economic parameters, based on a manufacturing cost of $644.4/ton. Curves refer to: 1 fumaric acid cost; 2 utilities cost; 3 chemicals and catalyst cost; 4 investment cost; 5 plant capacity cost factors; 6 sales volume cost factors.

plant depreciation charges of citric acid production.

In order to understand fully the potential of L-malic acid, a comparison between the processes of L-malic acid and citric acid should be made. Citric acid is produced by fermentation using molasses as the substrate. The citric acid produced is recovered, after broth filtration, by precipitation as the calcium salt. L-malic acid is produced by an enzyme catalyzed reaction and is also recovered as the calcium salt. The purification of both the citric and the L-malic acid then follows identical processing steps. It is evident that, if calcium L-malate can be produced at a lower cost than that of the calcium citrate, then the cost of production of the L-malic acid will be lower than that of the citric acid. The cost of raw materials and utilities to produce calcium citrate has been estimated to be $430/ton, while the cost to produce calcium L-malate has been estimated to be $385/ton. In this estimate 62% of the latter cost is that of the fumaric acid feedstock, which was assumed to be $271/ton. It should also be noted that the investment cost required for the production of calcium citrate is much larger than the investment cost for the production of calcium L-malate; and this again favors the L-malic process.

III. CONCLUSIONS.

If fumaric acid can be made available as a by-product derived from pollution control, it can be recovered at a price substantially lower than $271/ton. In this case, L-malic acid could be produced at considerably lower prices than that of the citric acid. Then the L-malic acid could compete with citric acid price wise, even if the investment cost of the citric acid plant had been fully amortized. The above considerations do not take into account waste disposal problems. Here again, the L-malic acid is favored since in the case of citric acid production by fermentation, there is the spent broth and the mycelium to discard. Further research work, however is still needed to assess the full feasibility of the L-malic acid process; and there are areas for improvement in respect to the results already obtained. These areas may include reduction of the catalyst cost and optimization of the operating conditions. The prospects however look very promising indeed.

REFERENCES

1. Anon. *Chemical Marketing Reporter*, Sept. 5, 1977.
2. ZAFFARONI, P., ODDO, N., OLIVIERI, R. & FORMICONI, L. *Agr. Biol. Chem.* 39: 1875, 1975.
3. DINELLI, D. *Process Biochem.* 7: 9, 1972.

APPLICATION OF PLANT PHENOL OXIDASES IN BIOTECHNOLOGICAL PROCESSES

G. N. Pruidze

Institute of Plant Biochemistry
Adademy of Sciences,
Georgian SSR, Tbilisi, USSR

Phenol oxidase (o-diphenol: O_2 oxidoreductase, E.C.1.10.3.1.) belongs to the copper-containing class of enzymes. It is widely distributed in plants, and plays an important role in cell metabolism. The enzyme catalyzes the oxidation of monophenols, polyphenols, aromatic amino acids, and amines. Plant phenol oxidases exhibit different substrate specificities, depending on their molecular weight and other physico-chemical properties (1-7).

The problems in the isolation and study of the properties of tea plant and grapevine phenol oxidases and the prospects for application of this enzyme in the food industry are discussed in this paper.

PHENOL OXIDASE USES

Phenol oxidases are widely utilized in the processing of various plant raw materials. The manufacture of many food products is based on the action of phenol oxidases on raw plant material. Thus, the oxidation of polyphenol compounds catalyzed by phenol oxidases is the basis for production of all kinds of tea, some types of wine, and many natural dyes. The products of the enzymatic oxidation of tea leaf polyphenol compounds largely determines the nature and quality of tea. The biochemical processes leading to the formation of various tea types, such as green, yellow, red, or black, may be directed by controlling the action of tea leaf phenol oxidases. As a result of the rational combination of a) enzymatic processes occuring under the influence of phenol oxidase and b) nonenzymatic transformations occurring at high temperature, high-quality soluble tea can be obtained from tea leaves.

Phenol oxidases are used in the manufacture of natural plant dyes. As is known, many synthetic dyes are harmful to human health. Therefore, extensive studies have been carried out recently to search for natural but harmless plant dye pigments, capable of replacing synthetic dyes. The technology of manufacture of natural yellow, green, brown, orange, red and black dyes rich in vitamin P from tea raw materials and from red beets, has been worked out on the basis of our studies (8,9). These dyes are harmless and contain various natural coloring substances, vitamins, and other compounds valuable for the human organism. The dyes are utilized in the food industry. Besides, they are finding applications in medicine, textiles, perfume, leather, and other industries.

Natural dyes are obtained by the control of oxidative conversion of polyphenol compounds under the influence of polyphenol oxidases. The rate of pigment formation depends on the enzymatic activity, substrate composition, amount of substrate, and chemical nature of the product. In the technological preparation of dyes, catechins with 3,4,5-trioxy substitution undergo mainly oxidative condensation, leading to the formation of theaflavin (Fig. 1) and thearubigin type dye products. It should be noted that the activity of phenol oxidases is inhibited by the products of catechin oxidation.

In this case the inhibitory effect depends on the chemical structure and the amount of the phenol oxidase catalyzed reaction product. In

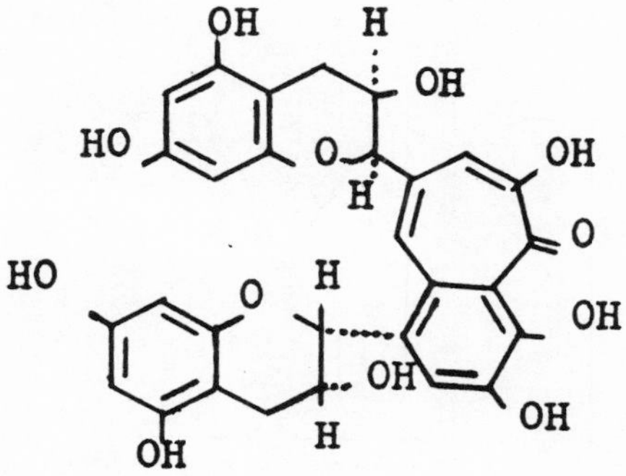

Fig. 1. Structure of theaflavin

accordance with the data given in Fig. 2, the total catechin preparation is intensively oxidized under the influence of phenol oxidase. The addition of theaflavines and thearubigins to the reaction medium results in a decrease of enzyme activity. With the increase in inhibitor concentration, the activity of phenol oxidase gradually decrease; and at a certain relation between catechin and inhibitors almost complete inhibition of enzymic activity is obtained.

It should be noted, that theaflavines are stronger inhibitors of phenol oxidase than are thearubigins. At the same time these compounds suppress the activity of the oxidizing enzymes more intensively than that of phenol oxidase, thereby regulating the rate of enzymatic oxidative processes in dye manufacturing.

Phenol oxidase activity is highly influenced by the pH and temperature of the medium and by the nature and concentration of the substrates and metabolites. Under certain conditions the state of phenol oxidase in leaf cells may be destroyed,

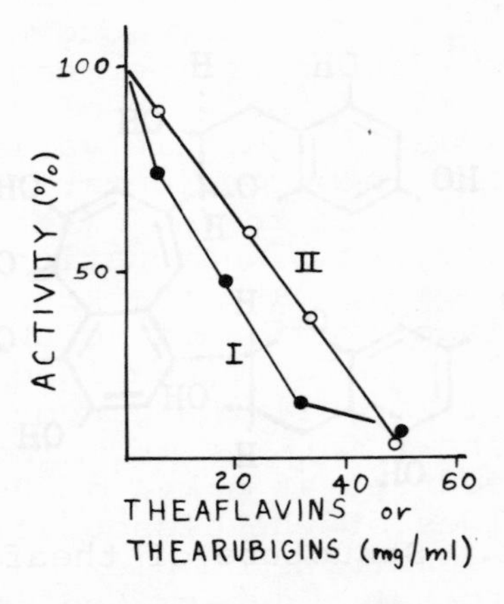

Fig. 2. The effect of theaflavins, I, and thearubigins, II, on phenol oxidase activity.

leading to a considerable change in the enzymatic activity. Tea plant phenol oxidase is activated by ethanol, acetone, and exposure to certain reagents. It should be noted, that the isolated enzyme preparation is distinguished for its higher activity as compared to that for the enzyme localized in the cells (Fig. 3). This supposedly is caused by the availability of natural phenol oxidase inhibitors in plants. As is seen from our studies, tea plants contain phenolic compounds that regulate the activity of oxidative enzymes. These compounds inhibit peroxidase more intensively than phenol oxidase (10). The quantitative content of the natural inhibitors in tea plants determines the rates and direction of enzymatic processes. The inhibitory activity of the compounds involved in phenol oxidase activity is partially removed by some amino acids and other metabolites. As is seen from Table 1, phenylalanine, glycine, and threonine are capable of removing the inhibition caused by the

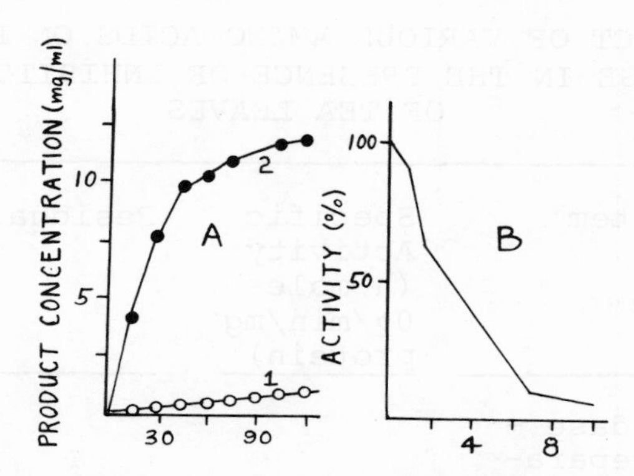

Fig. 3. Part A: Product concentration from phenol oxidase of tea leaves (line 1) and from total enzyme preparation isolated from the leaves (line 2). Part B: Inhibition of phenol oxidase by natural inhibitors from tea leaves.

tea leaf inhibitor; and phenylalanine can almost completely restore phenol oxidase activity.

The presence of natural inhibitors of oxidative enzymes in tea plants can explain the observation that the isolated phenol oxidase preparation catalyzed the oxidative process several times more strongly than did phenol oxidase in the initial leaf material (Fig. 3).

PHENOL OXIDASE PROPERTIES

We have worked on phenol oxidase isolation from tea and grapevine and have studied the most important characteristics of these enzymes. Phenol oxidases isolated from tea and grapevine differ substantially in their substrate specificity (Fig. 4). The purified total fraction of grapevine phenol oxidase highly catalyzes catechol oxidation

TABLE 1

THE EFFECT OF VARIOUS AMINO ACIDS ON PHENOL OXIDASE IN THE PRESENCE OF INHIBITORS OF TEA LEAVES

System	Specific Activity (μ moles O_2/min/mg protein)	Residual Activity (%)
Phenol Oxidase + Total Preparation of Tea Catechins	0.120	100.
The Same + Natural Inhibitor from Tea	0.009	7.5
The Same + Natural Inhibitor from Tea + Amino Acids:		
phenylalanine	0.103	85.8
glycine	0.085	70.8
threonine	0.063	52.5

but is a much poorer catalyst for pyrogallol. The enzyme does not oxidize (-)epigallocatechin gallate or (-)epigallocatechin. By contrast tea phenol oxidase intensively oxidizes (-)epigallocatechin, its gallate, and pyrogallol. The enzyme oxidizes catechol less intensively. The structural formulas of catechin and epicatechin derivatives are shown in Fig. 5.

The maximum action of grapevine and tea phenol oxidase on catechol is observed at pH 4.0-4.8. In

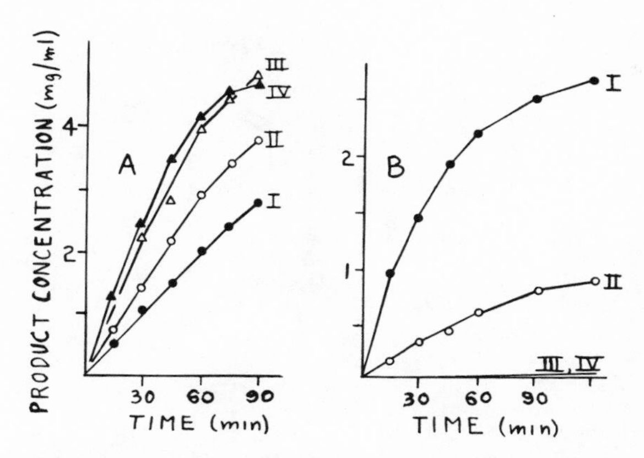

Fig. 4. Substrate specificity of tea (A) and grapevine (B) phenol oxidase; substrates; (I) catechol, (II) pyrogallol, (III) (-) epigallocatechin, (IV) (-)epigallocatechin gallate.

more acidic medium (pH 3.0-3.5) the enzyme still has high activity. In the case of gallocatechin as substrate, tea phenol oxidase has a pH optimum of activity at 5.4-5.7 (Fig. 6).

Studies of substrate specificity and some of the kinetic properties allow one to conclude that phenol oxidase from grapevine catalyzes mainly the oxidation of phenolic compounds containing catechol nuclei. While phenol oxidase from tea plants catalyzes the oxidation of phenolic compounds containing both pryogallol and catechol nuclei.

Total phenol oxidase preparations from tea and grapevine consist of multiple forms of the enzyme differing in molecular weight and electrophoretic properties, and changes. Themolecular weights of phenol oxidase multiple forms from grapevine are mainly 30,000-35,000 and 60,000-70,000 while those from grapevine shoots are in the 60,000 and

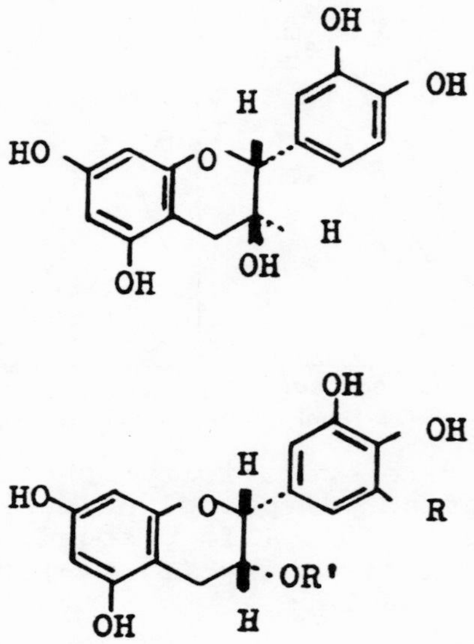

Fig. 5. Structure of (+)catechin (top) and epicatechins (bottom). For (-)epicatechin, R=R'=H; for (-)epicatechin gallate, R=H, R'=3,4,5-trihydroxybenzoyl; for (-) epigallocatechin, R=OH, R'=H; for (-) epigallocatechin gallate, R=OH, R'= 3,4,5-trihydroxybenzoyl.

70,000-80,000 ranges. Phenol oxidase of tea shoots is a mixture of multiple forms of enzymes with the monomer having a molecular weight of 30,000-35,000. Tea leaves contain enzyme with a molecular weight mainly 120,000-140,000, while that found in shoots is 70,000 and 30,000-35,000. Using protein fractionation by ammonium sulphate followed by ion exchange chromatography on CM and DEAE cellulose and hydroxylapatic columns, multiple forms of enzyme with high phenol oxidase activity have been isolated from the total tea phenol oxidase preparation

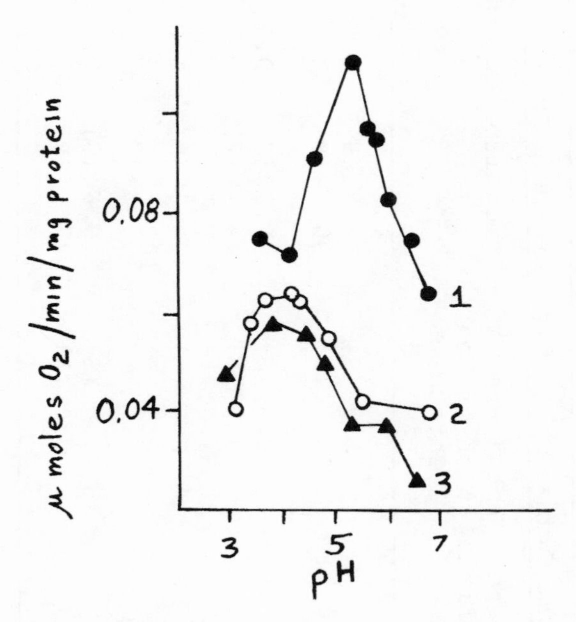

Fig. 6. The change of tea (lines 1 and 3) and grapevine (line 2) phenol oxidase activity with pH. Gallocatechin was the substrate for line 1 and catechol for lines 2 and 3.

The enzyme preparations obtained differ in substrate specificity (Table 2). Some multiple enzyme forms are capable of catalyzing the oxidation of polyphenol compounds containing pyrogallol groupings. Other multiple forms oxidize mainly phenolic compounds with 3,4-dioxy substitution, containing a catechol nucleus, the activity of which is controlled by triatomic polyphenolic compounds containing pyrogallol grouping. The inhibitory effect of these polyphenols on the enzyme activity depends on their concentrations. Oxidation of the total catechin preparation, which con-

TABLE 2

SUBSTRATE SPECIFICITY: MULTIPLE FORMS OF TEA AND GRAPEVINE PHENOL OXIDASE

Plants [(b)]	Enzyme Activity[(a)] for Substrate Shown				
	(-)Epigallo-catechin	(-)Epicate-chin Gallate	Total Prepara-tion of Tea Catechins	Pyrogallol	Catechin
Tea (I)	5.0 (S) 96 (R)	0.8 (S) 15 (R)	5.2 (S) 100 (R)	4.7 (S) 90 (R)	2.8 (S) 54 (R)
Tea (II)	0.30 (S) 43 (R)	0.7 (S) 100 (R)	0.0 (S) 0.0 (R)	0.10 (S) 14 (R)	0.5 (S) 71 (R)
Grape-vine (I)	0.0 (S) 0.0 (R)	-- --	0.0 (S) 0.0 (R)	0.70 (S) 100 (R)	0.37 (S) 53 (R)
Grape-vine (II)	0.0 (S) 0.0 (R)	-- --	0.0 (S) 0.0 (R)	0.64 (S) 29 (R)	2.2 (S) 100 (R)

[(a)] Specific activity (S) given as µmoles O_2/min/mg protein; relative activity (R) as %.

[(b)] Multiple forms.

tains 50% (-)epigallocatechin gallate, 22% (-) epigallocatechin, and 8% (-)gallocatechin, is strongly catalyzed by the basic multiple form of tea leaf phenol oxidase (multiple form I). Oxidation of a given preparation is not catalyzed by multiple molecular forms of the grapevine enzyme (Table 2).

SUMMARY

The enzyme preparation from tea plants represents a phenol oxidase complex consisting of multiple molecular forms. This allows the preparation of dyed products of various hues and shades. Phenol oxidase preparations also may be utilized in a number of fields in the food industry, including wine making, baking of bread, caucao, and tobaco. The use of immobilized phenol oxidases to obtain natural dyes and the use of soluble tea preparations are prospects for the future. A highly thermostable enzyme would be very desirable for us in column reactors for the above areas.

REFERENCES

1. GREGORY, R. & BENDALL, D. *Biochem. J. 101*: 569, 1968.
2. HAREL, E. & MAYER, A. M. *Phytochem. 7*: 199, 1968..
3. HAREL, E. & MAYER, A. M. *Phytochem. 10*: 17, 1971.
4. BUZUN, G. A., JEMUKHADZE, K. M. & MILECHKO, L. F. *Biokhimia*. (Russ. edition) *35*: 1002, 1970.
5. PRUIDZE, G. N. & CHACHUA, L. SH. In "Biochimia Rastenii" ("Plant Biochemistry", vol. I, Russ. edition) (S. V. Durmishidze, ed.) Metsniereba, Tbilisi, 1973, p. 36.
6. PRUIDZE, G. N., DURMISHIDZE, S. V. & KINTSURASHVILI, D. F. *Izves. Adad. Nauk. GSSR. Seria Biol.* (Russ. edition) *1*: 243, 1975.

7. SAMORODOVA-BIANKI, G. B., STRELTSINA, S. A. & ZAPROMETOV, M. N. *Biokimia* (Russ. edition) *42*: 443, 1977.
8. British Pat. No. 111079; 1969
9. BOKUCHAVA, M. A., PRUIDZE, G. N. & ULIANOVA, M. S. In "Biochimia Rastitelnix Krasitelei" (Biochem. of Plant Dyes", Russ. edition) (S. V. Durmishidze, ed.) Metsniereba, Tbilisi, 1976.
10. PRUIDZE, G. N. & GRIGORASHVILI, G. Z. *Prikladnaia Biochimia Mikrobiologia* (Russ. edition) *13*: 104, 1977.

IMMOBILIZED ENZYMES IN NUTRITIONAL APPLICATIONS

W. Marconi

Assoreni,
Monterotondo (Rome), Italy

Extracellular microbial enzymes play a fundamental role in food technology production of cheese, yoghurt, beer, wine, and so on. For this purpose it is generally much more convenient to use whole living cells instead of complicated multi-enzymatic systems extracted from the cells. Extracellular microbial enzymes also are used as additives, where they find application both in analysis and in the processing of food. The object of this paper is to discuss the role of enzyme additives in the processing of food materials, with special emphasis on enzymes immobilized to various supports.

In food processing enzymes can be added a) to improve a step in a traditional microbial process such as cheese production, b) to improve the feasibility or the yields of a process, and c) to improve the quality of the obtained products. Examples of each type are rennin and malt enzymes (case a), pectinases and cellulases in the extraction of fruit juices and oils (case b), and collagenase and protease for tenderizing meat, β-galactosidase for hydrolysis of lactose in milk, and invertase conversion of glucose syrups by partial isomerization to fructose and sucrose (case c).

The general characteristics necessary for successful application of enzymes in food processing are several. First, one must take into account that the operating conditions such as pH, temperature, substrate concentration, or presence and amount of inhibitors and cofactors, generally are far from the optimum for the enzyme. Second, the safety aspects must be considered. For example in the United States, only a few microbial sources are reconized as safe. Thus, enzymes from other microbial sources must be cleared through a very expensive food additive petition route. Other characteristics affecting the choice of the enzyme, and often the decision whether or not to use an enzymic additive or enzymic process, include the purity, availability, stability in storage, and ease in controlling the activity of the enzymes. Further economical and technological limitations must be taken into account when we envisage the use of enzymes in immobilized form.

Particulate enzymes, enzymes requiring addition of expensive cofactors such as NAD^+, and enzymes which have to operate for long times, such as the socalled flavor enzymes, generally cannot be considered for immobilization. Similarly, insolubilized enzymes generally cannot be employed in cases where the substrate is insoluble, thus ruling out of consideration important classes such as pectinases, collagenases, cellulases, and amylases for desizing clothes. Severe use limitations also are present with slurries or with soluble high molecular weight substrates. In addition immobilization should not be considered when the existing technologies employ soluble enzymes or resting cells of low cost and large production, as with some amylases and proteases, unless other important advantages can be envisaged.

Sometimes, the real cost of immobilization of the enzyme as well as the necessity for modifying and automatizing preexisting plants can discourage the adoption of an immobilized enzyme technology. Minor possible drawbacks can derive from bacterial contamination of enzymic reactors or from difficult control of the operating conditions affecting enzyme stability. Conversely, fewer problems in product treatment, such as decoloration or deionization, the possibility of realizing a more reproducible continuous process, reduced plant dimensions, and the absence of pollution from protein materials in the final product can make it preferable to choose an immobilized enzyme technology.

I. GLUCOSE ISOMERASE

Glucose isomerase is an intracellular enzyme which isomerizes glucose to a mixture of glucose and fructose. The product of the enzymic reaction has a sweetness comparable to that of invert sugar, which is obtained by acid or enzymic hydrolysis of the more expensive sucrose. Certainly, the enzymic isomerization of glucose to fructose is, at the present time, the largest process using immobilized enzymes in the world. Of course the main advantage of using immobilized glucose isomerase lies in the reduction of the operational costs. In fact, notwithstanding the very many efforts made in recent years, the fermentation costs of glucose isomerase are still too prohibitive to allow its use free in solution. However, even if the production costs of the enzyme can be reduced, companies will not give up the other intrinsic advantages of the immobilized enzyme process, such as the possibility to operate on a continuous basis, the absence of protein contamination of the product, and the low residence time at relatively high temperatures with scarce formation of byproducts. Table 1 illustrates the source and method of immobilization for some commercial preparations of glucose isomerase. It is interesting to notice that, among the most efficient preparations currently used in the industrial practice, no covalently bonded glucose isomerase appears. Even Corning Glass, that has developed excellent methods for covalent immobilization of enzymes, for this particular case has chosen the adsorption method. Also, most of the immobilization methods

TABLE 1

IMMOBILIZED GLUCOSE ISOMERASE*

Enzyme source	Immobilization method	Company
Streptomyces albus	Temperature fixing inside the cells	Agency of Industrial Science and Technology
Streptomyces sp. [(a)]	Adsorption onto DEAE cellulose	Standard Brands
Arthrobacter	Aggregation with a flocculating agent	Reynolds Tobacco Co.
B. coagulans	Crosslinking with glutaraldehyde	Novo
Streptomyces phaechromogenes [(a)]	Adsorption on phenol formaldehyde resin	Kyowa Hakko Kogyo

TABLE 1
(Cont'd)

Enzyme source	Immobilization method	Company
Streptomyces sp. [a]	Adsorption on special porous alumina	Corning Glass Works
Streptomyces phaechromogenes	Adsorption on special ion exchange resin	Mitsubishi Chemical Industries
Actinoplanes missouriensis	Occlusion in gelatin and crosslinking with glutaraldehyde	Gist-Brocades
Streptomyces olivaceus	Crosslinking with glutaraldehyde	Miles Laboratories
Streptomyces sp. [a]	Entrapment in fibers	Snamprogetti S.p.A

(a) Refers to enzyme from this source; entries without (a) refer to cells.

employ cells instead of a more or less purified enzyme extract.

The cost of glucose isomerase for the production of isosyrups is now so reduced that only cheap immobilization techniques can survive. Moreover, there are only very few immobilization methods that are so cheap and efficient as to afford the additional costs of purification of the enzyme from the cells.

Little information is available in the literature on the industrial performance and the activity of the afore mentioned glucose isomerase preparations. Of course there are some characteristics which depend on the enzyme source. With the exclusion of the Novo and Reynolds tobacco enzymes, the other sources are from *Streptomyces* sp. However some features, that are important from the industrial viewpoint, seem to be common and include the severe sensitivity of the enzyme to oxygen and the strong inhibition by calcium ions. The differences between the two classes of enzyme in regard to their pH profile, temperature dependence, and cofactor requirements seem to be less significant and mostly nuances suggested by the manufacturers with the aim of differentiating their product from others.

On the contrary, the immobilization method that is used causes remarkable differences as far as performance, stability, and catalyst morphology are concerned. Clinton Corn Processing Co. employs a series of shallow beds to allow a low pressure drop across the bed. The half-life of their glucose isomerase is probably about 20 days; and the activity is high. Novo claims a half-life of 40 days with a productivity of 1 - 1.5 tons per kg of immobilized cells. At Corning Glass, they were able to use an enzyme adsorbed onto a porous ceramic carrier and obtain a half-life of 40 days. The glucose isomerase included in gelatin by Gist-Brocades shows a productivity of about 0.6 tons per kg of enzyme-gel. Glucose isomerase entrapped in cellulose triacetate fibers under the proper conditions shows a half-life of about 70 days and a productivity in the range of 5-6 tons per kg of enzyme fibers.

Recently very good performance of the glucose isomerase fibers was achieved with a radial reactor which fit very well with the filamentous structure of the fibers. The radial reactor was prepared by rolling up the fibers around a perforated pipe in an ordered way like a bobbin. The reaction mixture flows from the holes of the central pipe, through the fiber layers where the enzymic reaction takes place, and outflow into a collection vessel. Packing densities up to 0.35 kg of dry fiber per liter of reactor volume were reached, making it possible to increase the efficiency of the enzyme fibers and to decrease the residence time. The radial reactors have the great advantage that they can be easily prepared, using standard equipment of the textile industry. Moreover, it is possible to use several of these reactors connected in series (Fig. 1) or in parallel, with great simplicity in changing each element when the enzyme activity drops below values too low for practical use. In such a way it is possible to guarantee a constant production rate and to make the most of the enzyme activity.

II. LACTASE

Another enzyme of importance is lactase. The potential use of this enzyme in the dairy industry offers one of the most promising commercial exploitations for immobilized enzymes. Lactase hydrolyses lactose, a sugar with poor solubility properties and relatively low degree of sweetness, to glucose and galactose. The hydrolysis of lactose offers some advantages in a dairy product. First, it improves the digestability of lactose. It is well known that some infants, adults, and some individuals of all ages of some ethnic groups are unable to digest lactose, probably because of the lack of lactase in the small intestinal mucosa. Another advantage is the possibility of preparing

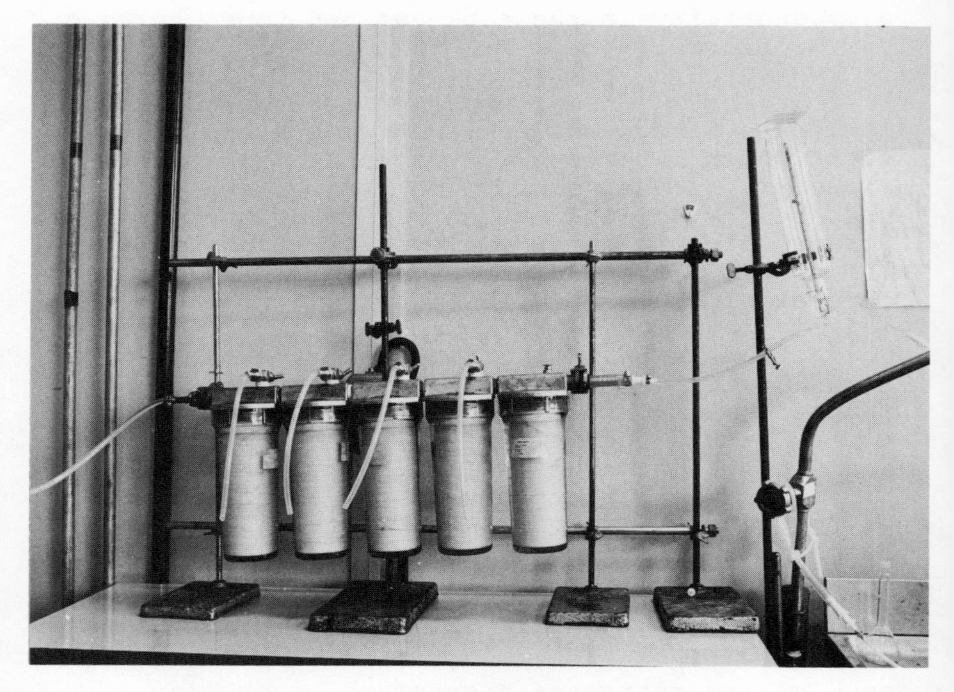

Fig. 1. Radial enzyme fiber reactors in series (7).

new food and dairy products with higher solubility characteristics and a higher degree of sweetness that can substitute, in some applications, for corn syrups. Moreover lactose is a by-product of cheese manufacturing, where many pollution problems exist. Lactose utilization also becomes more and more interesting because the cost of sweeteners is tending to increase.

The first company to commercially hydrolyse lactose in milk by immobilized lactase was the Centrale del Latte of Milan, Italy utilizing the Snamprogetti technology (1).

An industrial plant (Fig. 2) with a capacity of 10 tons per day is in operation in Milan. The

Fig. 2. Industrial plant of the Centrale del Latte of Milan, Italy, for producing low lactose milk using immobilized lactase (7).

entrapped enzyme is lactase from yeast; and the reaction is performed batchwise at low temperature. The processed milk, after having reached the desired degree of hydrolysis of lactose, is separated from the enzyme fibers, sterilized, and finally sent to packing for distribution. This process provides low cost, high quality, dietetic milk featured by: remarkable digestive tolerance, pleasant sweetness, absence of foreign matter, unaltered organoleptic properties, and good shelf-life characteristics. The following description is provided by the Centrale del Latte di Milano:

"Analisi percentuale:
Proteine (N x 6.38): min. 2.9%, max. 3.5%; Lipidi: min. 1.5%, max 1.8%; Zuccherl: min. 4.6%, max 5.2% (del quali lattoslo non plu di 1.3%); Sali minerall: min. 0.6%, max. 1%; Residuo secco: min. 10.2%, max. 11.0%.

"Many consumers do not tolerate milk because they cannot assimilate the sugar in it (lactose), which leads to intestinal disturbances. In this milk a process, new throughout the world and not involving additives, has been used to convert at least 75% of the lactose to its components, glucose and galactose, which are easier to digest. In this way a product has been obtained which is particularly indicated for all -- whether children or adults -- who have given up milk because of intolerance."

Many people have directed their efforts to the hydrolysis of lactose in whey. It is well known the importance of increasing whey utilization, which is a growing area of concern to industry also because of the pollution caused by the lactose. A typical composition of whey is shown in Table 2. It appears clear at once the value of the two major components: protein and carbohydrate. The protein fraction has considerable value for its aminoacids; and now there are available technologies for its separation and recovery using ultrafiltration and ion-exchange resins. However the advantages of the protein recovery have been to date outweighed by the high costs of processing; so that economical utilization of the carbohydrate fraction associated with the protein recovery must be taken into consideration by milk processors. At present this unfavorable economic picture is changing due to the cost increase for producing sweeteners. Moreover, there are now available a number of immobilization techniques for lactase that can reduce the costs of lactose hydrolysis.

TABLE 2

TYPICAL COMPOSITION OF CHEESE WHEY

Material	Concentration (kg/cubic meter)
Fat	2.0
Protein	7.0
Lactose	40 - 45
Mineral Salt	4.5

Several immobilization procedures have been described for lactase. Among these, the method proposed by Corning Glass seems to have reached the most advanced experimentation. The enzyme lactase from *Aspergillus niger* has been immobilized on porous silica supports (2). Activity of about 700 units per g, with a coupling efficiency higher than 50%, has been achieved. The half-life of the immobilized enzyme depends on the quality of the whey and ranges from 60 days for deproteinized and demineralized whey to a few days for whole whey. Microbial contamination, which is the major problem in this process, appears to have been solved by daily washing of the enzyme reactor. From pilot-plant experiments, an operating cost of about 0.03-0.04 $/lb of lactose, excluding the cost for demineralization and concentration, has been estimated. A total cost of 0.10 $/lb can be foreseen. When the degree of hydrolysis of lactose is in the range of 70 - 80%, the mixture so obtained has a degree of sweetness similar to that of 40 dextrose equivalent (D.E.) corn syrup, whose price is about 0.17 $/lb.

The immobilization of lactase in fibers also has been applied to the hydrolysis of lactose in whey. Most of the work has been done on a bench-scale. Attention has been paid to the hydrolysis of lactose both in sweet and acid whey; for this reason, lactases from yeast and from fungal sources, having their optimal pH values in the neutral and acidic ranges, respectively, have been employed. Very active enzyme fibers have been obtained, thus allowing very fast hydrolysis and minimizing the risks of bacterial contamination. Also, whole whey could be processed after procedures for very efficient washing of the fibers, that left unaltered the activity of the enzyme, were developed. Extended large-scale pilot plant investigations are now in progress by milk processors to establish the economy of the process and to sample the different products.

III. α-GALACTOSIDASE

α-Galactosidase has potentially large applications in the hydrolysis of raffinose in beet sugar syrup or juice and the hydrolysis of other complex carbohydrates present in vegetable meals.

Raffinose is present in considerable extent in sugar beets, especially in the beets of those countries having a cold climate. This carbohydrate, made of galactose and sucrose, interferes negatively with the crystallization of sucrose, thus lowering the yields. Moreover, raffinase accumulates in the molasses, whose utilization becomes more difficult. α-Galactosidase hydrolyses this trisaccharide, giving galactose and sucrose, and eliminating the negative interference in sugar crystallization and increasing the yield of sucrose. A Japanese company, Hokkaido Sugar Co., has patented and is using enzyme pellets of *Mortierella vinacea* for raffinose hydrolysis. A three chamber reactor is employed, with a stirrer passing through the central part of each chamber. The enzyme pellets, obtained by culturing *Mortierella vinacea* in a proper medium, are charged in equal amount to the chambers; and a

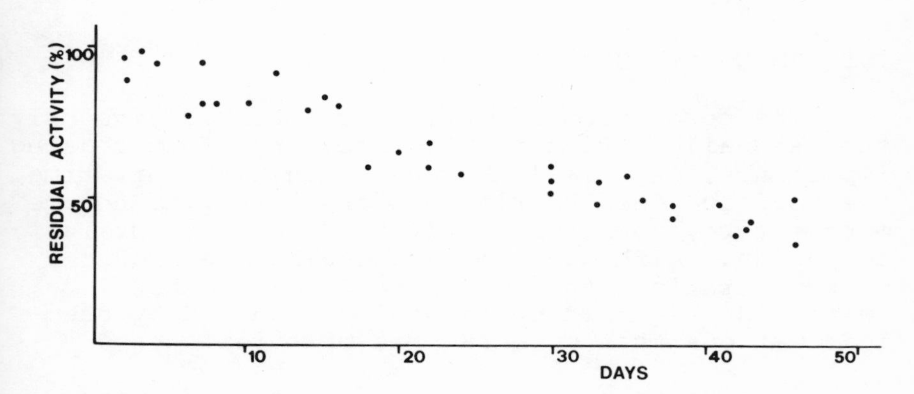

Fig. 3. Operational stability of fiber entrapped α-galactosidase.

sugar at 50°C is circulated through the reactors. After four days the size of the enzyme pellets becomes smaller and smaller; and the activity drops below 50%. The enzyme pellets are refrigerated.

α-Galactosidase also has been entrapped in cellulose triacetate fibers. Fig. 3 shows the operating stability of a sample of fiber entrapped α-galactosidase, working on raw molasses. The good stability enables absorption of the additional cost of purifying the enzyme. Moreover, it is not necessary, as in the Hokkaido process, to transport the immobilized enzyme in refrigerated form.

Another interesting application of α-galactosidase lies in the hydrolysis of those carbohydrates that cannot be split by animals into nutritionally useful metabolites. For example, the enzyme can be used fruitfully to treat soybean milk and whey allowing a more effective use of these by-products. α-Galactosidase hydrolyses raffinose and stachyose, reducing the flatulence that occurs in the lower intestine because of bacterial fermentation of undigested carbohydrates.

IV. AMINOACYLASE

The resolution of optical isomers to give only the naturally active form is becoming of increasing importance, especially because within a short time the food and feed regulations are likely to become more severe, permitting only the use of natural compounds. Although some aminoacids are produced by fermentations that give the L-form, other aminoacids are more successfully produced by chemical methods that give racemic mixtures.

Resolution of chemically synthesized aminoacids may be performed using aminoacylase, an enzyme which catalyses the deacetylation of the L-form of the N-acetyl aminoacids leaving unaltered the N-acetyl-D-aminoacid, that can be easily separated, racemized, and recycled. Aminoacylase, both from hog kidney and microorganisms, has been immobilized with various techniques; and this enzyme probably represents the first industrial application of an immobilized enzyme (3). Aminoacylase is immobilized by ionic adsorption onto DEAE sephadex. The enzyme reactor can be reloaded quite simply by passing in fresh enzyme solution when the previous enzyme is deactivated. The deacetylation of the L-form is performed continuously by passing through the reactor the DL acetyl aminoacid.

Aminoacylase also has been entrapped in cellulose triacetate fibers. The stability of the entrapped enzyme is very good, with the loss of enzyme activity only 25-30% after 50 days of continuous operation. The process has been applied to the resolution of the racemic mixture of tryptophan obtained by chemical synthesis. A bench-scale unit, employing 4 kg of fiber entrapped aminoacylase, was operated for some months. The activity of the aminoacylase fibers showed a 20%

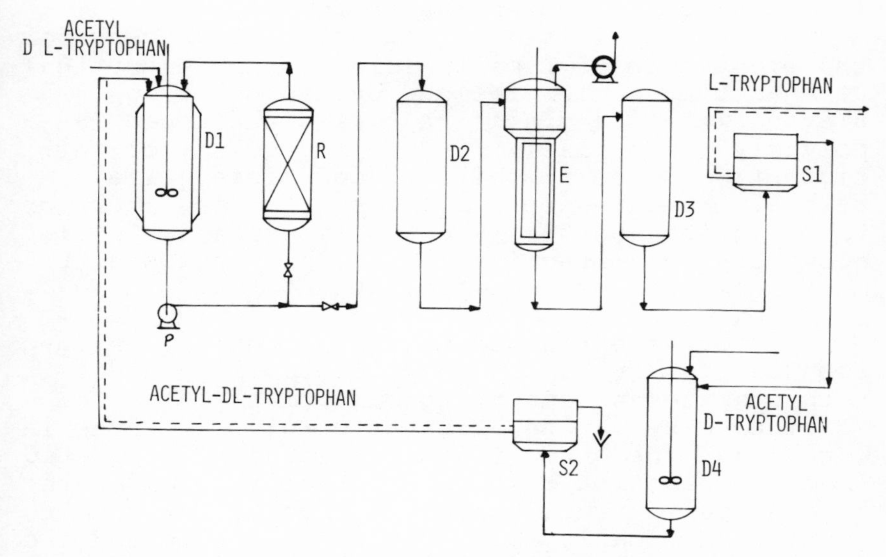

Fig. 4. Flow diagram for the pilot plant producing L-tryptophan by fiber entrapped aminoacylase. D1, reservoir containing the racemic acetyl derivative solution; R, enzyme reactor; P, recycling pump; D2 & D3, storage vessels; E, evaporator; S1, centrifugal separator; D4, racemization vessel, S2, centrifugal separator (7).

decrease during this period; and the productivity was about 400 kg of L-tryptophan per kg of fibers. The flow-sheet of the process is schematically diagrammed in Fig. 4.

V. FUMARASE

Malic acid is becoming of greater market interest as a food acidulant in competition with citric acid. At the present time the racemic mixture of malic acid is produced by a cheap chemical synthesis from maleic anhydride and water, whereas the natural form of malic acid produced by fermentation is too expensive to find its own market in the field of acidulants. A process for the production of L-malic acid for pharmaceutical use has been commercialized by Tanabe using *Brevibacterium ammoniagenes* cells entrapped in polyacrylamide gel (4). The process is very efficient; and the immobilized enzyme is quite stable. No data are available from the literature to establish whether or not this process also is economically attractive for the production of L-malic acid for use as an acidulant.

Fumarase entrapped in spun fibers of cellulose triacetate has shown good stability under both storage and operating conditions (5). The continuous hydration of fumaric acid was investigated; and the product L-malic acid was separated from the reaction mixture by precipitation as the calcium salt. This enzymatic process for the production of L-malic acid from fumaric acid and the comparative economics for production of acidulant grade L-malic acid and citric acid are presented in a separate paper in this volume (6).

VI. FUTURE POTENTIAL APPLICATIONS

Invert sugar is produced and sold in large quantities; so many studies have been done on the potential application of immobilized invertase. When one considers only the pure aspect of sucrose inversion and compares the enzymic to the chemical method, it is clear that the chemical way is by far economically more attractive. However, the chemical method gives only relatively limited degrees of conversion, which is not a big problem because for the maximum solubility of invert sugar one need only hydrolyse half the sucrose. Furthermore the chemical method does not work very well in the presence of inorganic salts; and if non-exchange resins are used, quite pure sucrose solutions are required. Invertase on the other hand does not suffer from these drawbacks. Furthermore, the ion-exchange resins must be regenerated at intervals; and the pollution caused by the regeneration effluents must be taken into account. The product obtained by invertase inversion is practically colorless and does not need any further treatment for refining. The reverse is true for the acid inversion. So, when the overall process is considered, the enzymic route is especially promising.

Significant research also has been done on the immobilization and use of glucoamylase, one of the most widely used free enzymes. This is another example of an immobilized enzyme that probably is not competitive with the free enzyme, but could become so, if one takes into consideration all the advantages due to its use in the immobilized form. The cost of free glucoamylase in the saccharification of starch is relatively low because many efforts have been made to improve the enzyme and the methods for its production. For this reason, a process based on the immobilized enzyme, must be very efficient and cheap. However, here again other advantages, such as the possibility of carrying out the saccharification of starch on a continuous basis, the significant reduction of the reactor volume, and the absence of foreign protein contaminant can push the starch manufacturers to adopt the immobilized method. In a study carried out at Iowa State University a column containing 16 kg of immobilized glucoamylase converted 600 kg of starch daily to a D.E. of about 92. Residence times in the range of 8-18 min were used as compared with 72 hr for the soluble enzyme process.

The clotting of milk by immobilized enzymes is also of great interest. Relatively little work has been done to date on the immobilization of rennin, probably because of the poor stability of the immobilized preparations. Immobilized rennin or other proteases or mixtures of them might allow the continuous coagulation of milk for cheese manufacture. The use of an immobilized enzyme is possible because the process can occur in two phases. In the first phase the enzyme displays its action on K-casein at low-temperature; and in the second phase the clotting of milk can be carried out by increasing the temperature. The advantage of using an immobilized enzyme is that the enzyme would not remain in the product. It is well known that if a protease or microbial rennin is used, the enzyme continues its action, leading to the formation of bitter peptides that negatively affect the organoleptic properties of cheese differently from calf rennin. This difficulty could be overcome by using an immobilized enzyme.

Another interesting enzyme that could be used profitably in immobilized form is catalase. Catalase has a potential use in milk processing for the destruction of hydrogen peroxide employed in the cold sterilization of milk. The treatment of milk with hydrogen peroxide is very interesting because, differently from the thermal sterilization, it leaves unaltered not only the organoleptic properties of milk, but mainly the natural composition without destroying the temperature sensitive compounds which are of great nutritional value. However, use of hydrogen peroxide has been rather limited due to the cost of catalase treatment and the difficulties of using the free enzyme. Thus, an immobilized, stable catalase might prove useful.

Another area of increasing interest in the near future will be the improvements to be done in starch manufacturing and in the obtaining of sugars different from the traditional ones. Immobilized pullulanase and beta-amylase are two enzymes that may find use in this area. The first enzyme is a debranching one; and its use might prove helpful as a convenient and economical means of increasing the yields of glucose production from starch in combination with glucoamylase. The second enzyme is now undergoing considerable expanded use for the production of maltose from liquified starch. Maltose is a sugar that has some nutritional advantages over the monomer glucose and other carbohydrates. The yields in maltose can be increased also by combined use of beta-amylase and pullulanase. It must be stressed that further research in the immobilization of these enzymes as well as in the simultaneous continuous use of their combination might lead to very compact industrial processes of extreme interest for the starch manufacturers.

A significant future is expected for multienzyme systems capable of performing high efficiency sequential reactions of the type found in cellular metabolism. For this reason, the immobilization of whole cells as well as of coenzymes bound to polymers has great potential. From further research in this field we expect the development of new processes for the conversion of agricultural renewable resources into feedstock materials for our chemical industry.

REFERENCES

1. PASTORE, M. & MORISI, F. *Methods Enzymol.* 44: 822, 1976.
2. WEETALL, H. H. & HAVAWALA, N. B. In "Enzyme Engineering" (L. B. Wingard, Jr., ed.) John Wiley, New York, 1972, p. 241.
3. CHIBATA, I., TOSA, T., SATO, T. MORI, T. & MATSUO, Y. In "Fermentation Technology Today" (J. Terui, ed.) Soc. Fermentation Technology, Japan, 1972, p. 383.
4. German Pat. NO. 2,450,137; 1975.
5. MARCONI, W., MORISI, F. & MOSTI, R. *Agr. Biol. Chem.* 39: 1323, 1975.
6. GIACOBBE, F., IASONNA, A., MARCONI, W., MORISI, F. & PROSPERI, G., this volume.
7. MARCONI, W. & MORISI, F. *Appl. Biochem. Bioeng.* 2: 219, 1979.

SECTION VI
FUTURE PROSPECTS

IMPACT OF ENZYME ENGINEERING ON SCIENCE POLICY

B. Adams, C. G. Heden and S. Nilsson

International Federation of Institutes
for Advanced Study
Ulriksdals Slott
Solna, Sweden

In view of the theme of this volume, it is interesting to take note of the history of the enzyme project developed under the International Federation of Institutes for Advanced Study (IFIAS) and especially its role in promoting international cooperation. A summary of this project for 1973 to 1977 has been issued (1). This project is an interesting illustration of the potential for the willingness of scientists in different countries and fields to work together on problems which are important for the long term survival of mankind. It is, however, also an example of the many obstacles for genuine international cooperation. Quick moves can be made by groups like IFIAS on subjects before the topic enjoys the status of an established academic discipline at the national research policy level. But, one runs the risk of not getting financial or other support when it is needed; and the whole initiative is killed. Or, in a few cases extraordinary efforts are made by the individual researchers involved, and inventive measures are taken to finance the component parts that constitute the necessary ingredients to test and develop the initial innovative idea. In the case of the IFIAS enzyme project, we were fortunate in having both dedicated leadership and the devoted cooperation of many researchers who believed in the ideals and the scientific potential of the project.

I. ENZYME PROJECT OBJECTIVES AND ORGANIZATION

The idea for the IFIAS enzyme project grew out of a series of conversations with Professor Ephraim Katchalski-Katzir of the Weizmann Institute of Science in Israel in 1971 and 1972. Of special interest was the enormous development potential of enzyme research for the fields of food, energy, medicine, and industrial production of various kinds. If IFIAS was going to have a project of a scientific nature but with great potential socio-economic impact, the field of Enzyme Engineering was an outstanding prospect. At the founding meeting of IFIAS in October 1972, the then President of the Weizmann Institute, Professor Albert Sabisi, made a proposal for a project on Enzyme Engineering. This later became the IFIAS project "Socio-Economic and Ethical Implications of Enzyme Engineering", with Professor C. G. Heden of the Karolinska Institute in Sweden as the Project Director. At about the same time Professor Katchalski-Katzir became the President of Israel.

The basic assumption behind the project was stated at the project design workshop held at Saltsjobaden, Sweden, in June 1974. The next few decades will see a continued increase in the population of the less developed countries, accompanied by serious social pressures caused by unemployment and rural-urban migration. Drastic measures will have to be taken to make the strains as small as possible and to provide for meeting the human needs of food, shelter, energy, health care, and education for as many as possible. The role of science and technology in the decentralized, self-reliant provision of food, energy, and health care to as many people as possible in different socio-cultural environments served as the focus of this IFIAS project.

The IFIAS project took the shape of a 4 year expert dialogue, revolving around the progress made in a few selected areas. The initiatives were aimed at a few areas which had enough research attraction to mobilize funds and to involve highly competent scientists. If those representing different specialties, were joined together by mutual interests, then cooperation, which otherwise might not have occurred, could be achieved to stimulate new important activities. The possible science policy implications of such an experiment of course had a general interest for IFIAS, which, on those grounds, deviated somewhat from its general principle not to sponsor monodisciplinary projects. At the project design workshop at Saltsjobaden the following main areas for the project were agreed upon: a) photosynthesis, its optimal uses and impact in less developed countries; b) laboratory medicine, food technology and environmental monitoring in poor countries; c) diagnosis of tropical diseases; d) biochemical fuel cells for decentralized energy production; and e) problems in biotechnology transfer. It should be pointed out that IFIAS does not view the field of Enzyme Engineering and Technology as the only one that could and should have been subject to such an assessment. It merely had been chosen as a case in point of very great interest and potentialities in many areas of socio-economic importance.

In all IFIAS projects one of the aims is to reach some policy impact. The outputs of the activity are therefore of equal importance to the inputs. A conceptual framework, describing the inputs, outputs, objectives, and receivers of the results, is shown in Figs. 1-4.

It is interesting to observe that as part of the Input, ten Foundations and Agencies contributed and more than 35 university departments and institutes from about a dozen countries participated. As output, about twenty workshop reports, articles and books were developed.

We will not go into any details here regarding the component studies since this has been described in the Summary report and in the partial studies referred to. Instead we will concentrate on the policy implications of the Enzyme Engineering project which is the subject of the present

A. Project Design Workshop
   - Pasteur Institute, Paris, France 1973

B. Annual Review Workshops
   - Saltsjobaden, Sweden 1973
   - Poona, India 1975
   - Philadelphia, USA 1977

C. Special Workshops
   - Physico-Chemical Aspects of Electron Transfer Processes in Enzyme Systems, Puschino-on-the-Oka, USSR 1975
   - Biotechnology of Electron Transfer Processes, Philadelphia, USA 1977

D. Commissioned Studies
   - 3 completed

E. Commissioned Study Workshops
   - Microbial Photosynthesis and Energy Conversion, Orsundsbro, Sweden 1976
   - Microbial Conversion Systems for Food and Fodder Production and Waste Management, Kuwait 1977

F. Participants:
   - IFIAS Members
   - United Nations Agencies
   - Government Agencies
   - Non-member Institutes
   - Corporations

G. Funding:
   - Sweden (private and government)
   - India, Department of Science and Technology
   - USSR, Committee on Science and Technology
   - USA, National Science Foundation
   - Corporations

Fig. 1. IFIAS Enzyme project inputs (1).

ASSESS THE SOCIO-ECONOMIC AND ETHICAL THREATS AND OPPORTUNITIES OF ENZYME ENGINEERING IN THE FOLLOWING AREAS:

A. Diagnosis of Tropical Disease
B. Photosynthesis
C. Environmental Monitoring
D. Enzyme Catalyzed Fuel Cells
E. Laboratory Medicine
F. Technology Transfer

Fig. 2. IFIAS enzyme project objectives (1).

A. Four IFIAS workshop reports (2, 3)
B. Impacts on national policies and programs: e.g. USSR, India, Sweden
C. Impacts on United Nations agency programs
D. Impacts on other IFIAS projects, e.g. soil, climate, energy, human settlement
E. TV & Radio
F. Articles for Journals (4-6, 9, 10)
G. Books (7, 8)

Fig. 3. IFIAS enzyme project outputs (1).

A. United Nations Agencies
B. Government Agencies
C. Scientific Community
D. Less Developed Countries
E. Corporations

Fig. 4. Receivers of the IFIAS enzyme project outputs (1).

paper and which, indeed, may serve as a valuable lesson for any future international collaborative effort in the field.

II. ENZYME PROJECT COMPONENTS

A. Tropical Disease Diagnosis

This project was based on the ELISA technique and was oriented to the detection of malaria and other endemic diseases. The special characteristics of the ELISA (enzyme-labeled antibody technique) are the high degree of specificity, rapidity, simplicity and automaticity. Large populations of people can be diagnosed quickly. It can provide survey data not only of medical significance but also for the ecological schemes in attacking some of the diseases. Special efforts were made to find a good antigen which can produce the appropriate antibodies to carry out the diagnosis by means of a simple kit. Special collaboration was established between research teams in India (Dr. Ramachandran) and in Sweden (Dr. Huldt) to use an Indian penicillinase to label the antibodies. This subproject has had a significant impact on research in Sweden, India, and the United Kingdom and on the WHO policy on tropical diseases. The research work will continue for development of small diagnostic kits under the new IFIAS Self Reliant Development program.

B. Photosynthesis: Its Optimal Use and Impact in Less Developed Countries.

This was the title of the commissioned study carried out by Dr. Russell Anderson. It was proposed by Dr. Viniegra of the Biomedical Research Institute in Mexico. Originally intended as a relatively short term survey supported by UNESCO and Swedish sources, this study grew to become a main part of the Enzyme Engineering project. The results of the study will be published (7). The study has made a policy impact in Sweden, India, Kuwait, and Mexico as well as on institutions such as UNESCO, UNEP, and the Board for Technical Development of Sweden. This study forms the basis for the new IFIAS Self Reliant Development program, which is to be seen as the continuation of the Enzyme Engineering project, but more strongly focused on integrated systems of biological and physical solar energy conversion for self reliant development.

C. Health Care Delivery in Less Developed Countries (LDC).

This was based partly on the development of diagnostic kits for tropical diseases. Also, a commissioned study was carried out by Dr. Sundaram regarding the costs and benefits of decentralized medical care in an LDC, with special regard to enzyme techniques. It was felt that enzyme engineering as it relates to health care and medical services would affect chiefly clinical chemistry, microbiology, and pathology. It was found that the main problems were resistance to change, lack of personal commitment to improvements, costs, and specifically in an LDC the dearth of servicing facilities as well as service personnel. Another substudy initiated a collaboration between Dr. Chang of McGill University in Montreal, Canada and the Pasteur Institute in Paris regarding a one shot vaccine. Certain diseases, such as small pox or cholera, require that vaccines be given in stages or repeatedly (boosters). Certain populations, where a dearth of medical personnel exists or where cultural aspects make constant health services impossible, would benefit greatly from such vaccines. A one shot vaccine containing enzymes that are mobilized on a timed basis would provide the necessary series of vaccines required over a span of weeks or months without the prolonged need for on the scene medical personnel to give booster shots.

D. Enzyme Catalyzed Electron Transfer or Biochemical Fuel Cells.

This part of the enzyme project was launched as a major multidisciplinary international research effort. It has generated a fast growing interest, even if many believe that the practical applications are rather far in the future. Under any circumstance the collaboration initiated by IFIAS has resulted in a self propelling activity by the parties involved and in some very intriguing research.

III. POLICY IMPACT AND THE FUTURE

It seems fair to say that the IFIAS Enzyme Engineering project has effected the science policy at the country level in Sweden, India, the USSR, and partly also in the U.S.A. It has, however, also influenced the research of institutions and individuals, which appears to be more important at this stage. After all, the selected problems must be solved by individual experts, some of whom claim that their work, for example on fuel cell problems, would not have come about if IFIAS had not served as a switchboard and coordinator. The Enzyme Engineering project also has increased the awareness and understanding of many institutions like UNEP and UNESCO of Enzyme Engineering and applied microbial genetics as tools in the development of small scale, decentralized technologies. It may also be worth mentioning that two UNESCO/IFIAS Fellows were engaged in November 1975 on the Enzyme Engineering project, directly under the Project Director in Sweden. They are now working on their Ph.D. degrees on the microbial production of methanol and ammonia and on simple health kits for an LDC, respectively. Apart from the limited policy impact at the institutional level, the Enzyme Engineering project seems to have had its main advantage in bringing many different experts and institutions together in a cooperative effort, which is both somewhat more goal oriented and long term than what is usually the case. One important conclusion from the Enzyme Engineering project needs to be emphasized. Thanks to the impressive voluntary participation and contributions, approximately ten times more money was put into research cooperation than was spent on administration.

The new IFIAS program on Self Reliance will still contain the energy and health components from the Enzyme Engineering project; but the new program will be more strongly focused on the socio economical and cultural aspects of self reliant development. The goal of the new program, under the direction of Dr. R. Anderson, is very practical and is designed to determine the maximal (optimal) energy budget a self reliant community can attain via an integrated system of biological and physical solar energy conversion.

Integration has a broad meaning in the new program. It includes a) the closed cycle integration of physical and biological processes in a dove-tailed manner for maximum energy use efficiency plus ease of recycling waste gases, nutrients, and heat; b) an integrated systems approach to satisfy simultaneously the basic needs for health, water, food and fuel; and c) integrating the biological aspects of hygiene, food, and solar energy conversion processes along with education and the enormous local agricultural and health practices experience.

Self reliance is defined as the ability to meet the basic needs locally. The program has four main component projects and two supplementary projects. The four component projects are: (a) energy and systems analysis of a solar community based on alternative integrated systems; (b) testing of the technical potential and practicality of the integrated systems, studied in component (a); to be carried out in a technically developed society; (c) stepwise integration of the items from component (b) which function well in a village community in an LDC; and (d) the social, cultural, environmental, and technical constraints along with the potential for enhanced bioproductivity of coastal zones (brackish water). The two supplementary research oriented projects are (a) laboratory aids for decentralized health care delivery systems and (b) the biochemical fuel cell.

In the new Self Reliance project, the role of IFIAS again is essentially integrative and catalytic, using the most flexible means for coordination and assistance in broad contacts, funding, and communication with policy makers. The integration of the various component projects into a conceptual whole will be promoted by annual workshops. Already a large number of researchers and institutions have been contacted and much interest in the program has been expressed. Some of the institutions concerned include the UN University, Dag Hammarskjold Foundation, UNESCO, IDRC in Canada, UNDP, UNEP, Board for Technical Development in Sweden, and Department of Science and Technology in India. If we succeed in doing a few of the things that the Self Reliant Development Program suggests, then after another four years we may be able to tell the world how the different sciences when integrated in the right way can contribute to a more self reliant development in the less developed countries.

REFERENCES

1. HEDEN, C. "Socio-Economic and Ethical Implications of Enzyme Engineering" Report No. 15, IFIAS, Stockholm, 1978.
2. IFIAS Report No. 13: Workshop on Physico-Chemical Aspects of Electron Transfer Processes in Enzyme Systems. Pushchino-on-the Oka, USSR, August 11-16, 1975.
3. PYE, E. K., ed, "Biotechnology of Electron Transfer Processes", IFIAS Report, in press.
4. WINGARD JR., L. B. *New Scientist 64*: 565, 1974.
5. HALL, D. & SLESSER, M. *New Scientist*: 15 July 1976.

6. ANDERSON, R. E. *Natural Science Research Council Yearbook* 1976/77: 177.
7. ANDERSON, R. E. "Biological Paths to Self-Reliance" Van Nostrand Reinhold Company, New York, in press.
8. SINYAK, K., HEDEN, C., LUNDBECK, H. & NEMEC, T. "Significance of Infectious Disease Control in Developing Countries - An Experiment in Military Redeployment" Stockholm International Peace Research Institute, Sweden, in press.
9. VOLLER, A., HULDT, G., THORS, C. & ENGVALL, E. *Br. Med. J.* *1*: 659, 1975.
10. HEDEN, C.-G. *Quart. Rev. Biophys.* *10*: 113, 1977.

SOME THOUGHTS ON THE FUTURE

L. B. Wingard Jr.* and A. A. Klyosov**

*University of Pittsburgh Medical School, USA and
**Moscow State University Chemistry Department, USSR

This volume is oriented to the future directions in which enzyme engineering is expected to play a useful role. Therefore, it is appropriate that we single out several of these directions and outline some of the key research problems that are relevant to the attainment of practical developments in enzyme engineering. No attempt is made to cover all of the directions or areas that may seem promising for enzyme engineering but merely to describe several that appear to us to be worthy of extensive future research and development efforts.

For the purpose of this discussion, we shall define enzyme engineering as the development of useful products or processes based on the catalytic action of enzymes isolated from their source of synthesis or intact within cells that are not growing. This definition omits fermentation and tissue culture systems, in which living cells are growing, but does not omit multi-enzyme systems in sub-cellular particles. Not all persons will agree with this definition of enzyme engineering; however, it provides a useful working definition for the purpose of this discussion. From this definition it is apparent that enzyme engineering is an applied area of research and development. However, it may not be apparent that enzyme engineering has such a strong dependence on the results of fundamental research in chemical enzymology, physical chemistry, microbiology, biochemistry, and polymer chemistry that often it is not possible to see a clear distinction between enzyme engineering and these fundamental sciences. For example, the immobilization of enzymes, the chemical modification of enzymes, studies into the mechanism of specific enzymes, and genetic manipulations to improve enzyme availability or process characteristics all are fundamental scientific topics which become part of enzyme engineering only when considered in light of a specific enzyme-based process or product. Similarly, the fundamental topics of heat and mass transfer, diffusional effects on enzyme-catalyzed reaction rates, and mathematical modelling of enzyme-catalyzed reaction kinetics or overall process economics become part of enzyme engineering primarily when applied to a specific enzyme-based process or product.

In the final analysis an enzyme-based process or product must provide something unique if it is to find practical acceptance. Examples of suitable unique contributions include: 1) produces a new and useful product or carries out a new and useful chemical transformation or an old transformation in a new environment, e.g. in vivo; 2) produces a better quality product; and 3) produces an existing product more cheaply.

I. FUTURE DIRECTIONS WITH HIGH POTENTIAL

Some of the future directions of enzyme engineering that appear to have high potentials for practical developments are listed in Table 1. A few comments on each of these directions is given below. Additional comments are found in the other papers in this volume that pertain to some of the topics listed in Table 1.

In biological systems one of the unique consequences of enzyme catalysis is the high degree

TABLE 1

SOME FUTURE DIRECTIONS OF HIGH POTENTIAL FOR PRACTICAL DEVELOPMENTS

- Multi-enzyme systems
- Enzymatic degradation of polymeric substrates
- Enzyme electrochemistry and energy transfer
- Immobilized cells
- Enzymatic amplification of weak signals
- Enzyme-catalyzed synthesis of fine chemicals
- Enzymes at aqueous-lipid interfaces
- Medical therapy and research techniques and clinical analytical chemistry

---

of specificity and high rate by which sequences of chemical reactions are carried out, with each reaction catalyzed by a different enzyme. Efforts to adapt such multi-enzyme systems for in vitro use require considerably more research; but the possibilities merit considerable effort. Work needs to be continued on the stabilization of biologically produced multi-enzyme systems, as found in chloroplasts and liver microsomes, for in vitro use. At present the non-biological preparation of multi-enzyme systems is limited to random distributions of different enzymes on the surface of solid supports or in solution. An experimental approach for the non-biological preparation of an ordered sequence of different enzymes immobilized on the surface of a solid support would constitute a major breakthrough with great potential for the in vitro application of multi-enzyme systems.

The enzymatic degradation of insoluble polymeric substrates, such as cellulose and certain proteins and carbohydrates, continue to present both fundamental and practical challenges for the controlled use of such processes on a large indistrial scale. The kinetics of the degradation of these substrates is complex, especially since multi-enzyme systems often are involved (e.g. the cellulase complexes); and the practical demonstration of economical processes must receive a major input of innovative research. Since cellulose, certain proteins, and carbohydrates represent replenishable substrates for energy, food, and other necessities of life, there is a large impetus to explore the possibilities for developing in vitro processes for the degradation of these polymeric substrates via enzymatic catalysis.

Enzyme electrochemistry and energy transfer is another direction where our present knowledge of biological energy transfer mechanisms suggests that there may be some unique in vitro systems capable of being developed. Studies so far have shown that enzyme cofactors can be incorporated into functioning electrodes and that enzymes can be used to catalyze electrochemical and energy transfer reactions. Various immobilization schemes and electron transfer mediators need to be explored; and the theoretical and experimental quantitative description of the rates of reaction and rates of energy transfer need to be developed for different electrochemical reactor configurations. The experimental demonstration of enzyme-catalyzed fuel cells, the enzyme-catalyzed biophotolysis of water, the microbial enzymatic conversion of carbohydrates by hydrogen, and the electrosynthesis of energy rich products are specific examples where a great deal of fundamental as well as practical research is needed.

Several industrial companies have developed commercial processes that use immobilized cells instead of immobilized enzymes for the conversion of substrates to desired products. The use of immobilized cells omits the expense and problems involved in the separation and purification of the desired enzymes; however, the repression or inhibition of unwanted enzymes that remain active within the cell is a problem that can be very troublesome. The quantitative description of the relative rates of substrate reaction and substrate and product diffusion need to be determined experimentally for cells immobilized by different methods, in various reactor configurations, and over a wide range of enzyme activities within the immobilized cells. The relative merits of immobilized cells versus immobilized enzymes warrants comparison both for single reaction (single enzyme) and multi-reaction (multi-enzyme) processes.

The in vitro enzymatic amplification of weak signals is a concept that has its in vivo counterpart in the cascade theory of blood clotting, wherein a very small concentration of clot-promoting tissue factor causes the eventual precipitation of a large quantity of fibrin particles. Both light and ultrasonic signals are suggested as inputs for the enzymatic amplifiers. This topic is discussed in considerable depth in an earlier paper in this volume.

The enzyme-catalyzed synthesis of fine chemicals is an area wherein alternative processes often exist and the selection of commercial processes is highly dependent on the processing economics. The synthesis of complex compounds having physiologic and pharmacologic activity, including antibiotics, prostaglandins, or neuropeptides, may be an appropriate avenue for enzyme-based processes; however, each product must be considered separately and in competition with alternative routes of fermentation, non-enzymatic chemical synthesis, or extraction from biological tissue or fluids. The development of economically practical processes for the synthesis or regeneration of high energy phosphate compounds, and the demonstration of ways to couple the hydrolysis of these compounds to the synthesis of organic compounds could lead to a wider scope of in vitro synthesis reactions where enzymatic catalysis might have an advantage.

In biological systems enzymes catalyze both the synthesis and degradation of lipid materials. The mechanistic role of how enzymes function in a lipid environment is not well understood, since enzymes are assumed to function primarily in an aqueous environment. With lipids, the enzymes may work at the aqueous-lipid interface rather than inside the lipid environment. This topic requires in depth fundamental research in order to learn how to adapt lipid enzymatic systems for use in vitro. Significant applications in food processing, waste conversion, chemical synthesis, and medical therapy can be envisioned.

In medical research enzymes have been used extensively for the development of experimental approaches in the fundamental medical sciences; and the incorporation of immobilization techniques promises to augment the usefulness of enzymes as tools in medical research. Enzymes have been much more slowly introduced into medical therapy; although this slowness is expected due to the major difficulty in developing any new therapeutic approach to specific disease states. Enzymes may prove useful in developing improved methods for the sustained release of drugs in vivo, for better targeting of drugs to specific body tissues, and for detoxification of body fluids under conditions when the normal systems are malfunctioning or overwhelmed. However, the actual clinical acceptance of such new enzyme-based techniques may not be achieved unless the new technique has major advantages and little risk to the patient. Detoxification with extracorporeal perfusion over an immobilized enzyme is a case in point. There will continue to be great clinical reservation against using such a method in the majority of cases because of the risk to the patient and because other methods may be available; however, in those cases where no other approach is available, then the clinical hesitation will be much less. Enzymes also are widely used in clinical analytical chemistry, mainly in soluble form and only in a few instances in an immobilized form. Economics plays a major role in the further acceptance of high-priced enzyme-immunoassays and special instrumented immobilized enzyme analytical devices. The development of enzymatic analyses for materials for which there is no current method of suitable sensitivity and for which there is a necessity to have such measurements has the best chance of gaining practical acceptance. Electrodes that function in vivo, devices for monitoring toxic constituents in the environment, and inexpensive methods for the determination of additional endogenous body fluid enzymes and metabolites are likely candidates.

II. FUNDAMENTAL RESEARCH TOPICS

It was pointed out earlier that although enzyme engineering is an area of applied science and technology, it is highly dependent on the results of a wide variety of the fundamental sciences. Since the future potential for enzyme engineering is tied so heavily to anticipated advances in the fundamental sciences, we have included a list of fundamental topics that need to be studied and much better understood if the future directions for enzyme engineering, discussed in the previous section, are to be realized. These topics are as follows:

1. Stabilization of enzymes against changes in environmental conditions, methods for the reactivation of denatured enzymes, kinetics of denaturation and its reversal.

2. Structure and function elucidation of enzyme complexes such as the cellulases and the redox electron transfer chains or enzyme-lipid complexes, relation of kinetics and specificity to ordered arrangement of the complex, structure of enzyme complex as it relates to the kinetics of

action on insoluble substrates, structural factors that enhance the coupling of free energy between the hydrolysis of high energy phosphate compounds and the synthesis of organic compounds via endogonic reactions.

3. Diffusional resistances and their influence on the overall rates of chemical conversion of substrates to product for immobilized cells, for multi-enzyme immobilized systems, and for lipid enzyme systems with a major emphasis on experimental verification or modification of theory.

4. Development of molecular models that simulate at least in part the substrate specificity and spatial orientation of substrate-cofactor complexes obtainable with specific enzymes, possibly using metallocomplexes or synthetic polymers and surfactant micelles.

ACKNOWLEDGMENT

A. Klyosov would like to thank I. Berezin and K. Martinek for many helpful suggestions and stimulating discussions. L. Wingard appreciates the input of these three individuals in putting this chapter together.

ADDRESSES OF AUTHORS

ADAMS, BARBARA
International Federation of Institutes for Advanced Study, Ulriksdals Slott, S-17171 Solna, Sweden.

BEREZIN, ILIA V.
Department of Chemistry, Moscow State University, Moscow 117234, U.S.S.R.

CERNIA, E.
Assoreni S. Donato Milanese, Milano, Italy.

CHAZOV, EUGENE I.
National Cardiology Research Center, USSR Academy of Medical Sciences, Petroverigsky Lane 10, Moscow 101837, USSR

GERASIMAS, VALDAS B.
Department of Chemistry, Moscow State University, Moscow 117234, U.S.S.R.

GIACOBBE, F.
Biochem Design S.p.A., via A. Bargoni 78, 00153 Roma, Italy.

GOGOTOV, IVAN N.
Institute of Photosynthesis, USSR Academy of Sciences, Pushchino, Moscow Region 142292, U.S.S.R.

GOLDMACHER, VICTOR S.
Present address unknown.

HEDEN, CARL GORAN
Bakteriologiska Institutionen, Karolinska Institutet, S-104 01 Stockholm 60, Sweden.

IASONNA, A.
Biochem Design S.p.A., via A. Bargoni 78, 00153 Roma, Italy.

IL'INA, ELENA V.
National Cardiology Research Center, USSR Academy of Medical Sciences, Petroverigsky Lane 10, Moscow 101837, U.S.S.R.

KAZANSKAYA, NOVELLA F.
Department of Chemistry, Moscow State University, Moscow 117234, U.S.S.R.

KLYOSOV, ANATOLE A.
Department of Chemistry, Moscow State University, Moscow 117234, U.S.S.R.

KÖSTNER, ADO I.
Laboratory of Enzyme Technology, Tallin Technical University, Ehitajate 5, Tallin 200026, U.S.S.R.

LARIONOVA, NATALY I.
Department of Chemistry, Moscow State University, Moscow 117234, U.S.S.R.

MARCONI, W.
Laboratori Ricerche Processi Microbiologici, Assoreni, C.P. 15-00015 Monterotondo (Roma), Italy

MARGOLIN, ALEXIS L.
A.N. Belozersky Laboratory of Bioorganic Chemistry and Molecular Biology, Moscow State University, Moscow 117234, U.S.S.R.

MARTINEK, KAREL
Department of Chemistry, Moscow State University, Moscow 117234, U.S.S.R.

MAZAEV, ALEXEY V.
National Cardiology Research Center, USSR Academy of Medical Sciences, Petroverigsky Lane 10, Moscow 101837, U.S.S.R.

MENYAILOVA, IRINA I.
Laboratory for Immobilized Enzymes, All-Union Scientific Research Institute of Biotechnique, Kropotkinskaya 38, Moscow 119034, U.S.S.R.

MORISI, FRANCO
Biochimica Ricerca, Laboratori Ricerche Processi Microbiologici, Assoreni, 00015 Monterotondo (Roma), Italy.

MOSKVICHEV, BORIS V.
Laboratory for Enzyme Immobilization, USSR Research and Technological Institute of Antibiotics and Enzymes for Medical Use, Ogorodnikov Prospect 41, Leningrad 198020, U.S.S.R.

MOZHAEV, VADIM V.
Department of Chemistry, Moscow State University, Moscow 117234, U.S.S.R.

NAKHAPETYAN, LEVON A.
Laboratory for Immobilized Enzymes, All-Union Scientific Research Institute of Biotechnique, Kropotkinskaya 38, Moscow 119034, U.S.S.R.

NILSSON, SAM
International Federation of Institutes for Advanced Study, Ulriksdale Slott, S-17171 Solna, Sweden.

PROSPERI, G.
Laboratori Ricerche Processi Microbiologici, Assoreni, 00015 Monterotondo (Roma), Italy.

PRUIDZE, GURAM N.
Institute of Plant Biochemistry, Georgian SSR Academy of Sciences, Tbilisi 380031, U.S.S.R.

RABINOWITCH, MICHAEL L.
Department of Chemistry, Moscow State University, Moscow 117234, U.S.S.R.

SAKHAROV, IVAN YU
Department of Chemistry, Moscow State University, Moscow 117234, U.S.S.R.

SIIMER, ENN
Laboratory of Enzyme Technology, Tallin Technical University, Ehitayate 5, Tallin 200026, U.S.S.R.

SINITSIN, ARKADY P.
Department of Chemistry, Moscow State University, Moscow 117234, U.S.S.R.

SMIRNOV, VLADIMIR N.
National Cardiology Research Center, USSR Academy of Medical Sciences, Petroverigsky Lane 10, Moscow 101837, U.S.S.R.

SVEDAS, VYTAS K.
A.N. Belozersky Laboratory of Bioorganic Chemistry and Molecular Biology, Moscow State University, Moscow 117234, U.S.S.R.

TERESHIN, IGOR M.
USSR Research and Technological Institute of Antibiotics and Enzymes for Medical Use, Ogorodnikov Prospect 41, Leningrad 198020, U.S.S.R.

TOLSTICH, PETER I.
I.M. Sechenov First State Medical Institute, Bolshaya Pirogovskaya 2, Moscow 119435, U.S.S.R.

TORCHILIN, VLADIMIR P.
National Cardiology Research Center, USSR Academy of Medical Sciences, Petroverigsky Lane 10, Moscow 101837, U.S.S.R.

VARFOLOMEEV, SERGEI D.
A.N. Belozersky Laboratory of Bioorganic Chemistry and Molecular Biology, Moscow State University, Moscow 117234, U.S.S.R.

VLADIMIROV, VLADIMIR G.
I.M. Sechenov First State Medical Institute, Bolshaya Pirogovskaya 2, Moscow 119435, U.S.S.R.

WINGARD JR., LEMUEL B.
Department of Pharmacology, School of Medicine, University of Pittsburgh, Pittsburgh, PA 15261, U.S.A.

ZHURAVLYOV, ANATOLE G.
N.I. Pirogov Second State Medical Institute, Malaya Pirogovskaya 1, Moscow 119864, U.S.S.R.

SUBJECT INDEX

Acetate removal, 408
Acetonitrile, 37
Acetylcholinesterase
  inhibitor, 363, 366
Acetyl tyrosine ethyl
  ester, 223, 227
Acrolein, 14
Acrylics, 227
Acryloylchloride, 14
*Actinoplanes
  missouriensis*, 469
Activation overpotential,
  340, 342
Adenosine triphosphate
  synthesis, 350-351
Adenosyl methionine
  transferase, 412
Adsorption,
  aryl-β-glucosidase, 154
  cellobiase, 142, 154
  on cellulose, 143-155,
    isotherm, 146-149
    reversibility, 151
  endoglucanase, 139
Aging, 55, 73
Albumin, immobilized, 224
Alcaligenes eutrophus,
  323
Alcohol dehydrogenase,
  348-350
Aldolase, 374
Algae, hydrogen evolu-
  tion, 327
Aminoacylase, 478-479
  immobilized, 479
7-Aminodeacetoxy
  cephalosporanic acid,
    258
6-Aminopenicillanic
  acid, 258
  production, 405-409
Ampicillin, 262, 285,
  406
α-Amylase, 374
β-Amylase, 482
Amyloglucosidase, 167-
    195
  immobilization, 168-
    193
    pore diameter, 173
  inactivation
    kinetics, 184-185
  kinetics, 184-185
  pH effects, 181
  preoxidation, 177,
    180, 181
  stability, 190-193
Anabaena cylindrica,
  324, 330
Antibiotics, 257-290
Antigenic, 219
  active carrier, 228
  polymer-enzyme
    adduct, 305-308
Arrhenius plot,
  chymotrypsin, 15, 24
  fumarase, 444

Arthrobacter, 57, 468
Aryl-β-glucosidase,
  activity, 88-91, 103
  ionization, 104
Aspartic acid, 4
Aspergillus, 140
  foetidus, 88-89, 97,
    116, 127, 130, 141
  niger, 90-91, 95, 127,
    130, 138, 141, 185,
    191, 200, 213, 475
  terricola, 296
Azo coupling, 174-175,
  178, 181
Azotobacter vinelandii,
  323

Bacillus subtilis, 408
B. coagulans, 468
Beets, sugar, 476
Benzyl penicillin, 32,
  260-270
  enzymatic synthesis,
    267-270
  hydrolysis, 405-409
  immobilized, 275
  kinetics of synthesis,
    270-280
  mechanism of synthesis,
    272
Bioactive carriers, 228
Biochemical fuel cells,
  494
Biocompatible polymer,
  221
Biodissolving enzyme
  particles, 221
Biphasic solvents, 34-36
Blue green algae, 325
Brevibacterium ammoniagenes, 411

Carbodiimide, 213, 244
Carboxymethyl cellulose,
  87, 102, 103, 113-117,
  119-125, 137, 141
Catalase, 482
Catechins, 458-463
Cell surfaces, 313-316
Cellobiase, 84, 140-143
  activity, 88-91, 94,
    99, 127
  adsorbed, 142, 147,
    149
  ionization, 104
  kinetics, 116, 144
  sonication, 102, 103
Cellobiohydrolase, 84,
  133, 140,
  activity, 96-99
Cellobiose, 85,
  steady state concentration, 112
  effect on endoglucanase, 135
Cellulase,
  description, 84-105
  initial velocity, 95
  ionization, 104
  kinetic parameters,
    144, 155
  mechanism, 131-132
  optimal, 160-161
Cellulose,
  conversion to glucose,
    83-165
  hydrolysis mechanism,
    105-135
  hydrolysis kinetics,
    108-125
Cephalosporins, 257, 405
Cephalothin, 260
Cephamycins, 257
Cheese whey, 474-476
Chemiosmotic theory, 350
Chemotrophes, hydrogen
  evolution, 322-323

Chlamydomonas
  reinhardii, 326, 330
Chlorella vulgaris, 326
Chloroplasts, 330, 332, 501
  inactivation kinetics, 77-78
Chymotrypsin,
  acryloylated, 15, 17
  acylation, 370
  alkylation, 8-9
  chemical inactivation, 363, 365
  crosslinked, 7-12, 231
  immobilized,
    on cellulose, 27
    in gel, 15, 19, 22-25, 42, 43
    on heparin, 229-230
    in membrane, 378
    on sephadex, 27, 223
  light effect, 387
  in octane, 37
  regeneration, 43
  substrates, 367
  thermoinactivation kinetics, 8, 15, 17, 19, 23, 27, 227
Citric acid, 439, 450
Clostridium,
  butyricum, 322-323
  kluyveri, 326
  pasteurianum, 323, 330, 332
  perfringens, 314, 322
Coenzyme A synthesis, immobilized cells, 413
Cofactors,
  overpotentials, 340
  immobilized, 344
Conformation, 3-12, 40-41, 207
  and crosslinking, 7
Copolymerization, 14
Copper enzyme, 453
Cost,
  of immobilization, 418
  lactase and whey, 475
  making malic acid, 446-450
Cotton hydrolysis, 125-135, 155
Crosslinking,
  glutaraldehyde, 9
  intramolecular, 7-12
  length of spacer, 9-12, 16, 231-233
Cyanobacteria, 325
Cyanuric chloride, 244
Cyclic voltammetry, 346
Cytochrome c_3, 331-332

Deacylation of acylenzyme, kinetics, 390
Denaturation,
  irreversible, 39-42
  SH groups, 39
Desulfotomaculum, 322
Desulfovibrio, 322, 331-332
Dextran, 226-227
Dextrose equivalent, 475
Diazo compounds, 358
Diffusional effects, 506
  and inactivation, 72
  in enzyme membrane, 348-350
  in enzyme photography, 387-388
Distribution, kinetics of,
  I-131 albumin, 224
  pancreatic inhibitor, 250-251
Dithiols, 23
Diurone, 328

E. coli, 4, 270, 322-
323, 406
Economic efficiency, 416
Effectiveness factor, 348
Efficiency, enzyme re-
actors, 415-435
Electroactivity, 340
Electrochemistry, 339-353
501-502
α, β-Elimination, 409-410
Endo-1, 4-β-glucanase, 84,
135-140
activity, 87-94, 127,
130
adsorbed, 139, 147, 149
influence of cellobiose,
135
ionization, 104
kinetics, 116, 122-125
Endomycopsis bispora, 168,
170, 184, 190
Energy transfer, 340-350,
494, 501-502
Entrappment, 18-28
Enzymes,
as amplifier, 501, 503
of light, 360-383
in biphasic solvents,
34-36
in electrochemistry,
339-353, 501-502
in energy transfer, 340-
350, 494, 501-502
in fine chemicals prepa-
ration, 405-413, 501,
503
immobilization, 4, 7,
167-170, 209-214, 223,
314
inactivation, 4, 100,
191-193
at interfaces, 501, 504
in nonaqueous solvents,
33-38
in nutrition, 465-483
in photography, 357-
403
photoinactivation,
376
polymer gel, 14
reactivation, 3-7,
38-44
stabilization, 3-54,
505
against pH, 28-32
against thermoin-
activation, 5-54
in surfactant organic
micelles, 36-38
in therapy, 219-237,
501, 504
Enzyme engineering in
science policy, 487
Epigallocatechin
gallate, 458
Ethylphenylacetate,
267-270
Exo-1, 4-β-glucanase,
84, 140
activity, 88-91, 94,
96-99, 127
ionization, 104
Exo-1, 4-β-glucosidase,
84, 88-91

Failure intensity, 60
Ferrodoxin, 326, 330
Fibrin, 228
Fibrinolysin, immobi-
lized, 225
Fine chemicals produc-
tion, 405-413
Flavor enzymes, 466
Food additive enzymes,
465-483
Formate, 329
Formate dehydrogenase
inactivation, 57
Free radicals, 100
β-Fructosidase, 425

Fuel cell, enzyme, 494
  approaches, 344
  free energy, 341-343
  schematic, 340-342
Fumarase, 412, 440, 480
  activity, 443, 445
  food acidulant, 479
  immobilized, pH effect, 442
  purification, 441
  in reactors, 443-444
Fumaric acid to malic acid,
  conversion, 439-450
  economics, 445-450

α-Galactosidase,
  hydrolysis of raffinose, 476-478
  immobilized, 476
β-Galactosidase,
  immobilized, 315, 425
  reactor, 423
G. candidum, 88, 96, 102-104, 113, 117, 120, 127, 130, 138, 141, 150
Geotrichum, 140
Gibbs free energy, 263-270, 340-343
Glucoamylase, 197-215
  immobilization, 202-206, 481
  inactivation kinetics, 198-206, 211
  stability, 197, 209-214
  substrate stabilization, 200-209
Glucose,
  from cellulose, 83-165
    production rate, 111-113, 115, 117, 130
    optimized from cotton, 156-160
  electrode, 350-353
  from starch maltodextrins, 207-209
Glucose isomerase, 467-471
  half life, 470
  immobilized, 468-469
Glucose oxidase,
  fuel cell electrode, 341, 343-344
  immobilized, 30, 315, 352
  kinetics of inactivation, 30
Glutaraldehyde, 170, 173, 179, 181, 352
Glutathione, 42
Glycoproteins, 313
Grapevine phenol oxidases, 457-464

Health care delivery, 493
Heparin as support, 229-230
Histidine ammonia lyase, 412
Hydantoinase, 412
Hydrogen peroxide, 352, 482
Hydrogen production,
  by algae, 327
  by chemotrophes, 322-323
  chloroplasts, 332
  mechanism, 328, 330
  by microorganisms, 322-327
  model systems, 330-334
  by phototrophes, 322, 325
  schemes, 329, 502
Hydrogenase, 62, 321-337
  models, 330

optimum temperature, 334
stability to oxygen, 333-334
Hydrolysis synthesis of β-lactams, 260-267
Hydroxynitrilo lyase, 412

IFIAS, 487
Immobilization,
by adsorption, 139, 142-155
amyloglucosidase, 167
Brevibacterium ammoniagenes, 411
carbodiimide, 213
to carbon, 346-347
of cells, 411, 476, 501, 503
to complimentary polymer, 13-16
covalent coupling to gel, 14
by crosslinking, intramolecular, 7-12
diazonium linkage, 316
by entrappment, 7, 18-28
in fibers, 440, 476, 478
in gel, 30, 42, 43, 309, 352
glucoamylase, 201-206, 211-214
glucose oxidase, 352
glutaraldehyde, 170, 173, 179, 209-214, 314, 348, 352
by ionic adsorption, 478
by light, 380-382
for medical therapy, 220
Mortierella vinacea, 476
multipoint attachment, 12
neuraminidase, 314
penicillin amidase, 275
to platinum, 352
riboflavin, 346-347
in semiconducting gel, 349
by silanization, 168, 209-214, 352
on silica, 475
spacer length, 9-12, 16-18, 231-233
terrilytin, 297, 309
trypsin, 297
by Wittig reaction, 347
Inactivation,
kinetics, 55-79, 102-103, 199-200, 302, 421
mechanisms, 5-54
due to microenvironment, 5
multi enzyme, 67-73
due to thermal effects, 5-54, 198
by ultrasound, 100-105
Insulin, 228
Invert sugar, 480
Ionization of cellulases, 104
Irreversible denaturation, 39-42
Isoalloxazine rings, 344

Kallikrein, 241-242
Kinetics,
aging, 55-79
inactivation, 55-79, 102-103, 199-200, 302, 421
multi enzyme, 105-135
see individual enzymes
Kinin inhibitor, 241

β-Lactam antibiotics, 257-290
  enzymatic synthesis, 270-290
  thermodynamics of hydrolysis, 259-270
β-Lactamase reactivation, 39
Lactase, 471-476
  immobilized, 472, 475-476
Lactoperoxidase, immobilized, 315
Lactose hydrolysis, 471
  in milk, 472-474
Lectins, immobilized, 315
Light alteration,
  of activated support, 379-383
  of membrane, 378
  of reaction medium, 377
Light amplification by enzymes, 360, 394
Light effect on enzymes, 372-377, 386
Light sensitive,
  effectors, 363-367
  inhibitors, 364
  substrates, 367-372
Lipids and enzymes, 501-504
Liposomes, 234-237
L-Malic acid,
  production, 439-450
  economics, 445-450

Maltose, 184
Maltodextrins, 197, 202
Mandels Weber method, 88-91
Manufacturing cost, 447
Mechanism of stabilization, 17, 25
Methyl viologen, 324, 327, 340
Micelles, 36-38
Michaelis constant,
  cellobiase, 97-98
  cellulase, 144-145
  chymotrypsin, 223, 227, 230
  glucoamylase, 205
  penicillin amidase, 281-282
Milk, commercial lactase treatment, 472-474
  clotting, 481
  peroxide removal, 482
Molecular models, 506
Mortierella vinacea immobilization, 476
Multi enzyme system, 501
  reliability, 66-73

NAD, 71
Nagase, 90-91
Neuraminidase immobilization, 314-315
Nitrogenase, 324, 333
Nocardia opaca, 323
Nonaqueous solvents, 33-38

Oligosaccharides, 85, 107
Optimization,
  glucose from cotton, 156-160
  immobilized enzyme reactors, 415-435
Organelles, 73
  aging kinetics, 74-78
Organic solvents, 4, 33-38, 44
Overpotentials, 340-343

Pancreatic inhibitor,

immobilized, 241-254
kinetics, 248
stability, 249
in vivo, 250-253
Papain, 372
Penicillin, 4, 257
Penicillin acylase, 405
Penicillin amidase, 260-270
benzylpenicillin synthesis, 270-280
immobilized, 32, 260-290, 425
inactivation and pH effect, 32
pH, 289
Peroxidase in micelles, 37-38
pH effect, 4, 28-33, 103, 247, 264, 269, 286, 289, 461
Phenol oxidases, 453-464
from grapevines, 458-463
inhibitors, 454-457
pH effect, 461
substrate specificity, 459, 462
from tea leaves, 457-463
Phenylacetic acid, 272
Phenylacetylglycine, 267-270
Phenylglycine acylase, 408
Phosphorylase reactivation, 39
Photochromic polymers, 373
Photography,
with enzymes, 357-403
methods, 384-393
with palladium, 359
with silver, 358-360
Photoinactivation, 376
Photoisomerization, 367
Photosynthesis, 492-493
Phototrophes and hydrogen evolution, 322, 325
Phthalic anhydride, 439
Plant enzymes, 453
Polymer substrates, 501, 502
Polyvinylpyrrolidone, 222
Pore diameter, 172-173
Potentiometric glucose electrode, 350-353
Prehydrolytic c-l factor, 84, 131-132
Probability of failure, 58
Proflavin, 365
Propargylpyridinium bromide, 349
Protease inhibitor, 241-254
Protein unfolding, 5-7, 29, 38, 44
Pseudomonas, 440
melanogenum, 408
Pullulanase, 482
Purple bacteria, 324-325
Pyrogallol, 38

Quantum yield, 358
amplification of, 359
enzymatic, 360, 383

Raffinose hydrolysis, 476-478
Rapidase, 90-91, 135, 138
Reactivation of enzymes, 3, 38-44, 372
Reactor,
batch, 155, 275, 443

column, 139, 150, 153, 187-189, 208
effective rate, 427
efficiency, 415-435
β-galactosidase, 423
optimization, 415-435
plug flow, 142
radial, 471-472
schematic, 479
stirred tank, 430-431, 444
temperature, 426
Reducing sugar method, 95
Reliability function, 59, 61

Reliability theory, 55-80
Reloading cycle, 420, 424
Rennin, 482
Rhizopus niveus, 207
Riboflavin, immobilized, 345-348
R. rubrum, 331

Science policy, 487-497
impact, 494
Self reliance, 495-496
Sephadex, 12, 26, 223, 478
SH groups, 39, 41, 372
Sialic acid, 314
Silanization, 168
Silica gel, 170, 173
Silochrome, 170, 175, 180-181, 189, 190
Silver photography, 358-360
S. maxima, 331
Sodium borohydride, 7, 296
Soluble supports, 226, 243-254, 295-309
Somogyi Nelson method, 95
Sorbose dehydrogenase, 412
Soybean,
milk, 477
trypsin inhibitor, 222
Spacer length, 9-12, 16-18
Spiropyran, 374
Stabilization,
against pH, 28-33
of enzymes, 3-54, 305
mechanism, 17, 25
by substrate, 198-207
Starch,
hydrolysis, 167-195
hydrolysis of non-reducing end, 197
liquified, 186-193

Streptomyces, 468-469
Structure and function, 505
Substrates, photo-excited, 369
Sugar beets, 476-478
Supports,
albumin, 297
acrolein, 303
acrylics, 227
carbon, 346-347
CM cellulose, 245
DEAE cellulose, 245
dextran, 226, 245, 297
for enzymes, 7
fibers, 440, 477, 478
glass, 171, 201-206, 211, 213, 316, 475
heparin, 230
hydroxyethylmethacrylate, 30
inorganic, 168-170
nylon tubes, 314
platinum, 352
polyacrylamide gel, 22-25, 42, 211, 309, 352
polyelectrolyte gel, 289

polyion complex, 32
polysaccharide, 25-27
semiconducting gel, 349
Sephadex, 12, 223, 225, 478
Sepharose 4B, 42, 314
silica gel, 170, 173
silochrome, 170, 175, 180
spatial considerations, 339-353
vinyl pyrrolidone, 297
Surfactants, 36-38

Tea phenol oxidases, 453-464,
in dyes, 454
Terrilytin,
antigenic properties, 308
immobilized on polymers, 297, 299
inactivation kinetics, 302, 306
inactivation by serum, 304, 307
stability, 305
Tetracyanoquinedimethane, 349
Tetracycline, immobilized, 316
Theaflavin, 454-457
Thearubigin, 454-457
Therapy,
enzymes, 219-237, 504
pancreatic inhibitors, 241-254
Thermodynamics of hydrolysis, 263-270
Thermoinactivation,
chymotrypsin, 15, 42
in gel, 15, 42
glucoamylase, 198-214
protein unfolding, 5-7, 29, 38
sulfur oxidation, 41
trypsin, 42
Thermostability, 6
chymotrypsin and alkylation, 8
in gel, 18-28
glucoamylase, 197-214
Thiele modulus, 348-349
Thiocapsa roseopersicina, 62
Thromboembolism, 225
Time to failure, 60, 66
T. koningii, 88-89, 127, 130, 137, 141, 149
T. lignorum, 88-89, 127, 130, 138, 141
T. longibrachiatum, 88-89, 95, 127, 130, 141
Transglycosilation, 137
T. reesei, 95-97, 135, 138, 141
s-Triazines, 244
Trichoderma, 140
T. roseopersicina, 331
Tropical disease diagnosis, 492
Trypsin,
activity, 299
immobilized,
in gel, 22
on polymers, 297, 299
on Sepharose 4B, 42
inactivation kinetics, 302, 306
inactivation by serum, 304, 307
inhibition kinetics, 248
modified, 395
pancreatic inhibitor, 241-254

pH effect, 247
soybean inhibitor, 222
stability, 305
substrates, 367
Tryptophanase, 409
L-Tryptophan production, 478-479
T. viride, 88-91, 141
β-Tyrosinase, 409

UNESCO, 494
Ultrasound and cellulase complex, 99-105

Velocity, maximum,
cellulase, 144
penicillin amidase, 280
Vibrio cholerae, 314
Viscometric method, 95, 103, 123
Viscosity of carboxymethylcellulose, 87, 93-94

Whey, lactase treatment 474-476